PRENTICE HALL
GRAPHIC COMMUNICATIONS
DICTIONARY

Daniel J. Lyons

AGAINST THE CLOCK
PERFORMANCE SUPPORT & TRAINING SYSTEMS

Prentice Hall, Upper Saddle River, New Jersey 07458

Library of Congress Cataloging-in-Publication Date

LYONS, DANIEL J.
    Prentice Hall graphic communications dictionary / Daniel J. Lyons.
    p. cm. – (Against the Clock series)
    ISBN 0-13-012226-2
    1. Printing–United States Dictionaries. 2. Graphic arts–United
States Dictionaries. I. Title: Graphic communications dictionary
III. Series.
    Z118 .L96 2000
    686.2'03—dc21   99-34860
                                              CIP

Acquisitions editor: *Elizabeth Sugg*
Development editor: *Judy Casillo*
Editorial / production supervision: *Barbara Marttine Cappuccio*
Director of manufacturing and production: *Bruce Johnson*
Managing editor: *Mary Carnis*
Manufacturing buyer: *Ed O'Dougherty*
Creative director: *Marianne Frasco*
Editorial assistant: *Brian Hyland*
Marketing manager: *Shannon Simonsen*
Cover design: *Wanda España*
Cover illustration: *Cooper*

© 2000 by Prentice-Hall, Inc.
Upper Saddle River, New Jersey 07458

Against The Clock and the Against The Clock logo are trademarks of
Against The Clock, Inc. registered in the United States and elsewhere.

Printed in the United States of America

10 9 8 7 6 5 4 3 2 1

ISBN 0-13-012226-2

Prentice-Hall International (UK) Limited, *London*
Prentice-Hall of Australia Pty. Limited, *Sydney*
Prentice-Hall Canada Inc., *Toronto*
Prentice-Hall Hispanoamericana, S.A., *Mexico*
Prentice-Hall of India Private Limited, *New Delhi*
Prentice-Hall of Japan, Inc., *Tokyo*
Prentice Hall Singapore Pte. Ltd.
Editora Prentice-Hall do Brasil, Ltda., *Rio de Janeiro*

DEDICATED TO

my wife, Jackie, for being
my best friend and companion
and
Jack C. Deibert
dedicated public educator, helper of youth, mentor
and always an advocate when one was needed.

CONTENTS

CONTENTS

PREFACE

This book began as a modest list of about 100 common printing terms used for a single occasion in a discussion group. As part of that exercise, conversations ensued about alternate, conflicting and sometimes confusing meanings of single words. Over time, more and more words came to be defined and itemized. Exciting developments in technology caused the list to expand from printing into the much wider realms of computing, the internet, prepress, video, audio, and, finally, general graphic communications.

Informally published each semester and updated biannually, it grew significantly over several years and generated requests for copies and suggestions for the addition of some new words and illustrations.

This edition is nearly five times the size of the last one, produced for my students in 1998, and represents an attempt to broaden its scope while improving the clarity of both the presentation and definitions.

It is not the great American novel. It is a practical offering to enable clear, concise and accurate interpretation of terms. Few people will, with the boundaries of a single profession or occupation, use all of the content. But as technology drives graphic communications forward, a reference with broader and more clear terminology is not only inevitable, but necessary. It is hoped that this offering will make the readers' successful transition easier and perhaps richer. It will be an aid both for new students and veterans of the graphic communications industry.

It has always been the my policy to encourage suggestions for new editions. If you think of a word that might help improve *The Graphic Communications Dictionary*, or whose definition you feel needs expansion, clarification or other help, please e-mail the author at dictionary@prof-lyons.com or the publisher, Prentice Hall, at www.prenhall.com.

ACKNOWLEDGMENTS

The author wishes to acknowledge the many people who have made it possible for him to enjoy a full career in Graphic Communications. The early influence of teachers, professors and later, professional associates, companies and associations cannot be over-emphasized.

Starting with a public school education in Williamsport and Carlisle, Pennsylvania as a student of David Hunter, Harrold Lesher and Jack Deibert and continuing through a college experience at Rochester Institute of Technology with Anthony Sears, Hector Sutherland, C. B. Neblett, Brian Culver, Alexander Lawson, and Peter Jedrezek, unimaginable positive influences have been exerted on my life.

Professional and management positions with William Penn Hastings of the Milton (PA) Standard Publishing Company and Dennis Schieck, Fred Smith and Donald W. Reynolds of the Donrey Media Group, and Robert Meibaum of Edgerton Germeshausen and Grier, recall heartfelt appreciation for both opportunity to grow and gain knowledge.

My years as a college professor and a department manager at the Community College of Southern Nevada has provided an opportunity to remain young at heart by having a never-ending stream of associates, employees and students who provide a breadth of curiosity, an atmosphere of enthusiasm and a confidence in the future. They, each in their own way, provided major stimuli for this compilation. During the past year I have had made new acquaintances and had considerable aid from the publisher's staff, notably Elizabeth Sugg, Judy Casillo and Pat Walsh. Their valuable assistance made it possible to provide a completed manuscript in true professional form. They are to be commended for their patience and persistence.

I am fortunate because all have reinforced the concept that if you find a job you like, you'll never have to work a single day but, rather, pleasure in the opportunity to participate.

Few people are as blessed as I with such extensive encouragement and support. My world would be less rich and certainly much poorer without professional friends like John Heiss, Bob Bay, Sherri Burgess, Bob Bremner, Terry Bratton, Mary Edwards and Jan Carr, several who helped forge a technically sophisticated production facility at the college, even in the face of overwhelming odds. They provided a cheering section when the work was daunting and never failed to provide the support and assistance needed to accomplish the objectives for which we all strived.

I would also like to acknowledge the following reviewers for their comments and assistance in the course of the development of this dictionary: Rich Eisenach, Fox Valley Technical College; Val LeFevre, Graphic Resolutions; Russell Rosener, St. Louis Community College; and March Schlueter, Fox Valley Technical College.

On a personal note, my life would have been untenable without the love and support of family. My parents, Theo "Dutch" and Clair E. "Red" Lyons, provided an encouraging environment to pursue all of my goals and did so at great sacrifice, often unknown to me as a youth.

With absolute surety, no greater void could be felt if I did not have the constant assistance, encouragement and the fun in my life provided by my best friend and wife, Jackie. She reads, writes, critiques and advises and is my personal pillar of strength in all that I do. Her contributions, both physical and spiritual, to the preparation of the manuscript are beyond measure and I will love her forever and always.

DJL

August 1999

GRAPHIC COMMUNICATIONS DICTIONARY

A guide through the words we use and misuse in nearly all areas of graphics, with delineation for the many multiple definitions for:

ANIMATION
ART AND DESIGN
COMPUTING
CONVENTIONAL AND ELECTRONIC PREPRESS
GENERAL PRINTING
INTERNET
PAPER AND PULP
PHOTOCOMPOSITION AND IMAGESETTING
PHOTOGRAPHY AND MOTION PICTURES
PRINTING CLASSES AND REPRODUCTION PROCESSES
TYPOGRAPHY
VIDEO AND AUDIO
WEB HOSTING AND PAGE CREATION

A valuable resource for both veterans of the industry and students of any area of Graphic Communications.

HOW TO USE THIS DICTIONARY AND WHAT TO EXPECT FROM IT

To assist you in using the Graphic Communications Dictionary, several points should be made.

• All of the listings are alphabetical and many of the words have multiple definitions. Every attempt has been made to make the explanations clear and concise without excessive technical references.

• The order of definitions, when there are multiples, is not formally structured. Generally, the first definition is the most used, from my experience, and subsequent ones are a bit more esoteric. Only the user can determine this.

• You will also find numerous related terms listed at the end of the definitions. These are useful as alternate words for exact-same or similar processes. Related terms, when listed, also may be antonyms of the word which is defined. All such terms are intended to provide another aid to clarify meaning and expand the scope of the Graphic Communication Dictionary definitions.

• Illustrations used throughout the book are simple and clear. They are not technical drawings. They are greatly simplified to show only the most rudimentary operating parts and principles of the subject. I have found that simplified drawings convey much more to a new user than complex ones. If the reader desires more detailed graphic explanations, there are many source books delving into the specific areas of interest within the various fields.

This compilation was prepared with an eye toward clarity and simplicity. The aim was to make a ready reference which can be used by anyone in the various graphic communications fields regardless of level of knowledge, experience or sophistication. One of the strengths of this dictionary is its scope. It provides a basic reference for people who are expanding their horizons into unfamiliar graphical areas. To professional educators, it provides a basic understanding and clarity for students.

Aa

A paper sizes — one of five categories of ISO paper sizes.

A/D — analog to digital conversion.

A/W — an abbreviation for artwork.

A2 envelope — an envelope used for note cards, as opposed to a #10 envelope, which is a business-size envelope used in the U.S.

A4 paper — an ISO paper size 21 x 29.7 cm (8.27 x 11.69 inches) used for letterhead. Closest international equivalent to 8 1/2 x 11 inch stationary.

AA — (1) in composition, author alteration. Any change or correction made by a customer. (2) proofreader's mark(s) showing such an alteration. (3) auto answer. An indicator light that tells you the modem is ready to connect to the phone line.

ABA (American Booksellers Association) — a trade association of publishers and booksellers.

ABI (Advance Book Information form) — a form filed by publishers with R.R. Bowker/Reed Reference Publishing, which is used to list the book in directories such as Forthcoming Books and Books in Print.

abnormal color vision — defective color vision ("color blindness"), which may take the form of protanopia (red and bluish green confusion); protanomalous (deficient in red response for some color mixtures); deuteranopia (red and green confusion); deuteranomaly (deficient in green response for certain color mixtures); tritanopia (blue and yellow confusion); and monochromatism (no discrimination of hue and saturation).

abrasion — markings on film, printed product, etc., caused by rubbing of one surface against another.

abrasion resistance — the ability of any substrate (foils, plastics, etc.) or ink to resist deterioration from rubbing. *Related Terms:* rub fastness; scuff resistance.

abridgement — alternate term for a digest format publication.

absolute address — a specific memory location or region in computer memory.

absolute insert — the nomenclature given to an accordion-folded mailing piece.

absolute resistance — the natural or artificially induced ability of a material to resist abrasive defacement.

absolute URL — an Internet address which includes in the protocol a complete network location of a file or server. Example: http://www.prof-lyons.com is an absolute URL. This is used instead of a relative URL which simply describes the path to a file within the same domain.

absolute value — a number without negative or positive deliniation; either a positive or a zero.

absorption — (1) regarding light, the extent to which paper diminishes light passing through it or reflecting from it. Selective absorption of the range of white light wavelengths produces colored light. (2) regarding fluid and vapors, the extent to which a material, such as ink, soaks into paper instead of drying on the surface. (3) in physics, the taking up of light energy by matter and its transformation into heat.

abstract — a brief summary of an article, book or other larger report or document.

accelerated graphics port (AGP) — a dedicated interface on the PC platform to enable high-performance graphics and full-motion video. Usually positioned between the PC's chip set and graphics controller, it significantly increases the bandwidth available to a graphics accelerator.

accelerator card — an expansion card (for computer's bus) that contains another processor that shares the work normally performed only by the computer's main microprocessor. *Related Term:* AGP.

accelerators — chemicals used to speed up the developing process of photographic film.

accent — a mark placed over, under, or through a letter to show a different pronunciation.

accent marks — *Related Term:* diacritical marks.

access — (1) to get into a computer system, dial-up service, or network, by dialing a phone number, logging on a network such as the Internet or retrieve information from a storage device (internal memory, disk, tape). (2) a relational database program from Microsoft.

access control — methods used to protect confidential computer data or logging on a computer network without authorization.

access time — the time required to either read from or write to a digital file. Technically speaking, it is the time elapsed between the instant a computer calls for data from a storage medium (like a hard disk or CD-ROM) and the instant the data is delivered. This can take minutes or microseconds if the data comes from a computer's DRAM.

AccessWatch — a UNIX utility for the World Wide Web that provides extensive daily usage activity reports for a Web site or an entire server. It is a regularly updated summary of server hits and accesses in both text and graphic formats

accordion envelope — *Related Term:* expansion or gusset envelope.

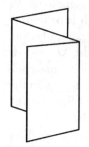

ACCORDION FOLD
A fold with too many names, it is one of the most utilized folds in industry; the number is optional and can be as extensive as required.

accordion fold — two or more parallel folds with adjacent folds in opposite directions. *Related Term:* concertina fold, Z-fold.

account — the standard PPP dial-up account of an Internet service provider. If your e-mail address is in the format yourname@ix.setcomx.com, you have such an account.

account holder — the person owns the credit card that an account is billed to.

accumulator — a computer mem-ory location where claculated results are retained.

ACDSee — a graphic image viewer and editor for Windows 95/98/NT which supports a number of graphic files formats including GIF and JPEG (JPG).

acetate — (1) in art and prepress, thin, flexible sheets of transparent plastic used to make overlays, allow the artist to write instructions or indicate where subsequent colors are to be placed or to provide a base for high-contrast film and peel-off masking materials. (2) a family of solvents also known as estars. In flexography, estars are used to clean the back side of printing plates and the mounting surface of plate cylinders. *Related Terms:* overlay; estar.

acetate proof — any one of a number of different proofing systems which use the overlay technique.

acetic acid — a common chemical used in dilute solution with water as a stop bath for photographic film. *Related Term:* short stop.

achromatic color — a color with no saturation, such as white, black, or gray additive color primaries red, green, and blue (RGB). These colors are used to create all other colors with direct (or transmitted) light (for example, on a computer or television screen). They are called additive primaries because when pure red, green, and blue are superimposed on each other, they create white. *Related Term:* subtractive color primaries.

achromatic color removal — alternate term for gray component replacement (GCR).

achromatic reproduction — a technique used in color separation to reduce the amount of additive primary inks used and emphasize black in the shadows and darker areas. *Related Terms:* gray component replacement; undercolor removal.

acid-free paper — a paper made from pulp having little acid so that it resists deterioration from age. Also called alkaline paper, archival paper, permanent paper, and thesis paper. *Related Term:* archival paper.

ACK (acknowledge) — an acknowledgment (approval) of a received data packet from its point of contact. When an unauthorized modification is received, the request will be rejected and a message will be sent back as a negative acknowledgment, or NAK. *Related Term:* NAK (No Acknowledgment).

acknowledgment — (1) a confirming receipt of an order, including specification, issued prior to production; (2) the front part of a book or publication expressing appreciation to people, organizations, etc., who gave the author support in the writing effort.

ACPI (advanced configuration and power interface) — a proposed power management standard from Microsoft, Intel, Toshiba and several others that would let the PC control power to its peripherals, as well as consumer devices to which they were connected. Also, the peripherals could use the interface to turn on the PC; it is a potential player in the home automation arena.

acquisition — the process of copying analog original data to a digital form.

acquisition editor — a person in a publishing house who is responsible for acquiring new titles.

Acrobat — a software solution from Adobe Systems to electronically communicate files with absolute definition, across platforms, networks and between dissimilar reproduction devices. Based on PostScript, it is becoming a standard for media communications between agencies, printers, customers, etc. It has several components including a freely distributed reader to enable non-Acrobat users to reproduce Acrobat files with actual fonts, colors, etc.

acronym — a word created from the initial letters or syllables such as EPROM (Erasable, Programmable, Read Only Memory.)

across the grain — *Related Term:* against the grain.

across the gutter — *Related Term:* crossover.

across the web — the reference to a web imposition where pages run at right angles to the direction of the web flow.

actinic — (1) describing the ability of light to produce changes in materials exposed to it, such as photographic emulsions. (2) the range of color of light emitted from vapor lamps, arc lamps, photo flood bulbs and other artificial light sources that will expose sensitized photographic films, paper or printing plates. *Related Terms:* actinic density; paper; archival paper; NAK (no acknowledgment.)

actinic output — energy that activates or hardens light sensitive coatings; mainly consisting of shorter wavelengths of the visible spectrum.

action paper — *Related Term:* carbonless paper.

activate — accelerating a chemical process or activity. *Related Terms:* accelerator; buffer.

activator — the term applied to the single solution developer for materials used in the diffusion transfer process.

active channels — a push delivery system from Microsoft when using Internet Explorer 4.+. The technology is based on developers who write and upload a channel definition file to their web site; any new content delivered automatically to users as often as the site is updated. Both developers and subscribers can control the update frequency of channels, subchannels and items subscribed to. DHTML as well as RealAudio and streaming video are often used to make the offerings more interactive and interesting. *Related Terms:* channel definition format; DHTML; push.

active matrix — in computer monitors, a type of LCD (liquid crystal display) that offers higher quality than a passive matrix display. Most flat panel displays for laptop computers are created by laying diodes over a superfine grid of wires that selectively activate the diodes by applying current to various points around the grid to make picture elements. If enough of these elements are activated, you per-

ceive an image. Active matrix displays use transistors to keep their diodes in an off or on state. This makes them brighter and produces better color, but they are much more complicated, unlike their passive matrix cousins which rely on the diodes' own persistence. The more extensive technology required to build active displays makes them far more expensive, but prices are declining as the industry masters the manufacturing techniques. *Related Terms:* diode; flat panel display; LCD; passive matrix; persistence; transistor.

active pixel region — on a computer display, the area of the screen used for actual display of pixel information.

active terminator — a terminator that can compensate for variations in terminator power supplied by the host adapter through means of a built-in voltage regulator.

active video lines — all of the video lines not occurring in the vertical blanking interval. True for all world formats including NTSC, PAL, and SECAM.

ActiveX — a set of technologies from Microsoft that provides tools for linking desktop applications to the World Wide Web. By using a variety of programming tools, including Java, Visual Basic and C++, developers can create interactive web content. ActiveX technology allows users to view files in ActiveX compatible software directly in a browser. *Related Terms:* Java; Visual Basic.

actual basis weight — the true basis weight of paper, which may differ by as much as 50% from the nominal basis weight stated on the packaging or in the price book.

acutance — the objective means of measuring a photographic files which shows a sharp line of separation between adjoining areas receiving different exposure. Acutance is referred to many times as the edge sharpness within the image.

ad banner — a Web page advertisement with links to an advertiser's site with a standard size of 468 pixels wide by 60 pixels high as set by the Internet Advertising Bureau (IAB). The cost is anywhere from free to in excess of $20,000 per month depending upon website visitor numbers.

ad hoc workflow — the least formalized workflow structure with no predefined outcome. Work is generally conducted in a narrowly defined group and there is little standardized documentation. Requests for task support usually come with specific direction and workflow in this environment is usually limited to tracking task assignments and due dates.

ad server — an extremely sophisticated program or a server which manages and maintains advertisement banners for a Web site or collection of sites. They are capable of keeping track and reporting site usage statistics on users. Advertisements can then be targeted toward specific types of individuals based upon cookie information queries. They also route banners so a user's repeated accesses will not repeat the same ad each time they visit.

adaptive compression — data compression software that continually analyzes and compensates its algorithm, depending on the type and content of the data and the storage medium.

ADC (analog-to-digital converter) — the conversion of data or signals generated or stored in analog format to the on-off digital format used by computers.

addendum — material supplementary to the main body of a book or other publication and printed separately at the start or end of the text or inserted as an additional sheet. Often used to correct errors or omissions.

additional servers — any name servers, aside from the primary and secondary name servers listed on the Domain Name Registration Agreement, that are used to connect a particular domain name to its corresponding Internet Protocol (IP) number(s). *Related Terms:* name server; primary server; secondary server; domain name registration agreement.

additive color — color produced by combining red, green, and blue light in varying intensities. Computer monitors use additive color, while the printing process uses subtractive color. This causes inconsistency between what a designer sees on the monitor and what comes off the prin-ting press. *Related Terms:* additive color process; additive primaries.

additive color process — *Related Term:* additive primaries.

additive plate — a presensitized printing plate, usually coated with a diazo emulsion; lacquer must be added to the surface to see the image.

additive primaries — red, green and blue light that produce white light when equally mixed. *Related Terms:* CYMK; subtractive primaries; secondary colors; color separation; video; RGB.

additives — (1) in papermaking, materials added to prepared pulp to improve the quality of the paper. (2) in printing, compounds that control such ink characteristics as tack, workability, and drying time.

additivity failure — a common condition of printing ink on paper where the total density of the overprinted ink films is not equal to the sum of the individual ink densities.

address — (1) in computing, a number that specifies the location of a single byte of ram data. (2) on the Internet, a method of directing data, the combination of letters, numbers, and/or symbols that will let you send e-mail to a particular person or organization. There are two types of addresses in common use: e-mail and IP or Internet Protocol addresses.

ADF (automatic document feeder) — a device which is attached to the top of a flatbed scanner or copier to process multiple pages in a single session without manually placing each page in the input position.

adhesive binding — a glue fastening of printed sheets or signatures. *Related Term:* perfect binding.

adhesive bleed — the activity when adhesive on self-adhesive labeling materials react with the material onto which it is mounted. Can take several forms, including lack of adhesion, color changes, etc.

adjacent color effect — the visual influence of a color area on an adjacent color. This effect is especially strong when the adjacent color area is relatively large and has high saturation.

adjusted exposure — the increase or decrease of exposure due to bellows extension, or the use of filters. Long exposures resulting in reciprocity effects require adjusted exposures.

ADDITIVE COLOR THEORY
The additive color theory primaries are red, green and blue. When combined equally, they make white light. Any two combinations make one subtractive color theory primary: cyan, magenta or yellow.

adjustments — using swings and tilts to the front and read standards of a view camera to align perspective and to bring subject or portions of subjects into proper focus.

administrative contact/agent — the administrative contact/agent is an individual or role account authorized to interact with the ISP on behalf of the domain name registrant. The administrative contact/agent answers nontechnical questions about the domain registration and the domain registrant. It is usually advised that the administrative contact/agent be the registrant or someone from the registrant's organization. *Related Terms:* role account; registrant.

Adobe PostScript — the preeminent computer imaging page description system used within the graphic arts and printing industries.

ADPCM (adaptive differential pulse code modulation) — a compression encoding method for sound data files that requires less storage space than pulse code modulation format used by WAV, AIFF and CD audio files. There are several varieties available. One is used on the mini discs to cram more data onto a smaller platter; another is used by Microsoft as part of Windows 9x's audio codecs. *Related Terms:* AIFF, codec, PCM, WAV.

ADSL (asymmetric digital subscriber line) — a standard phone line transmission system designed to provide high-speed data communications. ADSL technology can deliver upstream (from the user) data speeds of 640 KBPS or 10x ISDN rates and downstream (to the user) speeds of more than 6 MBPS. Since it uses the portion of a phone line's bandwidth not utilized by voice, simultaneous voice and data transmission are possible. Wide availability of ADSL is not expected before the year 2001. *Related Term:* ISDN.

advance — money paid to an author, usually at the time a contract is signed, that is a portion of expected royalties that will be paid to the author once the book is published. Originated from the phrase "advance against royalties."

advance copies — in publication and bindery work, a limited number of copies sent quickly for review before the total order is complete. Often used to provide copies to reviewers and critics in advance of public sale.

advanced bar code — an eleven number bar code used by the postal service. It is the ZIP+4 numbers, plus the last two numerals of the street address. It is designed to facilitate machine reading of mail to expedite processing and delivery.

advertisement — a commercial space. Areas in newspapers and magazines which are contracted for the sole purpose of promotion or a product, activity, event, etc.

advertising agency — a firm or company which designs, produces or generates information for public dissemination, most often in the form of advertisements.

advertising paper — *Related Term:* uncoated book paper.

afterimage — the sensation that occurs after the stimulus causing it has ceased. Because of cone fatigue, the colors of the afterimage may be complementary to those registered initially. *Related Term:* persistence of vision.

afterword — Part of a book's back matter in which the author or publisher offers parting remarks to the reader.

against the grain — the right angle to the grain direction of paper. Also called across the grain and cross grain. *Related Term:* grain.

PAPER GRAIN DIRECTION

AGAINST THE GRAIN

GRAIN
Grain direction is the direction of travel of the Fordrinier wire on the papermaking machine.

agate — a unit of measurement used in magazines to calculate column space. 14 agate lines equal 1 inch. (Agate was originally the name of a 5½ point type.)

agate copy — a general term applied to small type composition, such as classified advertisements, etc.

agate line — a measurement of column depth, used to calculate advertising space. There are 14 5½ point agate lines to the inch.

agent — A person who represents an author or other artistic talent by showing the author's manuscript or artist's photos and credits to prospective publishers or employers and who handles contract negotiations, helps to sell subsidiary rights, and manages the author's business and financial transactions. *Related Term:* artist's representative.

agitation — the movement of fresh solution back and forth over film during chemical processing.

AIFF (audio interchange file format) — an audio file format developed by Apple for storing high-quality sampled audio and musical instrument information. Played by a variety of downloadable software on both the PC and the Mac, it is also used by in several professional audio packages. *Related Terms:* ADPCM; PCM; sound; TrueSpeech; VOC; WAV; waveform.

AIIM (Association for Information and Image Management) — in production management, the industry trade association for imaging and micrographics technology.

AIM — Automatic Identification Manufacturers, Inc. The trade association of manufacturers of automatic identification systems.

aim point — usually associated with press makeready, the starting value of ink, prior to color adjustment.

air — an amount of white space in a layout. *Related Term:* negative space.

air brush — (1) a compressed air driven sprayer tool shaped similar to a pen and used to generate a fine mist of ink or paint. Traditionally used to retouch photographs and create continuous tone illustration, the technology is being replaced by computerized systems and photomanipulation software. (2) the term applied to retouching operations, either with the airbrush or computer. Sometimes spelled as a single word.

air bubble packaging — the use of double layered polyethelene or other polymer plastics in a configuration where individual cells of air are sealed between them, providing a cushion for delicate materials during shipping.

air knife coater — a device that applies excess coating to paper and then removes the surplus by striking the fluid coating with a flat jet of air, leaving a smooth, metered film on the paper.

air mail — the traditional fastest method of sending a letter. Domestic mail is routinely sent via this method today, but foreign rush mail is normally sent in special envelopes designed to highlight their importance.

AIR MAIL ENVELOPE
The air mail envelope is characterized by a border of parallelograms, front and back, usually in blue or red and printed on a pale blue stock.

airbrush drawing — an illustration made with an airbrush.

airbrushing — the retouching of prints or artwork by dyes or pigments sprayed on with high pressure air. *Related Term:* airbrush.

airmail paper — a somewhat antiquated term referring to any letter and envelope paper with a basis weight of 30# (40 gsm) or lighter and usually light blue to signify travel by airplane.

ALA — American Library Association. The largest library association in the United States.

albertype — *Related Terms:* collotype; artotype.

album format — an alternative term for a horizontally oriented project, i.e., the page width exceeds the depth.

alert box — a box that appears on the monitor screen that reports an "error message or warning message."

alias — (1) a name, usually short and easy to remember, that is translated into another name, usually long and difficult to remember. Commonly used in the UNIX realm to "abbreviate" verbose commands. (2) also used similarly in the context of electronic mail. Mail aliases are the basis of many electronic mailing lists.

aliasing — the unrealistic visual effect on a computer screen where images have jagged edges or stair-stepped appearances along what is supposed to be a smooth curved surfaces and/or diagonal lines on the screen. Sometimes called the jaggies. Anti-aliasing software techniques are used in imaging systems and graphic manipulation software such as Adobe PhotoShop to make these curved edges or diagonal lines look smooth and continuous. *Related Terms:* anti-aliasing; distortion; artifact.

alignment — (1) the horizontal positioning of characters. In base alignment, characters rest on a common horizontal line, excluding descenders and irrespective of aesthetics and design proportions. (2) to line up typeset or other graphic material as specified, using a base or vertical line as the reference point. (3) in bar code decoding, the relative position of a scanner or light source to the target of the receiving element. *Related Terms:* justify; right justified; left justified.

alkaline paper — *Related Terms:* acid-free paper; archival paper, considered to be long lasting and does not yellow with age.

all caps — the indication that type is to be set entirely in majuscules or upper case characters. *Related Terms:* upper case; majuscule; miniscule; capitals; caps and small caps.

alley — space between columns on a page. *Related Term:* gutter.

allocate — (1) in computing, the assignment of a resource, such as a disk file, or a diskette file, or time to perform a specific task. (2) in printing, specifying the quantity of a brand of paper that a manufacturer will supply to distributors and customers until a specified date. Paper on allocation is rationed to customers who may not receive more than they are allotted.

all-rag paper — a paper made using only fibers from cotton or linen, as compared to wood fibers. *Related Terms:* rag content; ground wood paper.

alpha blending — an additional channel in computer graphics, where each pixel has three values to enable interpolation of hue and shade.

alpha test — the preliminary evaluation of a project or product; often nonpublic. Usually followed by beta testing which will broaden the exposure of the product to a wider and often noncorporate user.

alt key — a key on PC computers which functions in conjunction with other keys to expand the command functionality of the keyboard. The activity initiated is often software-specific and can be used to access extended character sets, perform automated operations through scripting, etc.

alt text — the fleeting text that appears before an image is loaded on a Web page. The site author codes an HTML alt tag when building a Web page to identify the related graphic or say anything they want.

alteration — any change in a typesetting or printing job once specifications have been agreed on or production has begun. *Related Term:* AA.

alternate characters — multiple versions of characters, usually in display typefaces, to allow greater variety or personality to copy. Often refers to swash characters, which usually over or under-hang adjacent characters with curve-like flourishes. *Related Terms:* ligature; diphthong.

AM screening — amplitude-modulated screening. Another name for traditional halftone screening. *Related Term:* FM screening.

amalgamate — to combine two or more orders or specifications into a single work to increase productivity, lower unit prices, etc. Sometimes called consolidation.

Amberlith — Ulano trade name for its orange masking material. *Related Term:* Rubylith.

American National Standards Institute (ANSI) — an organization charged with coordinating standards development for measurements, terminology, safety, etc., for industry. Standards approved by this organization are often called ANSI standards. It belongs to the International Standards Organization, and has helped graphic communications by formulating standards for proper viewing conditions, electronic data interchange and instrument calibration.

American Standard Code for Information Interchange (ASCII) — a standard character-to-number encoding widely used in the computer industry. *Related Term:* ASCII.

amortization — the process of distributing equipment costs across time periods or jobs.

ampere — in electronics, a unit used in measuring current volume as it passes through a conductor.

ampersand (&) — the ampersand was originally a ligature for "et," the Latin phrase for and, expressed as "et cetera," which gradually evolved to its present form. Used in place of the word "and."

amplitude — the height and depth of the peaks and valleys of a wave form. By measuring the difference and then dividing by a factor of two, we get the amplitude or strength of the wave. The larger the amplitude, the stronger the wave. For instance, the brightness of the video signal at any given point in time. In audio, the overall volume. *Related Terms:* frequency; phase.

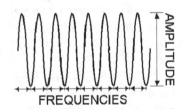

analog — a continuously varying electrical signal in the shape of a wave, transmitted electronically; a form of information which is represented by continuous wave forms that vary as the source varies. Uninterrupted information that is not sampled at a given rate, but is continuously variable. *Related Term:* digital.

analog computer — a nondigital computing device which represents data in a modulated state by monitoring voltages, amperage, cycles, etc.

analog controls — analog controls are dials, knobs, and similar mechanisms. They have a potentially infinite number of settings that cannot be stored in a profile and recalled. This means that analog controls require readjustment whenever you change mode (louder, brighter, etc.). *Related Term:* digital controls.

analog proofs — proofs created by photomechanical mechanisms, i.e., film positives or negatives. Most use nonelectronic or digital image generation.

analog recording — traditional nondigital recording. The process is based not on presence of bits and bytes, but on voltage changes caused by magnetic inducers, in the case of audio or video tape, or induced voltage flux caused by a needle moving in a record track.

analog transmission — AM radio; conventional TV, etc. A transmission system based on the varying amplitude of wave forms. *Related Terms:* FM; frequency modulation.

analogous — a color harmony that results from using two colors next to each other on the color wheel.

analog-to-digital — the conversion of an analog (continuous) signal into its digital (discrete) equivalent. Such a conversion measures the input voltage and outputs a digitally encoded number corresponding to that voltage.

anamorphic — (1) nonproportional enlargement or reduction on the X and Y axis. (2) a lens on a graphic arts camera that induces controlled distortion in negatives by reducing copy in one dimension while allowing the other dimension to remain unchanged. *Related Term:* proportional.

anchor — another name for hyperlinks—the underlined words or phrases you click on in World Wide Web documents to jump to another page. *Related Term:* hyperlink.

aniline ink — a fast drying ink compound often used in flexography.

aniline printing — the original name given to flexographic printing because alcohol based aniline dyes were used as inks. *Related Term:* flexography.

Anilox rollers — a steel or ceramic ink metering roller used in flexography. Its surface is engraved with tiny, uniform cells that carry and deposit a thin, controlled layer of ink film onto the plate. In flexographic press work, Anilox rollers transfer a controlled ink film to the printing plate (or rubber covered roller) which in turn prints to the web substrate. Anilox rollers are also used in remoistening glue units and to create "scratch-and-sniff" perfume ads.

annotation — in design and production, the ability to attach notes to graphics or text. Useful for clarifying documents or editing images.

annual — (1) a term often applied to a publication cycle of once each 12 months. (2) a yearbook.

anodized plate — an offset printing plate with a specially treated surface to reduce wear during printing.

anonymous FTP — the file transfer protocol that enables anyone to download files from a properly configured FTP server. Users without accounts can access files by entering the username anonymous, along with their e-mail address as a password. Anyone can access files that have been set up for anonymous FTP, so Webmasters need to use it only for files meant for general consumption. *Related Term:* FTP.

anonymous posting — any message posted to a newsgroup or e-mail discussion group without the identity of the person sending it.

anonymous remailer — an anonymous remailer is a computer that strips away identifying information (such as your e-mail address) before passing your message on to an e-mail address or a news group. When you send mail using a remailer, the From: field of your message becomes a made-up address like 12xyx@re-mailer.com. There are two kinds of services called anonymous remailers. The first is truly anonymous: no one anywhere knows your identity. The second, called pseudo-anonymous, knows your identity and can be forced in a court of law to reveal it. Truly anonymous services are often free, but are often difficult to use. Many pseudos charge a fee, but are a little more user-friendly.

anstigmat—a type of photographic lens from which all astigmatism errors have been removed. All high quality and professional lenses for 35mm cameras up to large format are most often anastigmatic.

antenna — a structure which transmits or receives electro-magnetic signals.

anthology — A collection of writings by one or more authors.

anti-aliasing — in computer graphics, the smoothing of the jagged, "stairstep" appearance of graphical elements. *Related Term:* jaggies.

antifogant — a chemical which is added to developing solutions to inhibit fogging of sensitized photographic film and paper, due to outdated materials, extended developing or when safelights are too intense.

anti-offset spray (or powder) — a misting device on some presses which places small amounts of a talc-like powder on finished sheets to assist in eliminating setoff of the image to the back of subsequent delivered sheets. *Related Term:* setoff.

antique finish — rough paper finish common in book and cover papers; handmade papers.

anti-skid varnish — in packaging, a special material sprayed onto cartons to prevent slippage in transit.

Aperture

aperture — (1) in photography, opening behind the lens whose size may be adjusted to control the amount of light reaching the film. Aperture settings are called f/stops. (2) in screen printing, the open space between the threads of screen printing fabric. *Related Terms:* iris; lens opening; f/stop; aperture percentage.

aperture grill — a series of wires stretched vertically down the inside of the CRT (cathode ray tube) or a perforated metal plate (shadow mask) used to direct the beams from the electron guns at the back of the tube to the appropriate phosphor on the inside of the face of the display tube. *Related Terms:* CRT; electron gun; shadow mask; Trinitron.

aperture percentage — in screen printing, the percent of the screen constituting the open or imaging area. *Related Term:* aperture.

aperture priority camera — a type of camera on which the photographer sets the aperture and the camera automatically adjusts the shutter speed.

apex — the point of a character where two lines meet at the top; an example of this is the top point of the letter A. *Related Term:* vertex.

API (application programming interface) — a broad and generic term for any software program that carries out a useful task; a series of functions that programs use to make the operating system perform as designed. Windows APIs, for example, open windows, files, and message boxes, as well as perform more complicated tasks, by passing a single instruction. Windows has several classes of APIs that deal with telephony, messaging, and other issues. *Related Terms:* 3D API; MAPI.

apochromatic — a lens used for color separation work. This lens will bring the red, green, and blue light bands of the spectrum to the same point of focus.

apparent color — depends upon which wavelengths reach the eyes. In printing, the color and finish of the paper have an important effect on the apparent color of the ink image. For example, if transparent red ink is printed on a paper that is a color other than white, the apparent color the ink will be black, not red.

apparent trap — *Related Term:* trapping.

appendix — that portion of a book appearing at the back which contains supplemental information for the main text including lists of resources, tables, or other reference material. *Related Term:* back or end matter.

Apple desktop bus (ADB) — an obsolete and relatively low speed serial port (bus) from Apple Computer Company to which the keyboard, mouse, tablet (digitizer), hand controls, or other specialized accessories are attached.

applet — a mini-application, referring to simple, single function programs that often ship with a larger product. Examples are calculators, file managers, notation pads, etc. *Related Term:* application.

application — an application is a major program that performs specific tasks such as graphics, word processing, drawing, page layout, database, disk maintenance, etc. This contrasts with an operating system, such as MacOS or Windows, which manages how your computer performs tasks, and "runs" these applications. *Related Terms:* applet; utilities.

application translation — in industrial computing, the process of changing the sequence and/or logic of an existing software package to facilitate use with smaller terminal displays, radio frequency ID sensors, or speech recognition, etc.

apprenticeship — (1) a specified period of time in which a person must work to learn a given technical trade. (2) the practice of learning a trade by working under the direction of a skilled worker.

appropriateness — the measure of how well a typeface is suited for a specific printed product. *Related Terms:* availability; readability; suitability; affordability.

apron — (1) additional white space allowed in the margins of text and illustrations when forming a foldout. (2) protective clothing which normally covers the front chest down to above the knees. Utilized in production areas where chemicals, inks and other materials may stain normal clothing.

aqueous coating — a water based coating applied as

an alternative to varnish by a printing press to a substrate to enhance and protect the previously printed image. May be applied in-line or as a post-press flood procedure. They are available in a variety of finishes from dull to gloss and, unlike conventional petroleum based varnishes, readily accept ink-jet addressing and other post press processing, and is fast-drying, durable and nonyellowing.

Arabic numerals — numerals used in modern languages which include ten figures: zero and numerals 1 through 9. They are so named because they originated in the Middle East. *Related Terms:* roman letters; roman numerals.

ARC (Augmentation Research Center) — a lab at the Stanford Research Center in Palo Alto, credited with developing the first practical Graphical User Interface tools such as the mouse, icons, and the first hypertext system. These tools were expanded upon by researchers at Xerox PARC and two decades later became the basis of the Apple Macintosh and its operating system. The center also developed ideas about teleconferencing, e-mail, and tools for workgroup processing.

arc lamp — a device that produces an intensely bright light from a sustained discharge of electricity across a gap in a circuit or between two electrodes (slightly separated carbons). The carbon arc lamp was once commonly used for exposing offset plates, engravings and photo resists, but, because it was a fire hazard, emitted toxic fumes and its light emissions fluctuated greatly, it has been replaced by quartz-halogen and pulsed-xenon-light sources.

Archie — an antiquated Internet file finder. Archie is a system that helps you find information anywhere on the net. Archie started life as an indexed directory of files from archives. Found files are retrieved using FTP (file transfer protocol). *Related Terms:* FTP; gopher; VSL; WAIS.

architecture — the basic configuration of a device such as a printer, computer, etc. How the parts work together to obtain desired results.

archival paper — paper produced without the use of acids and whose surface characteristics are nearly chemically inert. Designed to prevent yellowing and cracking and have a long life, up to one hundred years. *Related Terms:* acid-free paper; archival quality.

archival quality — (1) a long lasting, non-acetic paper used increasingly to print books and other important records and documents. (2) technique for printing books, photographic prints, testaments and records intended to last 100 years or more. (3) the long-term storage of image information on photographic, magnetic, or other media.

archival storage — inexpensive, long-term storage, kept in compressed form for reference but infrequent access.

archive — (1) in business, the process of creating archival storage. (2) the reference to the storage or its space, once created.

area composition — term typically applied to imagesetters which prepare data in such form that all or as many elements of the final page as possible are imaged in place. This reduces or eliminates paste-up. Area composition output falls somewhere between galley output (requiring extensive paste up) and full page makeup with all elements in place.

argument — reference or checking values used in analysis, typically to computer programs in the process of calculating results based on a formula or equation.

arm — that part of certain typographic characters which projects horizontally or slopes upward.

ARPA (Advanced Research Projects Agency) — the agency of the Department of Defense that developed the Internet. Originally called DARPA, it is sometimes referred to as ARPANet.

ARPANet — precursor to the Internet. The U.S. military was desperately afraid of a nuclear attack in the 1960s and computer scientists in the Advanced Research Projects Agency attempted to design a secure network that would connect military bases and other military agencies. They created a system based on linking distant computers with a newly developed set of protocols called TCP/IP. This new network became ARPANet. In the early 1980s, ARPANet began to evolve into what we now call the Internet. *Related Term:* TCP/IP.

array — a grouping of devices connected to act as a single entity. (1) storage devices, such as a hard disk array. (2) CCDs or CMOS elements arranged linearly or by area as receptors for digital cameras or scanners.

array scanner — a scanner which performs scanning operations using linear arrays, sensors which are in a fixed line and move across the object, and are uniformly activated in se-quence to ren-der a total image. Often called CCDs, CMOS or videocom devices.

array, area — a group of light sensitive CCD or CMOS elements arranged in horizontal and vertical rows to form a matrix or grid. Used in digital cameras as the image sensing device, all elements are activated simultaneously, resulting in a total image in one exposure cycle, making it ideal for moving objects. *Related Term:* array, linear.

COPY   SENSOR ARRAY

AREA ARRAY
Pictured is an imaginary flatbed scanner. In place of a moving linear (or row) array such as is most common today, a single snapshot would capture the entire image instantly.

array, linear — a group of heat generating or light sensitive imprinting or recording elements often arranged in a single line. Used as a scanner image sensing device and as a print head in some nonimpact printing architectures. The system functions or cycles continuously as either it or the subject matter is in motion, resulting in a series of exposures or small reactive printing areas. The CPU or other control device determines both the construction of the individual exposures to assemble a total picture and also the temperature of the individual elements when used in a printing device. *Related Term:* array, area.

arrow keys — cursor movement control keys such as those found on a typical computer keyboard. Usually up, down, right and left but occasionally diagonal.

art — (1) original copy, whether prepared by an artist, camera, or other mechanical means. Loosely speaking, any copy to be reproduced. (2) In general art usage, all matter other than text material, e.g., illustrations and photographs. *Related Term:* artwork.

art board — *Related Terms:* mechanical; base art.

art director — an employee, often of an advertising agency, who supervises creation and preparation of copy for reproduction.

art paper — a smooth coated paper obtained by adding a coating of china clay compound on one or both sides of the sheet.

artifact — a visible defect in an image, usually caused by bad analog signal quality or digital video compression or other limitations in the input or output process (hardware or software). They all have one thing in common: they impair picture quality and usefulness by the placement of random and unwanted stray pixels that don't belong in the image. *Related Terms:* bandwidth; pixel.

artist's representative — person who handles marketing and other business matters for designers, illustrators, and photographers. *Related Term:* agent.

artotype — *Related Terms:* collotype; albertype.

artwork — images, including type and photos, prepared for printing. Some printers include type in "artwork," others don't. *Related Terms:* copy; art.

ASA (American Standards Association) rating — a formalized measure of photographic film speed. Outdated terminology. *Related Term:* ISO.

ascender — The portion of a letter that rises above its x-height (the height of a lowercase "x" in a particular typeface) such as k, b, and d. *Related Term:* ascender height.

ascender or cap line

ascenders are letter parts which extend above the mean line

mean line

base line

CAP HEIGHT
Measured from the bottom of the x-height (base line) to the ascender line.

ascender height — (1) referring either to that part of a lower case letter which projects upward, above the x-height of the letter, or (2) to the actual letter itself (b, d, f, h, k, l, or t).

ASCII — the abbreviation for American Standard Code for Information Interchange: usually pronounced "Askee." An eight-level code for data transfer adopted by the American Standards Association to achieve compatibility between data services. An ASCII file contains only plain text and basic text-formatting characters such as spaces and carriage returns; it does not contain graphics or special character formatting. The ASCII character set of a microcomputer usually includes 256 characters or control codes. For example, the letter "A" is stored as ASCII 65, "B" as 66, "a" as 97, "b" as 98, etc. Some ASCII "characters" do not display as characters on the screen, but instead control the display in other ways. ASCII 8 is the backspace, 10 is the line feed, 13 is the carriage return, and 27 is escape. Other ASCII characters, consisting of letters from non-English alphabets and graphic symbols, fall in the range from ASCII 128 to 255. These "upper ASCII" characters will not always display or print in consistent ways. The most consistent ASCII characters are those that can be seen on the keyboard; they fall in the range from ASCII 32 to 127 and are called "plain ASCII." A plain ASCII file can be read by just about any program. Most operating systems use the ASCII standard, except for Windows NT, which uses the suitably larger and newer Unicode standard. *Related Terms:* American Standard Code for Information Interchange; ANSI.

ASCII art — a type of art which uses low-ASCII characters. Its sophistication runs the gamut from simple little symbols to complicated random-dot 3D stereogram images. *Related Terms:* smileys; emoticons.

ASCII-armored — a feature of Pretty Good Privacy (PGP), which "encases" an encrypted message in ASCII, allowing the message to be sent via e-mail as a regular message. *Related Terms:* encryption; pretty good privacy (PGP); public key encryption (PKE).

ASIC (application specific integrated circuit) — a custom microchip designed for a specific application. The circuitry does not begin as a concept which must be built. Most primarily involve integrating functions from a library and of pre-engineered plans.

as-is — (1) a term used to describe the current condition or process as in a road or a business process.(2) a sometime condition of sale: Sold "as is," indicating that the seller assumes no responsibility for functionality, completeness, durability or reliability of the product.

aspect ratio — (1) in computer graphics, the ratio of width to height of a screen or image frame, (2) in code reading, the ratio of bar height to symbol length. (3) in video, the true proportion of an image size given in terms of horizontal length (first number) versus the vertical height (second number). Standard NTSC television is an aspect ratio of 4:3; HDTV (high definition television) has an aspect ratio of 16:9; 4" x 5" sheet film is an aspect ratio of 5:4.

asphaltum — a tar-like material used as an acid resist in the preparation of many gravure cylinders and some lithographic image carriers.

assembly — (1) finishing operations that bring all elements of a printing job together into final form; common assembling operations are gathering, collating, and inserting. (2) assembly and manipulation on a computer screen.

assembly language — a unique language based on

computer type. Very low level of sophistication where each statement directs only a single action. Typically such things as store, erase, copy, etc.

asset tracking — in management, a bar coding system that gives a company the ability to consistently locate each asset.

associate — (1) in web printing, the overlaying of several webs on a web press prior to entering the former and nips. (2) in business, an employee; sometimes a minor partner. *Related Term:* marry.

asterisk (*) — (1) in publishing, reference mark used in the text to indicate a footnote. (2) in the digital world, a wild card entry in some computer operating systems to indicate missing letters or words. *Related Term:* wild card.

as-told-to — a book produced by a writer in collaboration with a nonwriter, usually a celebrity.

asymmetrical compression — a compression system which requires more processing capability to compress an image than to decompress an image. It is typically used for the mass distribution of programs on media such as CD-ROM, where significant expense can be incurred for the production and compression of the program but the playback system must be low in cost.

asynchronous — the characteristic of any operation that is independent of a master clock or time signals; also refers to information that is sent or exchanged independent of any specific time.

asynchronous communication — this term describes how your computer uses a modem to connect with other computers. In the days of teletypes and dumb terminals, computers sent data synchronously—they operated using a shared timer that marked the transmission of each character. This technique proved inadequate to meet the needs of efficiently sending large blocks of data over copper phone lines. Modems today use the asynchronous technique: instead of the character synchronization previously used, modern computers transmit a start bit, a stop bit and, optionally, an error-checking parity bit to indicate to receiving computers the boundary of each character. *Related Terms:* start bit; stop bit; parity bit.

asynchronous ink jet — *Related Terms:* drop on demand ink jet; continuous ink jet.

AT commands — a command set used to program a line of modems manufactured by Hayes Microcomputer

Products. AT commands program a variety of modem hardware settings and were adopted as a de facto standard by other modem manufacturers who wanted to market their wares with the coveted phrase "Hayes-compatible." At one time it was important to know that ATL0 turned your modem speaker down and ATM0 turned completely off. Now the commands are usually hidden under a menu option in your communication software. *Related Term:* Hayes-compatible.

ATAPI (attachment packet interface) — the Enhanced IDE standard usually works well for hooking up disk drives to PCs. People also need to attach tape drives and CD-ROMs to that same controller. EIDE's supports the ATAPI standard which makes it possible to use ATAPI-capable hardware and software drivers, mixing different types of drives on the same EIDE controller. *Related Term:* EIDE.

atlas — a type of book containing maps and other related information such as tables and charts, all of which primarily deal with geography, climate, currents, etc.

ATM (asynchronous transfer mode) — (1) in computing, asynchronous transfer mode, one method of file transfer across the digital network. (2) in desktop publishing, Adobe Type Manager. A program which allows PCs and Macs to display bitmap type from Type 1 outlines and lets PostScript fonts display on non-PostScript printers. *Related Term:* asynchronous communication.

attachment — any file linked to an e-mail message is an attachment. Many mail packages use MIME encoding to attach files. *Related Terms:* MIME.

attenuate — the action which reduces the strength of a signal. *Related Terms:* AM; amplitude modulation.

attribute — distinguishing characteristic of a sensation, perception or mode of appearance.

AU — a type of sound file used on Sun Microsystems or other UNIX computer systems. The Internet is dominated by UNIX systems and AU is used quite heavily. Regular browsers such as Netscape Navigator are usually able to play AU files, which have the extension .au. Sound files originating on a PC, however, are likely to be in WAV or MIDI format. *Related Terms:* MIDI; sound; WAV.

audio — in general terms, any type of sound which is transmitted. Specifically, sound for multimedia systems with frequencies from 15 Hz to 20,000 Hz.

audio graphics — a reference to the transmission of graphics and text information over a narrow band telecommunications channel, such as a telephone line or a subcarrier.

audio mixing — creating a custom audio track from several different sources using a microphone mixer or other sound mixing device.

audio stream — the sequenced frames of compressed audio.

audio-speaker-microphone unit — equipment that usually includes at least one speaker, multiple microphone and a telecommunications interface to accommodate a group of people in a teleconference.

audio-teleconferencing — two-way electronic voice communication between two or more groups, or three or more individuals, who are in separate locations.

audiotex — a telephonic connection designed to deliver audio information through dialing directly or through use of a modem or other access device.

audit trail — in business, particularly accounting, records of activity occurring in certain files, accounts or on computers which can be used as a basis to reconstruct fiscal activities, fund sources, histories, etc.

auditor — third party companies that track, count and verify ad banner deliveries or verify a Web site's own ad reporting system. Not to be confused with a "counter" which is a company that strictly counts ad and page deliveries.

authentication — (1) a method of verification for access. Authentication ensures that digital data transmissions are delivered only to the intended receiver. (2) rechecking data against source to ensure accuracy. Authentication also assures the receiver of the integrity of the message and its source. A common and simple form of authentication uses a password associated with a specific username to gain access to particular data. Sophisticated authentication protocols are often based on secret-key encryption or on public-key systems using digital signatures. *Related Terms:* DES; PAP; CHAP; authorization; encryption.

author — (1) in literature and book publishing, the original writer or creator of a body of work such as a novel, history, text book, etc. (2) in computing, to create a script, program, or document, usually with an authoring or scripting language such as C, C++, HTML, or Java.

author's alterations — changes in copy or artwork

after it has been typeset and sent to the printer, often called AAs. These types of changes frequently cost extra; the additional costs incurred by AAs are charged to the client, not the printer. *Related Term:* AAs.

author's corrections — *Related Term:* AA.

authoring system — software that helps developers design interactive courseware easily, without being expert at computer programming.

auto discrimination — in optical reading, the ability of code reading equipment to recognize and correctly decode more than one symbology. *Related Term:* auto distinguish.

auto distinguish —a scanner's ability to recognize a number of different symbologies and process the data without operator intervention; prerequisite for linear bar code scanners employed in open systems. *Related Term:* auto discrimination.

auto responder — an automated program that acknowledges receipt of an e-mail message and creates a tracking number for the message. It sends an e-mail acknowledgment to the requester with a subject line containing the tracking number. *Related Terms:* tracking number; domain name registration agreement.

Autochrome — the first commercially successful glass base film for color photography, introduced in 1904 by Auguste and Louis Luminere in Lyon, France.

automatic data collection (ADC) — direct entry of data into a computer system or other microprocessor controlled device without using a keyboard. It provides a quick, accurate, and cost-effective way to collect and enter data. *Related Terms:* POS; point of sale.

automatic image replacement — a process in which low-resolution FPO (For Position Only) images are automatically replaced by high-resolution images before outputting the final pages.

automatic white balance — a process in videography and digital photography where the device automatically interprets color ambient temperature and adjusts its sensitivity to accurately render the scene without color shift.

automation — the use of mechanical and electronic devices to perform human functions.

autonomous system number (ASN) — a group of Internet Protocol (IP) networks that adhere to a single routing policy. An autonomous system number (ASN) identifies the autonomous system, all of the networks subscribing to the specified routing policy,

etc. and enables the system to exchange information with other similar systems.

auto-parser — an automated program that extracts information from the fields in registration forms. It is designed to detect errors or incomplete information. Upon receipt of complete and correct data, it enters the appropriate information into a domain name database. *Related Term:* registration forms.

autopositive film — a film type which is designed to respond opposite of conventional film, i.e., it produces a postive image instead of a negative.

autoscreen film — a film that has a halftone screen built into the film emulsion. Not usable for line photography.

autotrace — the ability of some software to convert a bitmapped image into a vector rendering by delineation of differences in contrast, color, etc.

availability — one of the major factors in type selection, along with legibility, printability and readability. If the face is selected, can it be acquired in time and within budget.

available light — artificial or natural light falling on a scene to be photographed without the addition of lighting controlled by photographer.

avatar — a graphical handle or digital placeholder. Conventional chat rooms used a word "handle" to distinguish between participants. In 3D chats, avatars are a graphical representation that you select to stand in for yourself; it can look like anything—man, monster, object, animal or thing. Serious users suggest that the avatar you select should graphically represent the way you'd like others in the chat to perceive you. *Related Terms:* IRC; handle.

average background reflectance —the arithmetic average of the background reflectance from at least five different points on a sheet, expressed as a percentage. A critical factor in determining quality levels of bar code printing and their ultimate functionality, when printed.

AVI (audio/video interleave) — the file format used by Video for Windows, and one of three video technologies used on personal computers. (The others are MPEG and QuickTime.) AVI picture and sound elements are stored in alternate interleaved chunks in the file. *Related Terms:* .mov; MPEG; QuickTime; Video for Windows.

AVK (audio video kernel) — digital video interactive

(DVI) software designed to play motion video and audio across hardware and operating system environments.

AVSS (audio video support system) — software for MS-DOS platforms which enables the playing of motion video and audio.

AW — an abbreviation for artwork.

Bb

B paper sizes — one of five categories of ISO paper sizes.

back cylinder — *Related Term:* impression cylinder.

back flap — the back part of a dust jacket that folds inward and contains copy continued from the front flap and/or a photo and biography of the author.

back gauge — a calibrated and adjustable guide on a paper cutter against which paper is positioned to be cut at specific sizes. Its position relative to the knife drop point determines the length of the cut.

back light — light (natural or artificial) from behind the subject, directed toward the camera. The condition with automatic exposure systems, if not compensated for, may cause the object begin photographed to be underexposed due to the excessive brightness of the perceived background. Commonly used to cast shadows and/or create a visual separation between subject and background.

back-lighted copyboard — auxiliary equipment in some process cameras for photographing transparent copy; light is projected through the back of the copyboard (which has a translucent panel) and toward the camera lens on to the film. *Related Term:* transillumination.

back list — previously published books that are not new but still in print and available from the publisher, as opposed to front list, which are recently released books.

back margin — the inside side or gutter margin of a publication. *Related Term:* gutter.

back matter — the bibliography, glossary, appendix, afterword, colophon and similar elements appearing after the main text of a book. *Related Term:* end matter.

back printing — (1) printing only on the second side of a substrate; usually with no image on the front, as opposed to backup. Often used for sequenced pages, but also for instructions, contract terms and conditions, etc. (2) printing on the back or inner surface of a substrate as compared to printing on the outside or front. A required process when the image must be viewed through the substrate, such as in decal work, etc.

back spinner — the part of a perfect binding machine designed to remove excess glue before delivery.

backbone — (1) in bookbinding, the back or bind-ing edge of a case-bound book. (2) a main network that connects multiple smaller networks or the top level in a hierarchical network; the Internet's high speed data highway that provides the major access points where other networks connect. *Related Terms:* spine; backing.

backfile conversion — the act of converting existing paper and microfilm information, using scanning technologies, to create images that can be used in digital environment. There are two strategies, (1) scan on demand is a system where file conversion is performed only on each document as needed. (2) complete conversion refers to converting large amounts of the information prior to using the imaging system.

background — (1) activity(s) that occur in the computing process "behind the scenes" enabling the user to perform other tasks such as printing or active work in a different program simultaneously. (2) in optical scanning, the area surrounding a printed symbol, including the spaces and quiet zones. *Related Term:* multitasking.

background color — a general color area which is placed behind the major text, graphic, picture or illustration, which may be transparent or opaque.

backing — a part of the case binding method when the backbone is flared to receive the binder board covers.

backing up — *Related Term:* backup.

backslant — the name given to type which is slanted to the left, opposite of italic. *Related Term:* oblique.

back-to-back register — the precise alignment of images from one side of a sheet with those on the reverse.

backup — (1) in print graphics, to print on the second side of a sheet already printed on one side. Such printing is called a backup. (2) in computing, a duplicate of original data, software or printout made in case original is lost or damaged.

backup blade — in intaglio printing processes, the stiff support which assures proper pressure and alignment of the doctor blade against the image carrier.

bad break — any break in text which violates an established typesetting or grammar rule, or a line break that is visually jarring, as hyphenating the first line of a page, or a page that begins with the suffix "-ing" or a column that contains a single word. Generally, the incorrect end of line hyphenation. *Related Terms:* widow; orphan.

bad color — a printed sheet with too much, too little, or uneven ink.

bad copy — any manuscript or disk that is illegible, improperly edited, or otherwise unsatisfactory to the typesetter. Most typographers charge extra for setting type from bad copy.

bad lay — a sheet poorly centered during printing.

balance — the pleasing visual relationship of image elements on both sides of the vertical axis; can be formal or informal; a design principle of layout planning.

balanced process inks — a set of process inks of which the ratios of the blue and green actinic densities of the cyan and magenta are equal, enabling the use of only one color correction mask for the yellow printer.

Ballard shell process — special technique used by many gravure publication printers for easy removal of the copper layer after the cylinder has been printed, in preparation for replating and reuse.

balloon — the encircled copy in a cartoon. *Related Term:* bubble.

band — a range of closely sized electromagnetic frequencies which tend to cause similar phenomena, reaction or sensation, i.e., the visible spectrum, broadcast, television, x-ray, etc.

banding — (1) in computers and video, incomplete or inadequate pictorial information, i.e., insufficient brightness levels available to produce a complete continuous tone (CT) image. Often characterized by stripes of inappropriate densities in a gradient or toned areas. (2) in bindery, a method of packaging printed pieces using paper, rubber, or fiberglass bands. *Related Term:* artifact.

bandwidth — (1) in electronics and on the Web, the range of frequencies measured in Hertz or cycles per second a transmission line or channel can carry without undue distortion. In a general sense, this term describes information-carrying capacity or the amount of information that can flow through a channel. A measure of the information-carrying capacity of a com-munications channel; the higher (wider) the bandwidth, the greater the information carried. It is sometimes used as a measure of how fast a web page can be loaded. It can apply to telephone or network wiring as well as radio frequency signals, and moni-

tors. Bandwidth is most accurately measured in cycles per second, or hertz (Hz), which is the difference between the lowest and highest frequencies transmitted. (2) on a more human level, the term can describe a person's capacity for dealing with multiple projects ("I'd like to update this database, but I don't have the bandwidth."). *Related Terms:* band; memory bandwidth; video bandwidth.

bang (er) (!) — an exclamation point used to signify surprise in an on-line forum. *Related Term:* interbang; banger.

bank — (1) lightweight writing paper. The English term for onionskin. (2) in hand composition shop, the sloping surface upon which type cases rest during composition. Usually accompanied by a storage rack for additional type cases below. (3) one line of a multi-line headline. *Related Terms:* bank paper; tombstoning.

bank paper — *Related Term:* onionskin.

banner — (1) in publishing, a large headline, especially one that extends across a full page. (2) in advertising, a horizontally oriented sign with larger horizontal than vertical dimensions. *Related Terms:* nameplate; flag.

bar — (1) in typography, the cross horizontal piece between two strokes like on the H. (2) in optical reading, the darker element of a printed bar code symbol.

bar code — a series of parallel bar patterns and spaces of varying thicknesses used to encode data in a particular symbology so it is readable by optical mark recognition (OMR) equipment. Bar codes are used in printing as tracking devices for jobs and sections of jobs in production and in the paper industry for tracking individual

V805BVL2A615
BARCODE
Barcodes from consumer products are read via a low power laser beam which reflects the relative widths and spacing of each line for computer interpretation for inventory control, to determine sale price, discounts, etc.

rolls and pallets of finished paper and are arranged in a predetermined pattern following unambiguous rules defined by the symbology employed. *Related Terms:* Uniform Product Code; intelligent character recognition; barcode character; quiet zone.

bar code character — a single group of bars and spaces which represent an individual number, letter, punctuation mark or other symbol.

bar code density — the number of characters which can be represented in a linear unit of measure. Bar code density is often expressed in characters per inch (CPI).

bar code label — a label which carries a bar code and is suitable to be affixed to an article.

bar code reader — a device used to read a bar code symbol. *Related Terms:* OCR; OMR.

bar code symbol — *Related Term:* symbol.

bar height — *Related Term:* bar length.

bar length — the bar dimension perpendicular to the bar width. *Related Term:* height.

bar width — the thickness of a bar measured from the edge closest to the symbol start character to the trailing edge of the same bar.

bar width reduction — reduction of the nominal bar width dimension on film masters or printing plates to compensate for printing gain.

Baronial cards — a type of card stock that often has a beveled edge, used for announcements and invitations.

Baronial envelope — a near-square envelope with an oversized flap. Used for formal occasion announcements, and similar projects.

barrel or roll fold — a three-parallel fold to produce three distinct panels, one of which is doubled back.

base — (1) in photographic film, stencils, etc., the material used to support the emulsion. (2) in mechanical artwork, a term often applied to the bottom piece or mount board of artwork, made of a heavier stock for support and durability. (3) in statistics and mathematics, the number of digits comprising a system. The decimal system has 10 (0-9), binary systems have 2 (0,1), etc. *Related Terms:* support; base art.

base alignment — positioning of characters so the bottom of the x height lines up evenly on a horizontal line; in phototypesetting this alignment is used for the even positioning of different type styles on a common line.

base artwork — any artwork relating to mounting board, not overlays. Sometimes the related additional components such as halftones or line drawings, etc., to be added to artwork before final reproduction.

base cylinder — the supporting cylinder of a press which supports attached image carriers, i.e., in flexography, the rubber plates; in gravure, the copper cladding or sleeves.

base line — *Related Terms:* base alignment; baseline.

base negative — negative from main copy. *Related Term:* base.

base side — the side of paper or film which does not have an emulsion coating.

base size — standard size of sheets of paper used to calculate basis weight in the US and Canada. *Related Terms:* basic size; parent sheet; basic sheet size.

base stock — the underlying paper to which mills apply coatings to make coated paper. *Related Term:* body stock.

baseline — an imaginary reference line used to specify the desired vertical position where characters rest when printed on the same line. The approximate bottom of the x-height character. Some curved letters like S, C, or G may extend slightly below this imaginary line. *Related Terms:* base alignment; mean line; base line; ascender line; descender line; waist line.

baseline shift — a character attribute command in certain page layout programs that enables selected text to be raised or lowered relative to the normal baseline position.

BASIC (Beginner's All-purpose Symbolic Instruction Code) — this standard, high-level family of programming languages is simple to learn but creates programs that are typically slow to use. Many types of BASIC are available—your package might be called Turbo, Quick, or Visual Basic. The language is not generally used for industrial-strength applications, although Visual Basic has spawned a lot of shareware programs—and even some commercial applications such as MicroHelp's Uninstaller. *Related Term:* Visual Basic.

basic density range (BDR) — range of detail produced from halftone screen by main exposure.

basic exposure — camera aperture and shutter speed combination that produces a quality film image of normal line copy with standardized chemical processing.

basic sheet size — basis from which all paper weights are determined by the manufacturer; differs for each of the four paper classification: book paper (25 inches x 38 inches), writing paper (17 inches x 22 inches), cover paper (20 inches x 26 inches), and Bristol paper (22 inches x 28 inches).

basis weight — in the US and Canada, the weight, in pounds, of a ream (500 sheets) of paper cut to the basic (parent sheet) size. The ream weight and substance weight varies with paper grade. In countries using ISO paper sizes, the weight, in grams, of one square meter of paper. *Related Terms:* grammage; substance; substrate.

bastard cut — a bindery cut which is not at a 90 degree angle to one edge of the sheet.

bastard title — the half title of a book found on the page in front of the title page. A page with just the book name, usually before the main title page. *Related Term:* half title page.

batch — a technique for increasing the efficiency of the data transmission by organizing several files into a single group or unit.

batch file — the processing of groups of programs or data files to be output as part of a singular process.

baud — a rate of information flow, given in bits per second (BPS). In general usage, how many bits a modem can send or receive per second. It is the number of times per second that the carrier signal shifts value. A 1200 bit-per-second modem actually runs at 300 baud, but moves 4 bits per baud (4 x 300 = 1200 bits per second). Modern modems transmit more bits with fewer changes in sound, so baud and BPS numbers aren't equal. To be technically correct you should care about the distinction. *Related Terms:* BPS; KBPS.

Baumé — density scale used by Antoine Baumé, a French chemist, in graduating his hydrometers.

BBL (Be Back Later) — a written shorthand added to a comment in a chat room or on-line forum indicating temporary lack or participation.

BBS (Bulletin Board System) — (1) a computerized meeting and announcement system for discussions groups, uploading and downloading files, and generally obtaining on-line information and services through a dial-up connection. (2) an electronic gathering via a modem that allows the users to post messages. Originally informal communities but now organized into political, adult, commercial, activist, etc., categories. Still an active area, but it is being largely replaced by the Internet.

BCD — *Related Term:* binary coded decimal.

BCNU — Internet newsgroup and e-mail shorthand for Be Seeing You

beak — the serif at the end of a horizontal stroke.

bean cutter — a cutting tool designed to cut large circles in screen-printing film and stripping film.

beard — in hot metal or foundry typesetting, the beveled space below the printing surface of a type letter.

bearers — (1) in presswork on presses, the flat surfaces or rings at the ends of cylinders on a printing press that come in contact with each other during printing which serve as a basis for determining packing thickness. (2) in flexography, the strips of metal of varying thicknesses that are used to determine plate thickness in the manufacture of molded rubber printing plates.

beating — separating and modifying the wood fibers in the papermaking process so that the paper will be smooth and even.

bed — (1) in conventional prepress, the base on which the form is held when printing by relief printing processes. (2) in process camerawork, the element which supports and positions the copyboard, front case and back case. (3) in finishing, the metal working area on a cutter where the paper is inserted or removed during the cutting process.

bellows — a light tight area made of cloth or plastic material between the camera back and the camera lens; it permits the lens to be moved closer to or farther from the copy board to focus the camera.

bellows extension — the degree to which a bellows will extend. The shorter the lens, the less extension; a long lens requires a very extended bellows. Long bellows extensions require longer adjusted exposures due to the distance of light travel from lens to the film plane.

belly band — a wrapping sleeve which is narrower than the publication it holds.

belt press — A large printing press that prints several pages in one pass. *Related Term:* Cameron belt press.

benchmark — a basic standard for testing of systems, software, machinery, etc. A comparison to an ideal or best available known value.

Benday — (1) in mechanical art creation, a regular pattern of dots or lines used to add tonal variation to line art. (2) also, the dry transfer sheets manufactured for this purpose. *Related Terms:* cold type; Zipatone.

Bernoulli — a removeable magnetic storage system based on the Bernoulli principle that spinning objects tend to become lifting surfaces.

beta — a preliminary or testing version of a software

product. Commercial betas are actually works-in-progress; they are test copy releases of the full and/or final version, used as a method of testing under actual user conditions. They can be released to developers, selected users or to the general public. *Related Terms:* demo software.

BetaSP — the professional, widely used Betamax video format, not to be confused with the obsolete Betamax consumer format.

bevel — (1) in graphics, the use of as shading around the edges of graphics, such as buttons to give the illusion that they're jutting out from the page in 3D. Beveling is often used to enhance the user's experience on the Web, creating a sense of interactivity and realism. (2) the outward sloping of printing plates to facilitate their adhesion to a base cylinder by the use of wedges or adjustable clamps.

beveled edge — When the outer margins of a stock of paper (usually a card stock) have been embossed or blind embossed.

Bezier curves — in object oriented illustration programs, a curve whose shape is defined by mathematical formula representing anchor points set along its arc. Used in computer illustration for drawing and type rendering. These points set the shape of a curve.

BF — Sometimes "bf." The designation for setting type in boldface.

BFD (big f - - - ing deal) — often used in flame mail and posting, this particular acronym indicates contemptuous dismissiveness. *Related Term:* TFB.

BFT (binary file transfer) — one method of transferring files using fax modems using an extension to the fax protocol.

bible paper — a thin, opaque, very thin, high-tensile strength book paper used where low bulk is essential for bibles, insurance rate books, encyclopedias, directories, etc. Basis weights normally range from 14 to 30 pounds. *Related Terms:* India paper; directory paper.

bibliography — that part of a book's back matter in which other books or magazine articles are cited as resources or for the reader's further reference.

bidirectional — a bar code symbol capable of being read successfully independent of scanning direction.

bidirectional printer — a printing device that speeds hard copy production by printing in both directions as the print head moves across the platen.

bidirectional read — *Related Term:* bidirectional.

bilinear texture filtering — in 3D graphics, especially with games, you don't want a graphics card to simply grab texture maps from memory and write them on your computer screen because as the polygons are drawn bigger on the screen, they would appear blocky or chunky. Bilinear texture filtering is a sophisticated technique that averages or interpolates each four adjacent texels, and based on the calculations creates a new one. *Related Terms:* texel; trilinear texture filtering.

bill — (1) in business, an alternate name for invoice. (2) in production, an alternate name for a poster or flyer, i.e., handbill, etc.

billboard — a large public advertisement, usually in high traffic areas. Often printed in 4, 8, 16 or 32 panels.

billing contact/agent — the billing contact is the person designated to receive the invoice for domain name registration and renewal (re-registration) fees. The billing contact should be in a position to ensure prompt payment of fees. *Related Terms:* contact/agent; invoice; registration fee; renewal (re-registration) fee.

bimetal plate — an image carrier manufactured with two dissimilar metals; one forms the ink-receptive image and the other forms the solution-receptive area. *Related Term:* trimetal plate.

bin — in video, a grouping of clips to be used in a program. Derives from the industry's bins in which filmstrips were hung to dry after being developed.

binary — the basis for calculations in computers; a numeric system that represents all numbers using only two digits: 1 (the presence of) and (0) (the absence of) voltage. Anything made up of two units or parts may be referred to as binary. In computer systems, a base-2 numbering system using digits 0 and 1. Each of these digital switches is called a bit. An 8-bit computer processor represents data using sets of 8 bits, a 16-bit processor uses 16 bits, etc. The more bits a computer can process at once, the larger the numbers it can process. *Related Terms:* bit; binary code; binary digit; binary numeric values.

binary code — *Related Term:* binary.

binary coded decimal (BCD) — a numbering system using base 2 that represents each decimal digit by four binary bits, with the place values equal to 8, 4, 2, and 1, reading from left to right. *Related Terms:* binary; binary code; binary mode.

binary digit — *Related Terms:* binary code; binary.

binary file — any file that is not plain, ASCII text. For example: executable files, graphic files and compressed (ZIP) files.

binary mode — one subset of FTP file transmission. While FTP is a good way to transfer files over the Internet, not all files transfer equally. If a text (ASCII) file is sent as simply a stream of binary data to another machine, it will not transmit well. ASCII does not encode real end-of-line information into their data; it substitutes special characters in their place. FTP clients usually provide an option to send files as text or binary. In text mode, the data gets translated correctly. Sending an executable or ZIP file would corrupt that data because they utilize more than plain ASCII characters. In that case, you'd use binary mode or an auto detect feature that uses the files extension to determine how to process it. *Related Terms:* ASCII; FTP.

binary numbers — a numbering system with a base of 2, unlike the normal system, which has a base of 10 (decimal numbers). Binary numbers are preferred for computers because building an electronic circuit that can detect the difference between two states (high current and low current, or 0 and 1, on and off) is easier and less expensive than building circuits that detect the difference among 10 states (0 through 9). The word bit is derived combining parts of BInary digiT.

binary numeric values — a system of representing numbers using the characters 0 and 1 to represent any number. *Related Term:* binary numbers.

binary scanning — a mode where individual picture elements are identified as only either black or white as opposed to gray scale, continuous tone, or halftone scanning. *Related Term:* 2-bit scanning.

bind — (1) in publishing, to fasten sheets or signatures and adhere covers with glue, wire, thread, etc. (2) in mechanics, excessive tightness between adjacent moving parts caused by lack of lubrication or poor fit.

bind margin — the page margin next to the binding edge of a publication. Usually slightly larger than the trim edge margin to accommodate the binding apparatus, glue, etc.

binder — (1) a person who does bindery work. (2) a device equipped with metal rings for holding looseleaf sheets.

binder's board — the stiff, thick paper composite used to support the cover material in case-bound books. *Related Term:* case board.

binder's creep — the slight but cumulative extension of the edges of each inserted spread or signature in a saddle-stitched publication. *Related Term:* creep.

binder's waste — a precalculated amount of spoilage which may be generated in the finishing operations and still produce the appropriate number of products at a profit.

bindery — print shop department or separate business that does trimming, folding, binding, and other finishing tasks.

bindery marks — marks that appear on a press sheet to indicate how the sheet should be cropped, folded, or bound.

bindery service — a trade shop or company which specializes in performing bindery operations for commercial printers who may not have the necessary equipment, staff or other capabilities.

bindery work — all work with press sheets other than actual printing; cutting, jogging, collating, folding and stitching are some of the normal operations of a bindery.

bindery workers — the people who cut, fold, punch, perforate, stitch, assemble, pad and glue the finished printed jobs.

bind-in card — the attachment in a publication's binding by staples or glue of a inquiry or subscription solicitation card.

bind-in tabs — tab sheets which are permanently affixed in the binding of magazines, periodicals, etc. Usually the tab edge is folded inward for packaging, and the user must unfold it to expose and use the tabs.

binding — the fastening of assembled sheets or signatures along one edge of a publication. The binding process also includes folding, gathering, trimming, stitching, gluing, and/or casing.

binding dummy — the collection of representative signatures to determine final bulk or thickness of a finished product. Often used to determine creep as part of the prepress production process.

binding edge — the edge where binding blue, stitches, staples, coils, etc. are applied; usually the left side of a portrait-oriented project.

bingo card — *Related Term:* reader service card.

BinHex (BINary HEXadecimal) — a program routine for converting nontext files into ASCII, mainly because of the limited file type capability of the Internet e-mail system.

bio — short for biography. The brief description of an author's life and/or publication history that appears in the back matter of a book.

BIOS (basic input/output system) — the BIOS is the code in a PC's ROM that provides the basic instructions for controlling the system hardware. The operating system and application programs both directly access BIOS routines to provide better system component compatibility. Some makers of add-in boards such as graphics accelerator and video cards provide their own BIOS modules that work in conjunction with or sometimes replace the system BIOS. *Related Terms:* ROM; I/O.

bisynchronous transmission — sometimes referred to as synchronous or bisynchronous. A transmission process which allows for a steady stream of data. Faster than asynchronous transmission, which is controlled by stop and start bits between each character. *Related Term:* asynchronous transmission.

bit (binary digit) — a contraction of the term binary digit. (1) the basic unit of information in a binary numbering system. Computer circuits detect the difference between two states (high current and low current) and represent these two states as 1 or 0. These basic high/low, either/or, yes/no units of information are called bits. Eight bits comprise what is called an octet, sometimes referred to as a byte. (2) the smallest unit of information making up the digital or dot image of a character or graphic. (3) the smallest parts of a letter or graphic; little dots. *Related Terms:* binary; byte, gigabyte, KBPS, megabyte

bit depth — the number of digital bits used to represent midtone values, i.e., grays. Common depths are 4, 8, 16, 24 and 32. The higher the number, the closer the reproductive match will be to the shadings in the original copy.

bite — *Related Term:* tooth.

bitmap — any picture you see on the web or from a scanner, or a page created with a desktop publishing

BITMAP
Round edges become ragged or pixilated when enlarged.

application, is a bitmap. It is a map of dots, similar to what you see when you look at a newspaper photo under a magnifying glass. Bitmaps come in many file formats (GIF, JPEG, TIFF, BMP, PICT, and PCX, to name a few) and can be read by paint programs and image editors. *Related Term:* vector.

bitmapped — an image formed by a rectangular grid of pixels (binary digits) of equal density. The smallest unit of information in a computer. The computer assigns a value to each pixel, from one bit of information (black or white), to as much as 32 bits per pixel for full color images. It can define by itself one of two conditions (on or off, "0" or "1"). *Related Term:* bitmap graphic.

bitmapped fonts — typically fonts which cannot be resized; nonoutline fonts. When resized they will show edge pixels which become stair steps. *Related Term:* jaggies.

BITNet — an interactive electronic mail and file transfers service for the academic community. BITNet hosts are not directly part of the Internet although they are reachable by e-mail through Internet gateways.

bitonal — synonymous to black and white or line art, it means the data in an image is gathered from only one channel at 1bit per pixel.

bits per pixel — the description of the relative brightness or intensity of individual pixels in a bitmapped image.

bitstream — a sequence of bits.

EFFECT OF BLACK INK
Black ink absorbs all white color components (RGB).

black — (1) the absence of all reflected light; the absence of color. (2) an ink that appears to absorb all wavelengths of light. It is used as one ink in the four-color printing process. The letter K is often used to designate it. *Related Terms:* full scale black; skeleton black.

black colors — *Related Term:* wanted colors.

black duotone — a two-impression halftone reproduction in which the halftones are made to different specification but both are printed in black ink.

black generation — the amount of black generated on the black plate of a color separation.

black ink — ink which has no apparent color due to its total absorbption of RGB total spectrum illumination. The most used type of ink because of its high contrast with light colored substrates.

black letter — one of the six classifications of type. The Gothic type style popular in Germany in the 15th century.

black light source — *Related Terms:* ultra violet; UV.

black patch — artwork material used to mask the window area on a negative image prior to "stripping in" a halftone. May actually be made of red Rubylith or Amberlith when using orthochromatic film.

black plate — a plate that is used with the cyan, magenta, and yellow plates for four color process printing. Its purpose is to increase the overall contrast of the reproduction and, specifically, improve shadow contrast. Sometimes called the key plate. The latter K is used to designate this color. *Related Terms:* full scale black; skeleton black; key plate.

black type — a term sometimes used for ultrabold font face variants.

black, three color — the color that would be produced if solids of ideal magenta, cyan, and yellow process color inks were overprinted. Since each process color ink absorbs some of the two other components of white light that it should completely transmit, overprinting solids of three typical printing inks usually results in a brown color. Many color control bars include a three-color patch.

black-and-white — originals or reproductions in which black is the only color, as opposed to one-color (which can be any single color), two-color, four-color, or more.

blackletter — an alternate name for the Gothic classification of type characterized by its resemblance to medieval northern European manuscript characters. A heavy and ornate face design; sometimes called Gothic or text. The designs were patterned after a style of handwriting popular in the fifteenth century. This is blackletter or text type.

blad — a term used to indicate dummy book pages that appear in prerelease promotional materials in lieu of Greeking.

blade coater — (1) a device that first applies a surplus coating to a paper web and then evenly levels and distributes it with a flexible steel blade. (2) a tool made of nonoxidizing metal used to apply light sensitive liquid emulsion to the fabric of a screen printing. *Related Term:* knife coating; air coating.

blank — a category of board ranging in thickness from 15 to 48 points. Blanks may be C1S, C2S, or uncoated and are used for signs and posters.

blanket — a fabric coated with rubber that is wrapped around the intermediate cylinder of an offset lithographic press. It transfers the inked image from the image carrier (plate) to the substrate. As inked, a reversed image is transferred, or offset, from this blanket cylinder onto the substrate, resulting in a right-reading impression on the press sheet.

blanket cylinder — the cylinder that carries the offset transfer blanket, placing it in contact with the inked image on the image carrier and then transferring or "offsetting" the ink from that image to the substrate carried by the impression cylinder.

blanket to blanket — a setup on a perfecting press whereby two blankets, each acting as an impression cylinder for the other, simultaneously print on both sides of the paper passing between them. *Related Terms:* perfecting press; blanket.

blanking interval — a period in which the monitor receives no video signal while the video disc player searches for the next video segment or frame to play. *Related Term:* blanking level.

blanking level — in video , the signal level during the horizontal and vertical blanking intervals, typically representing zero output. *Related Term:* blanking interval.

blatherer — a chat room or usegroup user who answers a single question that needs few words to answer but takes multiple screens to say it.

bleaching — generally, the removal of stains and other discoloring from the wood fibers in the paper-making process. Specifically, the process where brown pulp is bleached before white paper can be made. This is done chemically, in a series of continuous bleaching stages, with chlorine as the primary chemical.

bleed — the area of plate or print that extends ("bleeds" off) beyond the edge to be trimmed. Applies mostly to photographs or areas of color. When a design involves a bleed image, the designer must allow some extra space beyond the trim page size for trimming. The printer often must use a slightly larger sheet to accommodate bleeds.

bleed allowance — the amount which a bleed must extend beyond a document's trim in order to allow for variations in cutting and folding.

bleed into gutter — an image which extends beyond the inside margins, across the gutter between facing pages of a book or magazine.

bleed to — *Related Term:* bleed.

bleedthrough — any image from the obverse side of a substrate which can be discerned on the reverse. *Related Term:* showthrough.

blend — an area in an image that merges from one color (or gray level) to another. Also known as a graduated tint, graduation, fountain, dégradé, or vignette.

blind embossing/debossing — a bas-relief impression made in paper (substrate) with a regular stamping die except that no ink or foil is used.

blind folio — nonprinted page numbers contained in a signature.

blind image — an image on a printing plate, particularly lithographic, that has lost its ability to hold ink and fails to print.

blind perf — a perforation designed to be invisible from the back side of a sheet to avoid defacement. *Related Term:* kiss cut.

blister — a raised area on paper which is caused by trapped moisture on its surface.

blister pack — a packaging system where products are affixed to a card placed between two plastic sheets from which the air is extracted, making the plastic form-fit the product. After extraction the two sheets are heat or sonic welded to assure package integrity.

BLOB (binary large objects) — in records administration, the ability to embed large binary objects (images) as part of a character database record.

block — a group of words, characters, or digits forming a single unit in a computerized system.

block in — to sketch in the main areas of an image prior to the design.

block out — liquid masking material used to cover the open areas of the screen printing fabric around the attached stencil.

block printing — one type of printing in the relief classification. Traditionally carved wooden blocks where the image to be printed remains at surface level, while the nonimage areas are carved below the surface. In modern usage, the solid wooden block has been replaced by a carefully machined wood block which has a thick linoleum surface. This facilitates the carving of much greater detail without the danger of splitting the wood.

blocked up — characteristic of shadow areas in a photo or halftone that lack detail because of overexposure or poor printing.

blockiness — artifact of compression appearing as misplaced color in rectangular areas of an image.

blocking — the descriptive term used to indicate drilled, punched or cut sheets which stick together.

blooming — the overflow of an illuminant into adjacent areas. Usually caused by too much energy being applied. For example, excess voltage in a CRT electron gun will cause adjacent, inappropriate, pixels to be activated, causing color distortion and/or reduction in sharpness and contrast.

blotting — in screen printing, the removal of water from the water soluble film when adhering it to the screen fabric.

blow up — (1) in photography, an enlargement, most frequently of a graphic image, photograph, or type. (2) the action of making an enlargement. *Related Term:* enlargement.

blue pencil — (1) pencil that writes in a light blue color that does not record on orthochromatic graphic arts film. (2) the process of writing instructions on artwork using a nonreproducing marker, pen or pencil. *Related Term:* nonreproducing blue (pencil).

blue sensitive material — photographic material that reacts primarily to the blue end of the spectrum.

blueline — (1) copy composed of blue lines on a white background. (2) prepress, photographic proof made from stripped-up film negatives or positives where all colors show as blue image on white paper and which can be used to check the position of page elements before printing. *Related Term:* blueprints.

blueline flat method — special method of preparing a flat as part of one multi-flat registration system; a blueline flat carries all important details, including register marks.

blueprint — in printing, the same thing as a blueline.

blueprint paper — inexpensive photomechanical proofing material; exposed through the flat on a platemaker, developed in water, fixed in photographic hypo, and then washed to remove fixer stains.

blueprints — *Related Terms:* blueline; white prints; brownline.

blurb — (1) in advertising, copy written with a sales angle, usually in brief paragraphs. (2) in book publishing, a summary of contents of a book presented as jacket copy, or a short commentary of a book or author on a book cover. Sometimes erroneously confused with callout or pull paragraph. (3) in periodical publishing, text that summarizes an article, usually set smaller than the headline and larger than the running text. *Related Terms:* breakout; pull quote.

blushing — an imperfect printed image caused by excessive moisture trapped between the ink film thickness and the paper.

BMP bitmap — BMP bitmap is a graphics file format developed by Microsoft and is the native graphics format for Windows. In the Windows program, the startup and closing windows and the wallpaper are all in BMP format. BMP files store graphical data inefficiently, compared to run-length encoded file formats. *Related Term:* run-length encoding.

BNC (Bayonet Neill Concellman) — developed in the 1940s, the BNC connector provides a well engineered, secure, easy-to-use means of connecting shielded cables to electronic equipment. BNC connectors are commonly used today in high-end video and computer networking applications.

board — (1) alternate term for mechanical. (2) a general term for paper over 110# index, 80# cover, or 200 gsm that is commonly used for products such as file folders, displays, and post cards. *Related Term:* paperboard.

body — (1) in composition, the metal block of a piece of type that carries the printing surface. (2) in printing, a term that refers to the viscosity, consistency, and flow of a vehicle or ink. (3) in publishing, main section of a brochure, book, article or other text material, not including headlines. (4) on the Internet, the part of an e-mail message you are sending which contains just the message itself without all the header and server information. (5) in Web publishing, the section that is designated by the use of a <body> tag.

body copy — (or body type) the majority of the copy in a book, magazine article, or marketing piece set in text type; the bulk of a story, not its headline or subheads. Generally 6 to 14 point in size. *Related Term:* body matter; body text.

body size — the height of the type measured from the top of the tallest ascender to the bottom of the lowest descender. Normally given in points, the standard unit of type size.

body stock —paper on which the text or main part of a publication is printed, as compared to the cover stock. *Related Term:* base stock.

body text — the paragraphs in a document that make up the bulk of its content. The body text should be set in an appropriate and easy to read face, typically at 10 or 12 point size.

body type — a particular font used for the main text of a printed piece, as opposed to headline type or caption type.

boiler plate — type which is repeatedly used and saved from use to use, to be ready to "fill in the spaces" allowed for it in composition. Often used for legal statements such as warranties, personnel practice statements, etc. The type file is always available for insertion into whatever document demands it.

bold — short for boldface.

bold type — *Related Term:* boldface.

boldface — a typeface which has been enhanced by rendering it in darker, thicker strokes so that it will stand out on the page. Headlines that need emphasis should be boldface. Italics are preferable for emphasis in body text. Generally, a heavier weight version of a typeface, used for emphasis. Indicated as "bf." *Related Terms:* bold; bold type.

bond paper — a writing and printing paper with a hard, smooth surface treated to take pen and ink well and have good erasure qualities, with a basis weight of 50 gsm or more. It was originally designed for printing stocks and bonds, and is now commonly used for writing, photocopying and printing of letterhead, stationery and business forms. Cheaper grades of bond paper are made from all wood fiber; the better grades are made from rag fiber (25%, 50%, or 100% rag content). Used where strength, durability, and permanence are required. *Related Terms:* business paper; communication paper; correspondence paper; writing paper.

bone folder — the most common type of hand folding device; consists of a long, narrow blade of bone or plastic which is pressed firmly against a fold in paper to make it sharp and well defined. *Related Terms:* buckle folder; knife folder.

book — a compiled work of individual sheets or signatures, bound together as a unit. The most common form of mass written communication for novels, poetry, nonfiction, educational texts and similar literary efforts.

book cloth — the cloth fabric used to cover case bound books.

book fair — an event or trade show where publishers promote their upcoming books.

book paper — the classification of paper that includes various grades and many finishes, coated and uncoated, but is used primarily for publishing bulk books and newsletters and for general printing needs. The most common type of paper found in the printing industry, with the exception of newspapers and pulp novels. A category or group of printing papers that have certain physical characteristics in common which make them especially suitable for the graphic arts. Noted for easy printability, book papers as a class include coated and uncoated papers in a wide variety of basis weights, colors and finishes.

book press — (1) a special clamping device used to press books together in the hand sewn case binding method of bindery. (2) a special type of printing press designed for the production of books and book signatures which generally utilize rubber plates. (3) the classification given to those who have book printing presses and equipment. *Related Term:* Cameron belt press.

bookbinder — *Related Term:* trade bindery.

booklet — usually, any publication of less than 48 pages. Generally the cover will be of a heavier weight stock than the inside signatures.

bookmark — (1) on the Internet, an anchor. It is a location within a page that can be targeted by a hyperlink and/or stored for later recall by an Internet browser. You can easily return later with a simple mouse selection rather than typing in the correct URL, if you can even remember it. Web sites often have hot lists which are really just collections of bookmarks. (2) any device used to mark, for later recall, a page or pages within a reference.

Boolean — an English mathematician, George Boole (1815-64), founded a field of mathematical study called symbolic logic. His name is now used to describe one type of refinement technique used in constructing database queries. The most common operators are AND if you're looking for all terms, OR if you're looking for at least one of the terms, and NOT if you're excluding a term. These are referred to as Boolean operators.

Boolean operator — *Related Term:* Boolean.

boot — the permanently installed computer power-up routine; getting into the program you're going to use. Contains basic instructions for initiating some utility programs and prepares the computer to receive application software input. When you shut down a system then restart, you are rebooting. *Related Terms:* warm boot; cold boot.

border — (1) in page makeup, a continuous or intermittent decorative design or rule surrounding the matter on the page or within a layout. (2) in artwork, an enclosing boundary for a graphic or other element. Usually, but not necessarily, appears on all sides of the element.

boric acid — used in fixing baths, it has a hardening property which helps prevent sludge on films.

bot (robot) — in general use and at its most general level, a bot is any type of automated software that runs on a computer 24 hours a day, 7 days a week, that automates mundane tasks for the owner, even if the owner is not logged in. This covers a pretty wide range of programs, but the term applies specifically to two areas: chat and web-cataloging software. (2) in the Web world, they often are called spiders and crawlers. They explore the Web by retrieving documents and following all the hyperlinks in them, and then generate catalogs that can be accessed by search engines. They are the basis for popular search sites like Alta Vista, Lycos, etc.

bottling — the process of skewing pages to compensate for paper thickness as it is folded. Primarily used on signatures designed for large web- or sheet-fed presses.

bounce — (1) in hand or machine composition, an uneven baseline. (2) the return of a piece of mail because of an error in the delivery process. Mail can be bounced for various reasons. (3) an undesirable phenomenon in which the reproduction of book or magazine pages is off by as much as 1/16 of an inch. (4) also refers to the message indicating the error (informal usage). *Related Term:* off the line.

bounding box — traditionally, computer programs have dealt with on-screen objects, such as images, by placing them in an invisible rectangle called a bounding box. You can see an example of a bounding box by clicking an image inside a word processor such as Microsoft Word. The outline that appears around the image is the bounding box. *Related Terms:* frame; box.

bowl — the interior part of a letter in a circle form such as in a, B, b, c, d, P, o and the upper part of g.

box — (1) a white space or page text area framed by a decorative border, usually presented separately from the main text and illustrations. (2) a rigid container made of corrugated paper board, chipboard or wood. Sometimes called a carton or a case. *Related Terms:* bounding box; box.

box board — the inexpensive or cheaply made cardboard used to manufacture box containers.

bozo filter — a program that filters e-mail from or posting by individuals who are on your b-list (bozo list).

BPI (bits per inch) — in bar coding, the linear density of encoded information (number of characters encoded per linear inch of the bar code symbol.

BPR (business process re-engineering) — the radical restructuring of the business processes, organizational boundaries and management systems of an organization. Also known as re-engineering.

BPS (bits per second) — a standard unit for measuring the rate of transmission for digital data. *Related Terms:* bit; baud rate.

BPSK (biphase shift keying) — a type of FM digital signal modulation that is less efficient, but also less susceptible to noise, than other similar modulation techniques. It is used for sending data over a coaxial cable network.

braces({ }) — a character used to connect or embrace lines. Used to set off matter extraneous to the context. *Related Term:* brackets.

bracket ([ ]) — (1) in editorial work and composition, the marks used to set off extraneous or incidental copy or data. (2) in type design, the rounded area between the serif and the main stroke. *Related Term:* braces.

bracketed exposures — making one or more exposures on both sides of the exposure calculated to be correct; done to be sure of obtaining one good exposure.

break for color — to separate the parts of a piece to be printed in different colors.

breakout — a sentence excerpted from the body copy and set in large type, used to break up running text and draw the reader's attention to the page. *Related Terms:* blurb; pull quote.

brightness — (1) generally, the reflective quality or brilliance of a piece of paper. In color, the amount of light reflected by a particular color. The variation of a color from dark to light. (2) in computers and video, the relative level of each pixel from 0-255, i.e., from white to black.

brilliance — in a lens, the ratio of transmitted light to incident light. All lenses transmit light to different degrees. Two lenses set at the same f/stop may require different exposures due to brilliance.

Bristol — in papermaking, a heavyweight paper used for printed materials that will be frequently handled such as posters, folders, announcements, direct mail pieces, and invitations.

BRM (business reply mail) — return postcards or envelopes in which the postage has been prepaid by the original sender.

broadside — a large, tabloid-size advertising circular.

brochure — a pamphlet of two or more pages that is folded or bound.

broken type — a face defect in a typeset or reproduced character glyph.

bromide — a photographic print made on bromide paper.

bronzing — an effect produced by dusting wet sizing ink with a metallic powder after printing.

brownline — an inexpensive photomechanical proofing material; is exposed through the flat on a platemaker, developed in water, fixed in photographic hypo, and then washed in water to remove fixer stains; the longer the exposure, the more intense the brownline image. *Related Terms:* blueline; white print.

brown-print paper — a silver nitrate based photographic proofing paper.

browser — a browser is a text or graphic software program which interfaces a computer to the World Wide Web. It usually includes a hypertext interpreter and enables viewing of sites and navigation between nodes or sites. *Related Terms:* hypertext; HTML; DHTML.

browser sniffing — a software routine which lets a web site detect which versions of various browsers are running, and determines whether or not they can have access site features. Browser sniffers are usually written with JavaScript and can also be used to detect if the person seeking access to a site has the proper media plugins for site features such as streaming audio, video, shopping, etc. *Related Terms:* download; plug-ins.

BTW — an Internet, e-mail and discussion group shorthand for By The Way. It is used mainly to cut down on keystrokes

bubble — *Related Term:* balloon.

bubble jet — an ink jet technology in which a drop emission is produced on demand by a thin film of boiling ink in a tubular chamber. *Related Terms:* asynchronous ink jet; drop on demand ink jet; valve jet, ink jet.

buckle folder — a bindery machine in which two rollers push the sheet between metal plates with positionable stop guides. When the leading edge

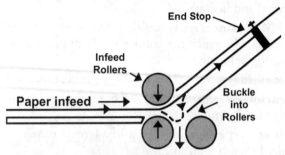

BUCKLE FOLDING
The sheet is driven to a head stop which causes a bulge at the place of the intended fold. At that point, the driven sheet is forced between two rollers, which create the fold.

of the sheet encounters the guides it causes a buckle at the entrance to the folder. A third roller working with one of the original rollers uses the buckle to fold the paper. Buckle folders are best suited for making folds which parallel the paper grain.

buckling — an undesirable effect that occurs when a sheet of paper has been improperly printed or folded, causing wrinkles.

buffer — (1) in computing, a portion of computer memory that is designated to store input until the CPU is able to process it. A buffer is an interface between system elements whose data rates are different. (2) in graphic arts, a chemical used in the developer to control the action of chemicals accelerators in photographic film processing.

bug — (1) in computing, a defect, often in the program software or hardware that causes a computer system to perform erratically, produce incorrect results, or crash. Corrective action is referred to as debugging. The term comes from a real insect which caused a circuit failure in ENIAC, the original electronic computer. A bug is not a glitch, which refers specifically to hardware problems. (2) in industry, a union or trade organization identifying mark often placed on printed matter.

buildup — a flat piece of material, such as frosted glass, plastic, etc., placed under the fabric when attaching a stencil to a screen printing unit.

bulk — the thickness of paper relative to its basis weight; the thickness of paper expressed in thousandths of an inch, and measured by the caliper of pages per inch. Sometimes also expressed as pages per inch. Thickness varies from thin Bible sheets, which can measure 1300 pages per inch through thick card stock measuring as few as 10-12 pages to the inch. Bulk is affected by both weight and finish. Most often, a heavy paper is thicker than a light one. A rough, toothy paper is thicker than a smooth one.

bulk mail — A class of mail sent by the U.S. Postal Service at a discount rate for business mail of at least 200 pieces that has been sorted by zip code. *Related Term:* third class mail.

bulk pack — to pack printed pieces in boxes without prior wrapping or bundling.

bulking book paper — a high bulk or thicker paper which results in a thick book relative to the number of pages. *Related Term:* novel paper.

bulking dummy — a paper dummy assembled from the actual paper specified for a printing job and used to demonstrate the final product thickness after printing and binding.

bulking number — the actual number of sheets of a given paper in a one-inch stack. Because one sheet has two pages, multiplying the bulking number by two yields pages per inch (ppi). *Related Term:* bulk.

bull's eye — (1) in prepress, a term which refers to a circle and cross hair register mark. (2) in print production a slang term for a hickey.

bullet (•) — not a dingbat or sort character. A common font character usually used for emphasis in place of numbers to draw attention to listed items. *Related Terms:* dingbat, wingdings; webthings; webdings; Monotype sorts; Zapf Dingbats.

bulletin board system (BBS) — at its simplest, a BBS consists of a computer and software that provides electronic mail and discussion groups, archives of files and any other services or activities of interest to the bulletin board system's operator subscribers. The system is a precursor to the World Wide Web. BBSs, ave traditionally been the domain of hobbyists, but many BBS's are operated by government, educational and research institutions. The web has replaced most BBSs, although some sysops and companies still maintain their systems or even connect them directly to the web. The majority, however, are still reachable only via a direct modem-to-modem connection over a phone line. *Related Terms:* e-mail; Internet; Usenet; newsgroup.

bump — (1) in camerawork, a shorthand term for a supplemental, no screen exposure sometimes made when making halftone exposures, in addition to a main and flash exposure. (2) ink applied from a fifth or higher plate in four-color process printing to strengthen a specific color. *Related Terms:* touchplate; bump color; bump exposure.

bump color — a touch plate. Adds a special color or accent color within a specific image area, for reaching optimal color match. Commonly used to achieve bright reds.

bump exposure — a brief no-screen exposure that supplements the main exposure when making a halftone. The effect is to compress the screen's density range without flattening tonal detail. The technique produces a full range of halftone dots from short density range copy and can also be used to expand highlight or shadow tonal separation, depending on whether negatives or positives are being made. *Related Term:* bump.

bundled software — software provided as part of a computer package. Bundling is the process. The bundle is all of the software included.

burn — to expose a plate when making printing plates.

burn in — (1) in lithography, to expose a blueline proof or printing plate with light. (2) in photography, to give extra exposure to a specific area of a print.

burner film — a piece of transparent film that allows full passage of light to the cylinder when preparing a gravure mask.

burnish — to rub dry transfer lettering or waxed galleys with a smooth and blunt instrument or burnisher (a pen-like device) in order to adhere it firmly to the art base. *Related Term:* burnisher.

burnisher — (1) in conventional art preparation, a tool used to apply pressure to a small surface; usually has a long handle and a hard, smooth ball or point. Such a tool is used in applying dry transfer letters and Benday or Zipatone shading tints.(2) also, the person performing a burnishing operation. *Related Term:* burnish.

burst — an art design, die cut, etc., which appears as an explosion.

burst perfect bind — to bind by forcing glue into notches in spines of signatures, and then adhering a paper cover. *Related Term:* perfect binding.

burst tester — a testing device designed to determine a quantifiable point at which paper strength is exceeded, causing tearing or bursting.

burst transmission — messages stored for a period of time then released at a much faster speed for transmission. The received signals are recorded and then slowed down for the user to process.

bursting strength — the ability of paper to resist rupture as measured by Mullen test. *Related Term:* pop strength.

bus — in the broadest terms, a bus is a common connection between electrical devices. In computing, bus most commonly means the data pathway that connects a processor to memory and to other "peripheral" buses, such as VESA and PCI. *Related Terms:* PCI, VESA; Universal Serial Bus (USB).

business cabinet — *Related Term:* stationery.

business envelope — an envelope with opening along the long edge, such as the #10 in the North American sizes and the DL in ISO sizes.

business paper — a stock classification that includes bond, duplicator, ledger, safety, and thin papers. *Related Term:* bond paper.

busy — (1) in advertising, excessively active content. Too many points, too many typefaces, borders, splashes, etc. (2) in telephony and computing, an incomplete circuit transaction due to the receiving party being occupied with another call or data stream.

butt — (1) in printed images, adjacent colors without spreads or chokes. (2) in stripping to join negatives and/or positives without overlapping or space between. *Related Terms:* kiss fit; butt fit.

butt end — the unused remainder of a supply roll on a web-fed press. *Related Term:* white waste.

butt fit — in presswork, printed ink colors that touch but not overlap. (2) in stripping, two negatives that are cut to fit in a precise manner, one to the other, without overlapping. *Related Terms:* trap; trapping.

butt waste — paper too close to the core of a roll to yield a satisfactory product when printed. *Related Terms:* stub waste; butt end; white waste.

buy out — to subcontract a specialty service for production operations such as die cutting, foil stamping, camera, platemaking, engraving services or bindery operations. Sometimes called farm out or job out. Often procured without customer knowledge. The operations generally are ones which are not especially economical for a single plant, but which can serve many operations. Typically, areas where this occur are service bureau operations for color scanning, engraving plates, etc., die cutting, foil stamping, camera, platemaking, engraving services or bindery operations. *Related Term:* trade shop.

byline — author credit in a newspaper or magazine article. Usually adjacent to the story or article, as opposed to editorial credit which may appear in the contributing writer area elsewhere in the publication.

bypass — in telecommunications, a private network which works external to normal transmission systems.

byte — (1) a unit of measure equal to eight bits of digital information, (also called an octet or a word). On personal computers, one character. The number of bits used to represent a character. (2) the standard unit measure of file size. *Related Terms:* megabyte; kilobyte; gigabyte. Pronounced "bite." Abbreviated as B.

bytecodes — a Java compiler created platform-independent codes that run inside of a Java Virtual Machine (VM).

Cc

C — (1) the Roman numeral for one hundred (100). (2) in paper specifications, CWT stands for hundred-weight (100 + weight).

C paper sizes — one of the five categories of ISO paper sizes.

C print — color photographic print made from a negative on Kodak C color print paper.

C&lc — the symbols for setting type in which the first character of each word is capitals. *Related Terms:* U&lc; C&lc.

C&sc — capitals and small capitals. In composition, used to specify words that begin with a capital letter and have the remaining letters in small capitals which are the same height as the body of the lowercase letters. THIS IS A SAMPLE OF CAPITALS AND SMALL CAPITALS. *Related Terms:* C&lc; U&lc.

C/N or C/NR — carrier to noise ratio. The ratio of signal power to noise power in a system, usually expressed as a power ratio in dB. *Related Term:* S/N.

C1S — the abbreviation for coated 1 side. Paper that has a coating, or finish, on only one side; often used for book covers and postcards.

C2S — paper coated on 2 sides. Paper that has a coating, or finish, on both sides.

cab — special film mask used in gravure printing; is punched and marked to be used as a guide for stripping negatives prior to making a gravure cylinder.

cabinet – (1) to paper purchasers, a standard 100 package count for blank announcements and embossed cards. (2) to paper merchants, a collection of paper swatch books and unprinted samples from many mills, usually organized into a box or other container and supplied to printers, agencies, and other major customers.

cable — an assembly of one or more insulated wires in a common protective sheath.

cable modem — a modem attached to a coaxial cable television system, and that can transmit data at 500 kilobytes a second, or up to 80 times faster than an ISDN line or about 6 times faster than a dedicated corporate T1 line. They are expensive, between $200 and $250, plus miscellaneous fees from the local cable company including a $50 to $100 installation fee, plus an additional $40-$90 per month, not includ-

ing TV program selection. *Related Terms:* modem; ISDN.

cable release — flexible device that attaches to a camera shutter button so exposure may be made with minimum vibration of the camera body.

cable television — transmission system that distributes broadcast television signals and other services by means of coaxial cable. Most cable systems have the potential for two-way communication in addition to broadcast television.

cache — an area to store information that can be retrieved quickly. Web browsers cache URLs, pages, graphics, and sounds on your hard drive. When you return to the page, the material is not downloaded again. It is already on the local disk where data access is much faster. Disk access is slower than RAM access, so there's also random access memory (RAM) caching, which stores information you might need in the much faster RAM. *Related Terms:* L1 cache; L2 cache; primary cache; secondary cache.

CAD (computer-aided design) — a computer system used by used by architects, engineers and 3D artists. Where they were previously forced to put their ideas on paper, by drawing out precisely scaled views of their project and making precise calculations for angles and distances, the computer now takes much of the drudgery out of the design process. They can now visualize and manipulate the objects or spatial environments they are designing and all work is saved digitally. Revisions entail adjusting the formulas and dimensions for a totally new drawing. Redraws are minimized. Revisions and design adjustments can be envisioned and turned into electronic representations quickly. CAD originally required high- powered workstations which were expensive, slow, and not very easy to use. Newer computer processing speeds and technology have moved more and more work directly onto the desktop and the modern systems have become a standard part of the design process.

café (cybercafé/Internet café) — a late 20th century phenomenon; a public eating and drinking establishment where the principal entertainment is in-line access through terminals at individual tables.

calendar — a printed product showing months and days by year.

calender — (1) a shortened term for calendering, the process in paper manufacturing for producing a very smooth, high-gloss surface on paper stock by passing the sheet between a set or stack of heated rollers at the end of a paper machine. (2) also, the heated, horizontal cylinders at the end of a paper machine used to produce the calender finish on paper. *Related Term:* calendering.

calender rolls — a set of horizontal rolls at the end of a paper machine to increase the smoothness of the paper.

calendering — in papermaking, the process of passing paper between the calender rolls to increase the paper's smoothness.

calf paper — a paper colored and embossed to resemble the leather that is occasionally used in book binding. *Related Term:* leatherette.

calibrate — to adjust an input device such as a scanner or an output device such as a monitor, imagesetter, or printing press to more accurately reproduce color.

calibration — a method of setting equipment to a standard measure for reliable and predictable results.

calibration bars — on a negative, proof or printed piece, a strip of color/tone values used to check printing quality. *Related Terms:* color bars; control bars; calibrated gray scale.

California job case — a specially designed compartmented tray in which handset type is stored and from

CALIFORNIA JOB CASE
Individual compartments are available for each of the characters, spaces, numbers and punctuation marks of a font. Their size varies based on the frequency of use of individual characters, i.e., "e" was larger than "x," etc.

which it is set. The cubicles are arranged for a minimum of hand motion and are sized to accommodate letters in quantities related to frequency of use. *Related Term:* type case.

caliper — (1) in paper specifications, the thickness of paper or other substrate expressed in microns (millionths of a meter), mils or points (each is 1/1000 inch), pages per inch (ppi), or pages per centimeter (ppc). (2) also, the name of the tool used to make the measurement.

calligraphy — (1) highly decorative writing. Often associated with hand addressing of invitations and other socially important documents, as well as personalization of diplomas, etc. (2) the art of producing such handwriting. *Related Term:* engrossing.

callout — descriptive label as part of an illustration; text that explains or amplifies. It is often accompanied by a line pointing to a particular area. *Related Terms:* blurb; pull paragraph; pull quote.

camera — a light-tight photographic device that records the image of an object formed when light rays pass through an aperture and fall on a flat, photosensitive surface. In addition to the aperture and a lens, other camera components may include automatic or manual focus and size adjustments, a film, or paper-holding mechanism, and an area for previewing the final image. A light source and metering device may also be included. *Related Terms:* darkroom camera; camera, process; camera, gallery.

camera back — the part of a process camera that holds photographic film in place with a vacuum or other material-holding system while an exposure is made.

camera back masking — a single stage color correction process where the processed masks are placed, in turn, over unexposed separation film in the camera back before producing the separation negatives.

camera body — light-tight structure into which film, lens and other photographic control components are mounted. The camera shutter is located on the body of some cameras.

camera copy — anything that will be photographed by a process camera in preparation for graphic reproduction. *Related Terms:* camera ready copy; camera ready.

camera flare — any unwanted light that reaches the film in the camera. Light leaks in the camera case, the bellows, misadjusted lights, etc., can all cause the condition.

camera lens — the part of a camera that passes light into the camera body and focuses the photographic image. Most often composed of several highly purified glass elements.

camera lucida — a mechanical aid used by illustrators to assist in enlarging and reducing subjects to be copied by hand. *Related Term:* Lucy.

camera ready — artwork or paste-up material that is ready for photoconversion and reproduction.

camera ready art (or copy) — (1) in conventional work flows, any artwork or type that is ready to be prepared for mechanical reproduction; the mechanicals, photographs and art fully prepared for printing. *Related Term:* camera ready.

camera service — business using a process camera to make diffusion transfers, halftone negatives, printing plates and other photo-related elements for commercial printing.

camera, aperture priority — a type of camera on which the photographer sets the aperture and the camera automatically adjusts the shutter speed based on film speed.

camera, darkroom — a process camera in which the film can be loaded in the darkroom and the camera extends from the darkroom into the area of the photography studio where "room light" operations can be performed.

camera, gallery — type of process camera normally situated entirely in a normally lighted room; film is carried to the camera in a light-tight case or film holder, the exposure is made, and the case is carried to the darkroom for processing. *Related Term:* camera, darkroom.

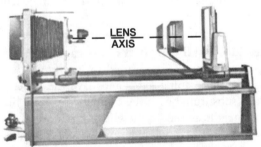

HORIZONTAL PROCESS CAMERA
A typical horizontally configured process. The film is placed at the extreme left, which would be inside the darkroom. All components to the right of the vertical support on the left are outside of the darkroom, in the gallery. Note the front case which houses the lens, and the copyboard on the right with attached lights. Both lens board and copyboard positions can be controlled in the darkroom to cause enlargement or reduction to an exact percentage of the size of the original.

camera, horizontal — most common quality graphic arts camera, where the exposure axis runs along a horizontal plane. A greater degree of enlargement and reduction is possible with this arrangement than can be practically done with a vertical camera which is physically limited as to how far the back case, front case and copy board may be separated.

camera, large format — camera that makes negatives 4 in. x 5 in. or larger.

camera, process — a large camera with a highly corrected flat field photographic lens, especially designed for sharp definition and having an accurately calibrated reduction and magnification scale. Used for creating negatives and positives used in the photomechanical reproduction of line, halftone and full color photography. *Related Terms:* camera, gallery; camera, horizontal; camera, vertical.

camera, shutter priority — a type of camera on which the photographer sets the shutter speed and the camera computer automatically adjusts the aperture based upon programmed film speed or ISO settings.

camera, single lens reflex (SLR) — a category of camera that lets the photographer view the subject through the same lens that is used to take the picture to eliminate parallax. Opposite of a range finder camera. *Related Terms:* camera; twin lens reflex; parallax.

camera, small format — camera making negatives 35mm or smaller.

camera, stat — obsolete type of camera which traditionally photographed directly onto photographic paper to produce a reversed print. No negative was generated and each copy was an "original." First court-accepted facsimile. Largely displaced by xerography and similar technologies which produce true positive duplications.

camera, studio — *Related Term:* camera, view.

camera, technical — *Related Term:* camera, view.

camera, twin lens reflex — a category of camera with two lenses. One lens is used to view the subject and focus the image. The second lens is used to make the exposure. *Related Terms:* parallax; SLR.

camera, vertical — a space-saving graphic arts camera that is usually housed entirely in the darkroom or an adjacent light tight-room. The copyboard and film holder are perpendicular to a vertical optical axis of the camera lens (i.e., the copyboard is at the bottom and the camera back at the top). Because of this arrangement, it has limits to its reduction/enlargement sizes without the use of a secondary or changeable lens. Not generally considered the best choice for quality reproduction, but favored by small shops and agencies with limited floor space. *Related Terms:* camera, horizontal.

camera, view — a camera with a ground glass on the film plane to allow a photographer to preview the subject prior to film exposure. Generally they are of medium or large format and most often are used in studio settings. Also referred to as a field camera, although the structures are somewhat different. View cameras are usually on a monorail of which the front and rear standards move from front to back. A field camera folds in a more compact fashion. It can be compared to a range finder camera or a single lens reflex camera, although the view and field have many adjustments for focus and perspective and can accommodate a large variety of lenses. *Related Terms:* studio camera; technical camera.

camera, viewfinder — normally, a small viewing square or rectangle to the side of the lens. This arrangement makes for a simpler and less expensive device, but at closer distances the parallax between the actual taking lens and the view seen by the small window do not coincide. While many amateur cameras use the technique, it is more typically found on disposable, one-use cameras. *Related Terms:* camera, single lens reflex; camera, twin lens reflex; camera, range finder.

cameraman's sensitivity guide — usually a piece of photographic film or paper with several shades of gray from white to black used in test exposures of photographic film. Each shade has calibrated density value. *Related Terms:* gray scale; photographer's sensitivity guide; step tablet; step guide; step wedge.

Cameron belt press — a type of belt press that both prints and binds a book, it is a web-fed printing machine used for book production which prints from rubber plates imposed on a flexible belt and fastened in place with double sided tape; designed to mass produce an entire book rapidly. Particularly popular in paperback and soft cover book publishing.

cancelbot — monitoring programs which are left running on an Internet server to automatically scan for and delete any Usenet postings from persons (or containing subjects) the system administrator deems inappropriate. Cancelbots are primarily used these days to filter out incoming spams on news servers from notorious e-mail addresses. *Related Terms:* cancelbunny; cancelpoodle; cancelmoose.

cancelbunny — *Related Term:* cancelpoodle.

cancelmoose — any person on the Internet who wages war against spamming.

cancelpoodle — the nickname for people who delete Usenet messages posted to newsgroups, claiming copyright infringement.

canned format — the specifications for composition and/or makeup of type kept on magnetic or paper tape for repeated use to command a typesetter.

canned text — *Related Term:* boiler plate.

cap — a contraction of "capital," meaning an upper case character. *Related Terms:* majuscule; capitals.

cap height — height of the capital letters of a typeface, expressed in points, usually about 2/3 of the type size or the height of a point size from the base line to the top of the upper case letter. *Related Terms:* x-height; cap line.

cap line — one of four typographic lines of reference. The uppermost limit of capitals and ascenders. An imaginary line across the top of capital letters. The distance from the cap line to the baseline is the cap size.

capillary film — used in screen printing to create the printing stencil. *Related Term:* presensitized direct film.

capitals — majuscules. The large characters, the original form of Latin inscription characters also known as caps or upper case. *Related Terms:* majuscules; miniscules; caps.

caps — an abbreviation for capital letters.

caps and small caps (C&sc) — capitals and small capitals. In composition, used to specify words that begin with a capital letter and have the lower case letters substituted with the capital letter form but of same height as the "x" height of the normal lower case letters. They look like this: THIS IS A SAMPLE OF CAPS AND SMALL CAPS. *Related Term:* C&lc.

capstan design — a film/paper transport mechanism used in many imagesetters. Photographic paper or film is fed through the machine by being "pulled" off a roll by passing between a series of rubber friction rollers. Output quality often considered inferior to the drum design. *Related Term:* drum design.

caption — identifying or descriptive text and the traditional term used in magazine layout for the explanatory matter accompanying art; usually called a cutline in newspaper editing and layout, but now more generally an explanatory text accompanying illustrations in all types of publications. *Related Term:* cutline.

CAR (computer assisted retrieval) — in records archiving, a microfilm system that is operated and controlled by a special computer which has an index of the stored materials. When a request for a specific record is entered, the system automatically scrolls the microfilm to the correct frame.

car sign board — the name given to board paper for signs inside buses, trains, etc. Car sign board is approximately .020 inch thick (20 points) and coated on one side.

carbon arcs — obsolete technology. Most often two copper-clad carbon rods to which current with high amperage was applied. The current jumped or "arced" between the rods, creating electromagnetic emissions which were heavy in blue and violet light; was used commonly as a light source for the graphic arts industry but has been abandoned due to toxic emissions and unpredictability of illumination quality.

carbon black — the pigment commonly used in black inks. Toners are usually combined with the pigment in the ink formulation to make the black ink more neutral.

carbon dioxide conditioning — one method of making a photopolymer plate very sensitive to ultraviolet light by displacing oxygen with carbon dioxide in a special pressurized chamber. *Related Term:* heat sensitizing.

carbon dioxide laser — an infrared laser used for cutting paper and other lightweight materials. Used in rubber plate engraving for flexography.

carbon paper — one type of paper coated with black wax-like material that transfers to an underlying sheet with pressure from writing or typing.

carbon printing — a process of transferring the positive image to the cylinder mask in conventional gravure.

carbon tissue — gelatin-based material coated on a paper backing and used in gravure printing. *Related Term:* carbon printing.

carbonate — salts formed by the combination of some metal with carbon. In photography, the most common is sodium carbonate.

carbonless paper — a specialty paper that produces duplicate copies of handwritten, typed, or otherwise impact-printed sheets without the use of carbon paper. To achieve this, the paper is coated with two different micro-encapsulated chemicals, one on the face and the other on the reverse side of the sheet. When pressure is applied to two attached sheets of paper, the encapsulated chemicals break and mix, producing a visible image similar to that formed by carbon paper. Originally manufactured by National Cash Register Corp. *Related Terms:* carbonless; NCR (no carbon required) paper; action paper; impact paper; NCR paper; self-copy paper.

card — (1) in electronics, a printed circuit board and the electronic circuit components on these boards. (2) in printing, any paper used to make products such as postcards and index cards.

cardboard — a term for stiff, bulky paper such as blanks, index, tag, or Bristol.

cardbus — a specification that allows PCMCIA (PC cards) to transfer data at rates exceeding 100MB/sec. Older, 16-bit cards transfer data at a rate of 20MB/sec. *Related Terms:* PCMCIA; PC card.

cardinal numbers — normal sequence of numbers, one, two, three, as compared with ordinal numbers, first, second, etc.

caret — (1) in typography, proofreaders' mark indicating where changes are to be or to indicate where corrections or additional copy are to be inserted. (2) in press make ready, cuts in the shape of inverted V's made in the packing of letterpress presses and used to align the make-ready sheet with the type form. (3) in computing, a "control" symbol for MS-DOS computers. *Related Term:* platen dressing.

carload (CL) — the selling unit of paper weighing approximately 20,000 or 40,000 pounds, depending on which mill or merchant uses the term.

carousel carrier — multicolor screen printing machine which has a stations for each color screen about a central axis. Impressions are made and the substrate is rotated to the next adjacent position about the axis for subsequent color application. *Related Term:* carousel screen press.

carousel screen press — equipment that holds individual screen frames, one for each color; rotates around a central axis and moves each screen into position sequentially over the object to be printed. *Related Term:* carousel carrier.

carrier — (1) in computing, another name for a phone connection through a modem. A flashing modem light labeled CD indicates that the modem is receiving a carrier detect (CD) signal and is hooked up to another computer. (2) in printing, an incorrect short version of image carrier or plate. (3) in broadcasting, a high-frequency radio signal which is modulated to carry information long distances through space or via cable. *Related Terms:* CD; FM; AM.

carrier sheet — (1) an alternate name for release sheet, and the name often applied to the base for pressure- sensitive label materials, designed to facilitate easy removal for use. (2) an insertion in bagged goods, such as publications and periodicals, which has the address label and postal indicia or other markings. (3) the material onto which dry transfer (rub-off lettering) is attached. *Related Term:* release sheet.

cartography — the skill of drawing or making maps and navigational charts.

carton — the shipping and selling unit of paper weighing approximately 150 pounds (60 kilos). A carton contains from 500 to 3,000 sheets, depending on the size of sheets and basis weight.

cartouche — (1) an elongated frame used in hieroglyphic symbolism. Characters enclosed were usually of royal family or retinue. (2) in modern typography, merely an oblong border used to enhance and add sophistication to a design or printed piece.

cartridge — (1) in art preparation, a thick general purpose paper used for print drawing and wrapping. (2) in imaging, the name given to the replaceable toner, imaging drum assembly and/or film supply in desktop publishing, imagesetting and copiers.

case — the covers of a hard-bound, or case-bound, book.

case bind — binding signatures with glue to a case made of binder's board covered with fabric, plastic, or leather, yielding hard-cover books.

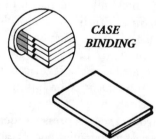

CASE BINDING

Printed signatures are gathered, sewn and glued into a heavy cover of binder board and cloth, paper or leather.

case binding — an expensive but durable type of binding in which a hard cover is placed on both sides of the bound sheets to form a useful book. Generally assoc-iated with most high-quality book-binding. Cases are usually covered with cloth, vinyl or leather. *Related Terms:* case bind; casing in.

case board — *Related Term:* binder's board.

case sensitive — the situations where it matters if letters are typed in uppercase or lowercase. Some computer programs, systems, and network services are case sensitive: "XYZ" is considered different data than "xyz."

case-bound — another name for a hard-bound book. *Related Terms:* hardback; hardcover.

casing in — part of the case-binding method in which the finished book bodies and book cases are united.

cast — (1) generally, in all graphics, an unwanted tinge or shade of color present in an image. (2) a cast, in computer language such as C or Java, is a program action that converts an object from one type to another: for example, changing a floating-point to integer. Different computer languages have unique and specific rules defining how a cast may occur. Programmers can optionally perform a cast directly or the language can perform the cast at processing time. (3) in hot metal composition, the act of injecting molten metal into a matrix for the purpose of forming relief letters from which to print. *Related Term:* Java.

cast coated — a high-gloss, ink-absorbent paper with an exceptionally glossy enamel finish, most often only on one side of the sheet. Cast coated paper is dried under pressure against a polished cylinder during manufacture.

cast coater — a device that applies a wet coating to a paper web before it contacts a highly polished, heated cylinder drum.

casting — (1) in typesetting, a process in which molten metal is forced into molds (matrices) to produce relief letter forms. Type can be cast as single characters or as complete lines. (2) in stereotyping, the casting of a full page, or components of pages, into metal printing plates (stereotypes) from matrices (mats) for newspapers or books. *Related Term:* hot metal composition.

A SINGLE LETTER OF CAST METAL

casting box — a mechanical device used in the process of casting a stereotype plate from a paper matrix.

casting off — a process of calculating and determining the amount of space manuscript copy will occupy when set in a given typeface within a given area. *Related Terms:* copyfitting; casting; castoff.

castoff — a calculation determining how much space copy will take up when typeset, based upon known alphabet character counts for a given type size and face. *Related Terms:* casting; casting off; copyfitting.

catalog — a purpose-specific publication designed to advertise and sell a wide or specific variety of goods and/or services, usually offered by a single company. They can be anything from a small pamphlet or brochure to a several hundred page book.

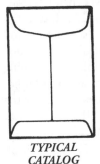

TYPICAL CATALOG ENVELOPE

catalog envelope — a large envelope with the opening on the short side or end. Originally designed to hold catalog or magazine publications.

catalog paper — coated paper rated #4 or #5 with basis weight from 35# to 50# (50 to 75 gsm).

catalog sheet — a single sheet, usually standard letter size, designed to advertise and sell good or services. Often they include an order form or similar ordering information.

catching up — a term that describes a condition in lithography in which the non-image areas of a press plate begin to take ink. *Related Term:* scum.

catchline — (1) in editorial layout, a line of display type between a picture and cutline. (2) in composition, a temporary headline for identification on the top of a galley proof or on imagesetter output.

Catch-UP — one browser-based service that provides software update information. It finds and downloads the latest software releases and keeps your software up to date with ease, as long as the manufacturer co-operates and your credit is good.

cathode ray tube (CRT) — (1) in third generation phototypesetting, an electronic tube used to display letter images, in the form of dots (computer logic character formation) or lines (character projection), for exposure onto film, photographic paper, microfilm, or offset plates. (2) in computing and television, an electronic tube used to project images on a screen for direct viewing off of the phosphor coated face. *Related Terms:* video

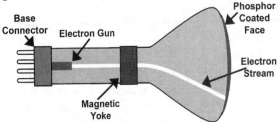

CATHODE RAY TUBE (CRT)
An electron stream fired onto phosphors on the interior of the face plate cause them to glow and produce visible light. The magnetic yoke can alter the path of the electrons, causing them to sweep across the face in a raster (lines) of information.

display tube; visual display unit; picture tube; VDT.

cause and effect chart — in production management, a tool used to outline all elements of a process or problem.

CAV (constant angular velocity) — a term used to describe the characteristics of the most standard type of laserdisk where the disk spins at a constant rate. *Related Term:* Red Book.

CB — carbonless paper abbreviation indicating back side only coating. *Related Terms:* NCR; CF; CFB.

C-band — frequencies from 4 to 6 gigahertz (Ghz) used to transmit and receive signals to and from satellites.

CC (color correction) filter — (1) in graphic arts photography, filters used to assist in retaining or dropping selected portions of color copy on the process camera or (2) in commercial photography, filters used to adjust color balance in color photo printing.

CCD (charge coupled device) — an optical chip; a light-sensitive solid state semiconductor that captures image elements (photosites). A CCD consists of a set of image elements (photosites) arranged in a linear or area array. Images are digitized by an external light source that illuminates the source document, which reflects the light through optics onto the silicon light sensors in the array. This generates electrical signals in each photosite, proportional to the intensity of the reflected illumination. Movement of a linear array and sequenced activation produces raster images as the array passes over the object. Much more expensive area arrays capture the entirety of the object at one time. Used in digital cameras, bar code and optical scanners, etc., to detect relative brightness of images is scans. In 1-dimensional bar code reading, multiple numbers of light-sensitive areas can be arranged linearly. 2-dimensional bar codes require a two-dimensional array for proper reading of the symbols.

CCD array — a group of light-sensitive solid state recording elements arranged in a line (linear array) or x-y matrix (area array) and used as a image importing device in scanners and digital photography. *Related Terms:* charge coupled device; CMOS.

CCD camera — a digital imaging device which utilizes charge coupled devices as sensors in place of conventional film. Easily portable, captured images are directly downloaded into a computer system and the magnetic or solid state storage medium (film) can be reused.

CCITT (Comité Consultatif Internationale de Telegraphie et Telephonie) — the international committee based in Geneva, Switzerland, that recommends telecommunications standards, audio compression/decompression standards (codecs) and the standards for modem speed and compression (V.34, V.90, etc.). The organization has officially changed its name to ITU-T (International Telecommunications Union-Telecommunication), but the old name persists.

CD — (1) in telephony, carrier detect, a communications signal or "handshake" created when two modems connect. CD remains present as long as the two modems are connected, though it can be interrupted by a call-waiting signal if one of the phone lines receives a second call. (2) in computing and multimedia, a compact disk. Many variants have been developed including audio CD, CD-ROM, CD-I and CD-R. These formats' standards are described in published specifications which are described by a color: the Red Book standard describes audio CDs: All must be 5.25-inch plastic-coated metal platters with a specified track configuration, etc. Most CDs of any format can fit even into players that won't recognize them, but many CD-ROM drives will recognize and play audio CDs. A new and more highly sophisticated version of the technology has now appeared in the form of DVD, which has (potentially) 30 times the data storage capacity of conventional CDs. It, too, is used for audio, video and computer data. Most DVD players are compatible with CD audio and computer disks, but conventional CD players of any type cannot read the DVD information. *Related Terms:* CD-DA; CD-I; CD-ROM; CD-R; DVD; Green

Book; Orange Book; Red Book; White Book; Yellow Book; AA; carrier; HS; RS-232.

CD-DA (compact disc digital audio) — another name for the Red Book audio CD format. This standard allows for up to 74 minutes of digital sound that's transferred at a data rate known as "single-speed," 150 kilobytes per second (K/sec) because the original CD-ROM drives transferred other data at the same rate; very slow when compared with conventional hard disk speeds. Subsequent developments resolved this issue with the introduction of 2x (double-speed), 4x (quad-speed) 8x and now 32x and faster CD-ROM drives. *Related Terms:* CD; CD-I; CD-ROM; Green Book; Orange Book; Red Book; White Book; Yellow Book.

CDF (channel definition format) — popular "push" technologies that automatically deliver channels (Web pages) to browsers or network clients. Push technology uses a separate file that specifies structure and lists the web pages, images, and controls to be pushed to the user. The file also includes headlines describing content, download scheduling information and other options. Developers use HTML-like tags to create the CDF file. *Related Terms:* push; HTML.

CD-I (compact disc interactive) — a proprietary CD-ROM format. It has its own CD Green Book standards, but it is not a widely supported platform: the standards are published by Sony and Philips. *Related Term:* Green Book.

CDMA (code division multiple access) — a digital spread-spectrum modulation technique used primarily in personal communications devices. It digitizes the conversation and marks it with a special frequency code. The data is then scattered across the frequencies of the band in a random pattern. The receiver is instructed to decipher only the data corresponding to a particular marking code to reconstruct the signal. *Related Terms:* spread spectrum; frequency spectrum; S-CDMA.

CD-R (compact disc recordable) — a compact disk format that enables recording of data onto compact discs that can be read by regular CD-ROM drives. Data can be recorded on different occasions (multi sessions).

CD-ROM (compact disc, read-only memory) — a compact disc used to store and play back computer data. It is not intended for audio. They can contain up to 650MB of data. The format has become a favorite medium for installing programs; they are extremely cost effective as they cost only slightly more to manufacture than a single disks and modern programs commonly contain up to 39 floppy disks. *Related Term:* CD-ROM drive.

CD-ROM drive — a computer peripheral that plays back CD-ROMs and, with the right software, audio CDs. The configuration includes a spindle upon which the disc rotates, an unmodulated laser that shines onto the disc's surface, a prism and optics that deflect and direct the laser beam. The surface is covered with "pits" arranged on a spiral track. As the laser shines onto the surface, the pits deflect the laser away from the optical path. A light-sensitive diode detects the light/no light condition. Potentially very fast as a data retrieval system, basic audio CD standard transfers data at only 150 kilobytes per second (KB/sec) and does not utilize the potential. Data disks contain much more material which must be accessed faster to satisfy expected computing speeds. The original (1x) speed has been replaced by speeds up to 48x. This is great for computing, and conditions are changing, but unfortunately most present CD-ROM titles still transfer data only within the 2x/4x bandwidth and, typically, there is not much advantage to drives with higher transfer rates than with a 4x drive. *Related Term:* CD-ROM.

CD-XA — an extended architecture for CD-ROM. Billed as a hybrid of CD-ROM and CD-I, promoted by Sony and Microsoft. The extension adds ADPCM audio to permit the interleaving of sound and video data to animation with sound synchronization. It is an essential part of Microsoft's multimedia computers.

cedilla (ç) — accent beneath the letter "c" indicating that it should be pronounced with a short sound.

cell — (1) in data base computing work, this is the smallest component of a table. A table may have one or more cells within any one row. (2) in gravure printing, the tiny etched depressions on the cylinder which will hold ink for reproduction. (3) in flexographic printing, the tiny engraved depressions in the anilox roll which meters and transfers ink to the plate.

cell padding — space between the contents of a cell and the inside edges of that cell. A table with no cell padding will wrap tightly around its contents.

cell spacing — the distance between cells within a table. A table with a cell spacing of 5 will push each cell 5 pixels away from each other.

center lines — lines drawn on a paste-up in light blue pencil to represent the center of each dimension of the illustration board.

center mark — (1) in stripping operations, a small triangle cut in a masking sheet to show the center of the masking sheet and to help identify the gripper edge of the plate. (2) generally, lines on a mechanical, negative, printing plate or press sheet showing the center of a page or press sheet. *Related Terms:* crop marks; corner marks; edge marks; trim marks.

center point — the dot that appears between letters of words, generally in a dictionary, to indicate syllable breaks. Should not be confused with a bullet, which has a different base alignment.

center spread — Two facing pages located in the centerfold of a signature or magazine. *Related Term:* spread.

centered type — lines of type set centered on the line measure. *Related Terms:* quad center; quad left; quad right; quad; center justified.

centerline — the vertical axis around which character elements are located for letters, numerals, or symbols.

centipoise — a measurement used for vicosity of flowing materials.

central impression cylinder press — a press that contains a large water-cooled central cylinder which acts as a substrate carrier and has the printing units positioned around the outer edge. The web of substrate is in continuous contact with this cylinder and printing units are moved against it, around its circumference. *Related Term:* McCall press.

central processing unit (CPU) — the " brains" or circuitry section of the computer that controls the interpretation and execution of instructions from a program or an input device.

centricleaner — a special cleaner used in the papermaking process that separates unwanted materials from the prepared wood fibers.

Centronics — named after the company that invented it, a brand trade name usually used to refer to the standard 36-pin connector used for parallel printers.

Century Schoolbook — a popular serif typeface used in magazines and books for text setting which has a large x-height and an open appearance.

CEPS (color electronic prepress system) — (1) an industry-standard production environment designed to facilitate production in pre-press operations. (2) a high-end computer system that is used to correct colors and assemble images into final pages.

CERN (The Conseil Européen pour la Recherché Nucléaire or the European Laboratory for Particle Physics) — the birthplace of the World Wide Web. A particle physics laboratory in Geneva, Switzerland, CERN is one of the world's largest scientific laboratories and an outstanding example of international collaboration of its many member states.

certificate — a document confirming completion, ownership status, etc. Often ornate and designed to impress.

CF — the abbreviation used with carbonless paper indicating coating on its front side only. *Related Terms:* CB; CFB; NCR; carbonless.

CFB — the abbreviation used with carbonless paper indicating coating on its front and back sides. *Related Terms:* CB; CF; carbonless; NCR.

CGA (color graphics adapter) — a generally obsolete low resolution video display standard, invented for the first IMB-PC. CGA pixel resolution is 320 x 200.

CGI (common gateway interface) — a standard mechanism used to extend Web server functionality, allowing the Web server to take external data from an Internet user, process the data and return results based on that data. These external programs are called gateways because they open up an outside world of information to the server. *Related Term:* HTTP.

cgi-bin (cgi binary) — a name that appears in a Web browser's URL window and indicates that you're running a common gateway interface (CGI) program, such as a search tool. *Related Term:* CGI.

CGM (computer graphics metafile) — a standard format that allows for the interchanging of graphics images between platforms.

chain gripper delivery — press delivery unit technique in which sheets are pulled onto the delivery from the impression cylinder to the delivery stack by a pair of bicycle-type chains with gripping fingers between them. The grippers are timed to open and close in conjunction with the printing press cycle. The result is a delivery pile of identically placed sheets which can be stacked higher than with a conventional chute delivery. Often the delivery is equipped with side and front jogging mechanisms to minimize the need to further handle the sheets prior to additional presswork or bindery.

chain lines — the widely spaced watermark lines (usually about 1" apart), caused by chaining marks made when smaller fragments of the paper machine wire are sewn together to form a whole. The marks usually run with the grain in laid papers. Chain lines are often natural in handmade papers, are considered attractive and are (artificially) imitated in machine made papers. Chain lines usually run with the grain on laid paper when cut to letterhead size.

chalking — (1) ink that is smeared or easily removed from the sheet or does not stick to the printed substrate. (2) a powdering effect left on the paper surface after the ink has failed to dry satisfactorily due to a fault in printing. (3) incorrect term for printed sheets which have been powdered to excess during the delivery operation.

chalky — the extreme harshness or hard contrast of a print in which the highlights are pure white with poor or no detail.

chambered blade system — a type of enclosed press ink fountain which automatically meters ink as required by the image carrier.

changing bag — a large bag, usually black, made of a number of thicknesses of material, with holes for arms to enter. Used for loading and unloading film in small cameras, and for loading film holders out in the field, on location, when no darkroom is available.

channel — in broadcasting, (1) a half-circuit. (2) a radio frequency assignment (which is dependent upon the frequency band and the geographic location), i.e., channel 1, channel 2, etc.

CHAP (challenge-handshake authentication protocol) — an authentication protocol used to verify a user's name and password for PPP Internet connections. CHAP is more secure than PAP; it performs a three-way handshake to establish the link between computers, and can reestablish the authentication anytime during the connection. *Related Terms:* PPP; PAP.

chapbook — a small book or booklet, often part of a series.

chapter heads — chapter title and/or number of the opening page of each chapter.

character — (1) in computing and data processing, any symbol which may be assigned, for information processing purposes, a numeric or symbolic representation in a coding scheme. (2) in printing, any printable symbol, including letters of the alphabet, numbers, punctuation, spaces and special symbols. (3) in optical reading, a single group of bars and spaces which represent an individual number, letter, punctuation mark or other symbol. *Related Term:* font.

character alignment — the vertical or horizontal position of characters with respect to a given reference line.

character compensation — *Related Term:* tracking.

character count (cc) — a process used as a first stage in type calculations. (1) a predetermined measurement of the approximate number of characters of a specific type size and design which can be set in one linear pica, or the actual count of each character and space in a piece of copy. (2) the number of characters in a line, paragraph, or piece of copy. *Related Term:* characters per pica (cpp).

character encoding — character encoding is a table in a font or a computer operating system which maps character codes to glyphs in a font. Most operating systems today represent character codes with an 8-bit unit of data known as a byte. Thus, character encoding tables today are restricted to at most 256 character codes. Not all operating system manufacturers use the same character encoding. For example, the Macintosh(R) platform uses the standard Macintosh character set as defined by Apple Computer, Inc. while the Windows(TM) operating system uses another encoding entirely, as defined by Microsoft. Fortunately, standard Type 1 fonts contain all the glyphs needed for both these encodings, so they work correctly not only with these two systems, but others as well. *Related Terms:* character, glyph, keyboard layout.

character generation — (1) the formation or projection, by the formation of scanning lines, of typographic images on the face of a CRT or by laser in high speed phototypesetting systems, as in third and fourth generation photoimager systems. (2) con-ventionally, the projection of character images onto film, paper or plates via a photo negative such as used in the first and second generations of phototypesetting. *Related Terms:* digital typesetting; phototypesetting.

character recognition — the act of the computer converting printed information into computer-readable text. *Related Terms:* optical character recognition; OCR; intelligent character recognition; ICR.

character set — those characters that can be encoded in a particular bar code symbology.

character string — a collection or sequence of digital bytes which are treated singly by a computer, such as dates, social security numbers, zipcodes, etc.

character-based display — a non-WYSIWYG terminal, without graphics. The only displayed characters are ASCII and/or some proprietary symbols pertinent and appropriate to the specifics of the machine's function.

characteristic curve — visual interpretation of a light-sensitive material's exposure/density relationship, most commonly via graphs or charts. *Related Term:* DlogE curve.

characters per pica (cpp) — a system of copyfitting that utilizes the average number of characters per pica of a given type face in a given size as a means of determining the length of the copy when set in type.

charge corotron — in electrostatic copying, a charging device (usually wire) which places an electrical charge on the photoconductive drum or belt to produce images in the imaging process. *Related Term:* charge transfer.

charge coupled device (CCD) — *Related Terms:* CCD; area array; linear array.

charge transfer — (1) the process where the toner particles are conveyed from the photoconductor to the substrate through a transfer corona. (2) the movement of electrical charges from one surface to another in an imaging system (e.g., from photoconductor to paper). *Related Term:* charge corotron.

chase — (1) a metal frame for preparing and holding letterpress forms in which metal type and blocks (engravings) are locked into position to make up a page for reproduction in a relief class printing press. (2) a prepared screen printing frame and stencil.

chaser lockup — a method of locking up type forms with dimensions of furniture that are larger than the width of the form and top and side furniture overlap at the corners.

chat — a generic term to describe real time interactive computer input between individuals sharing a common interest. Most chat rooms have a particular topic (which you are expected to discuss) but there are some that are purely for meeting other people. Communication is by typing messages which appear on all other screens in the area. More advanced products assign avatars (2D or 3D characters) to each participant or may even have expressions selected by the chatters. In addition to typing, some of the most advanced software also let users with sound cards speak to each other. *Related Term:* avatar.

chat room — a virtual "place" offered by an online service provider in which one or more people can participate in live chat. *Related Term:* chat.

check character — a character included within a symbol whose value is used for the purpose of performing a mathematical check to ensure the accuracy of the read.

check digit — *Related Term:* check character.

check paper — *Related Term:* safety paper.

checkbox — (1) in computing, a form field used to select an item by simply clicking inside a box. This is typically used to represent a selection among a number of choices. (2) in typography and composition, one form of dingbat or sort which usually is comprised of an outline box with a check mark inside, indicating a selection.

checking for lift — a procedure that checks all elements within a locked chase; determines whether the elements are fixed tightly in position and will not move or fall out during handling or printing.

chemical pulp — paper pulp made by using chemicals to remove lignin and other impurities from wood chips and separate them into fibers. *Related Term:* groundwood pulp.

Cheshire labels — addresses printed on wide computer paper in a format that can be cut into labels and affixed to newsletters by machines built by the Cheshire Company, a part of Xerox Corporation.

China clay — a general term for clay used to make the coating for coated paper.

Chinese white — a very light hued opaque mask or paint used in design and pre-press operations.

chip — (1) a thin silicon wafer on which electronic components are deposited in the form of integrated circuits; the basis of digital systems. (2) in papermaking, the resulting product of the process of chipping.

chipboard — an inexpensive, single-ply board, usually brown or gray. Chipboard is commonly used as the backing for note pads and point-of sale displays. *Related Terms:* cardboard; fiber board.

chipping — reducing wood logs to small chips in the pulp paper making process.

chloride paper — a very slow printing paper for contact prints. The light-sensitive emulsion contains chloride and silver.

chlorobromide — a type of printing paper which give warm, black tones. It is the intermediate between chloride and bromide papers.

choke — a photographic masking technique in which one color area is made slightly smaller, used in conjunction with another trapping technique called a "spread," in which another color area is made slightly larger to allow for misregistration on press. *Related Term:* spread.

chopper fold — in web publishing, the final fold made at the delivery.

chroma — the brightness or purity of a color. It is sometimes called the grayness of the color.

chroma check — a peel-type color proofing system manufactured by E. I. Dupont.

chroma key — image overlaying (keying) technique where one video signal is placed over another. The areas of overlay are defined by a specific range of color, or chrominance, on one of the signals. TV meteorologists, for instance, often stand in front of a blank blue or green wall to create the primary signal; the secondary signal (weather maps) are overlaid electronically.

chromatic — perceived as having a hue; not white, gray or black.

chromatic attributes — those attributes associated with the spectral distribution of light: hue and saturation.

chromaticity — that part of color specification which does not involve illuminance. Chromaticity is two-dimensional and is specified by pairs of numbers such as dominant wavelength and purity.

chromaticity diagram — the graphical representation of two of the three dimensions of colored light sources. Not intended to represent surface colors. Often called the CIE diagram.

chrome — an end term used to describe color transparency materials. It is a contraction of brand names such as Kodachrome, Fujichrome, Ektachrome or Agfachrome, etc.

chrome plating — the process of placing a thin coating of chrome metal over the surface of an engraved gravure cylinder to make it durable for long printing runs.

chrome yellow — an inorganic pigment that is primarily lead chromate, used for making opaque yellow inks.

chrominance — in broa2dcasting and computing, the color portion of a video signal.

Cibachrome — a direct positive photographic dye bleach process paper which is exposed through a color transparency and has very stable color characteristics.

Cicero — the typographic unit in the Didot system. Equal to 4.55mm or .1776 inches. Each Cicero contains twelve Didot points, and is approximately 7% larger than a pica.

CIE (Commission Internationale de l'Éclairage) — translated as the International Commission on Illumination, the main international organization concerned with color and color measurement. The international group that first developed a universal set of color definition standards starting in 1932 and made revisions since. Has developed a set of color definition standards endorsed by Adobe for PostScript Level 2.

CIE 1976 L*a*b* color space — a uniform color space utilizing an Adams-Nickerson cube root formula, adopted by the CIE in 1976 for use in the measurement of small color differences.

CIE 1976 L*u*v* color space — a uniform color space adopted in 1976. Appropriate for use in additive mixing of light (e.g., color TV) and when an associated chromaticity is desired

CIE chromaticity coordinates — the ratios of each of the tristimulus values of a color to the sum of the tristimulus values. In the CIE systems they are designated by x, y, and z.

CIE diagram—*Related Term:* chromaticity diagram.

CIE luminosity function (y) — the 1924 CIR plot of the relative magnitude of the visual response as a function of wavelength from about 380-780 nm.

CIE standard observer — a hypothetical observer having the tristimulus color mixture data recommended in 1931 by the CIE for a 2 degree field of vision. A supplementary observer for a larger 10 degree field was adopted in 1964.

CIE tristimulus values — the amounts of the three reference or matching stimuli required to give a match with the color stimulus considered, in a given trichromatic system.

Cinepak — developed by Apple and SuperMac Technologies, Cinepak can compress video files to 1/

25th their original size. While the codec (compressor/decompressor) was originally developed to play QuickTime movies off CD-ROMs, it is now also widely used to shrink QuickTime and AVI movies for transfer over the Internet. It's even built in to Windows 95/98. *Related Terms:* Codec; AVI; QuickTime.

circle cutter — a cutting tool designed to cut perfect circles in screen printing film and conventional stripping film. *Related Term:* bean cutter.

circuit — in electronics, a means of both-way communication between two or more points.

circular — an advertising flier inserted into a newspaper.

circular screen — (1) in process camera work, a circular-shaped (usually glass) halftone screen mounted on a revolving frame at the film plane of a process camera that helps the camera operator to obtain proper screen angles for halftones by rotating the screen. (2) in design, a mechanical contact screen with (usually) evenly spaced concentric circles about a midpoint. When used with illustrations it causes the eye to gravitate toward the center point, which is usually placed at the desired impact area of the art.

circulation — the number of readers of a periodical such as a magazine or newspaper.

citric acid — a tribasic organic acid used as a preservative in developers and the acid constituent in some fixing and clearing baths.

CL — in paper manufacturing and sales, abbreviation for carload.

clasp envelope — generally, a white or kraft envelope which can have a glued flap, but which always has a reinforced receiving hole and a metal prong on the envelope body. By closing the flap and bending the prongs through the hold, temporary closure is made.

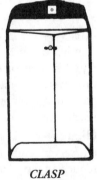

CLASP ENVELOPE

class — basic divisions in type classification such as sans serif, serif, square serif, gothic, script and decorative.

classified ad — an advertisement that uses only text, as opposed to a display ad, which also incorporates graphics.

clay — the fine-grained natural material, usually aluminum silicate, used to make coating for coated paper. *Related Term:* China clay.

clean copy — copy that is ready to be typeset, or copy that has already been typeset and contains no further corrections.

clean floppy disk — a magnetic floppy or flexible disk containing error-free typeset copy.

cleanness — synonym for high saturation.

cleanup sheet — an absorbent, blotter-like paper sheet used on a lithographic duplicator to remove ink from the ink rollers.

clear areas — the portions of a prepared offset plate that have no ink. Also known as the nonprinting (hydrophilic) or ink-repellent areas. *Related Term:* quiet zone.

clear key — a terminal, computer or workstation key used to cancel a function.

clearance merchant — *Related Term:* job lot merchant.

clear-text password — the plain text version of an encrypted password. *Related Terms:* encryption; encrypted password.

cleat bind — *Related Term:* side stitch.

click — the singular punch of a mouse key for computer input. *Related Term:* clicks.

click here — an overused and all too common term to try and get a user to press the mouse button. The words "click here" can be either hypertext or a clickable graphic which links to another Web page.

click rate — the percentage of impressions that resulted in users clicking on an ad banner.

clicks — (1) in computing, a term referring to the number of times a user presses the mouse button on an ad banner or to execute a software function. (2) otherwise, the number of Web pages a person must go through (by pressing a mouse button) in order to reach a certain destination.

client — the customer or workstation side of a client/server setup. A computer system or process that requests a service of another computer system or process. When you log on to a server, the word client can refer to you, to your computer, or to the software you are running. For example, to download something from an FTP site, you use FTP client software. *Related Term:* server.

client-side image map — a locally residing image map that encodes the destination URL of each hot spot directly in the page. These types of image maps are processed on the client side, do not require processing from the server to respond to clicks on the image map, and are faster and more efficient. Not a universal approach as not all browsers support the technique. *Related Term:* client-side program.

client-side program — *Related Term:* client-side image map.

cling — the property or tendency of one material to affix itself to another without adhesive. Often caused by static electricity.

clip — a short snippet; an abbreviated portion of video, audio or printed source material.

clip art — (1) in mechanical art preparation, originally art limited to high contrast drawings printed on white, glossy paper supplied in camera-ready form; copyright given with purchase. Could to be cut out and pasted to a mechanical. Available from a number of companies that specialized in providing art to printers. (2) in modern usage, an electronic collection of black and white or color photos, drawings, icons, buttons, and other generally useful graphics files that can be inserted into page layouts or web pages. Often supplied on CDs and having several resolutions of each image to enable use in publishing in different sizes or video, etc.

clipboard — a temporary storage area on the computer for cut or copied items. It can store text, graphics, or a group selection for later use.

clipping service — a company that collects articles of interest from newspapers and periodicals for its clients.

clock — the timing device which is part of a computing system. Used to synchronize the internal operations and functions and control processing speed. Generally, the higher the clock "speed," the faster its computer will perform operations.

clogging — ink which dries in the pores of a screen stencil, causing poor ink tansfer on to the substrate. *Related Term:* drying-in.

clone — a deceptively similar product usually designed to be confused with an established one. Very common in the computer world.

cloning — pixel manipulation technique used in image processing programs such as PhotoShop to add or remove detail in a picture through duplication of selected pixels from elsewhere in the same or another file.

close formation — in papermaking, the relatively uniform distribution of fibers in a paper sheet.

closed circuit television — a transmission system that distributes television programs, live or on tape, both audio and video, to a limited network connected by cable. The telecast cannot be received by other television sets outside the selected network, and the signal does not have to meet FCC commercial specifications.

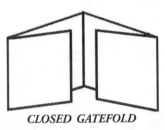

CLOSED GATEFOLD

closed gatefold — a 3-parallel folded sheet on which the outer panels are folded inward and then the entire product is folded where they meet.

closed loop system — any type of automated system which performs self diagnosis of trouble, may attempt automatic corrections and monitors usage, wear, hours of operation and maintenance scheduling.

closed shop — shop that requires craftspeople to join a union to maintain employment.

closeout — a paper merchants' term for sale on papers needing to be cleared from inventory.

close-up — (1) a proof correction mark to reduce the amount of space between characters or words. (2) in photography and graphics editing, the tightness by which a subject is cropped by framing only a portion of it in the viewfinder or on the ground glass of a view camera. Moving the camera closer to the subject, isolating it from the remaining scene. Whereas a group shot would be general composition, moving in on a single individual would be a close-up.

clustering — connecting multiple computers and making them act like a single machine. Cluster servers are often used to distribute computing-intensive tasks and risks. In the event of failure, some operating systems can move processes between servers, allowing work to continue uninterrupted while repairs are made.

CLUT (color lookup table) — a palette of colors within image editing applications and system software. *Related Term:* LUT.

CMOS (complementary metal-oxide semiconductor) — (1) in general computing, chips made with this low-power semiconductor technology used to hold basic start-up information—such as the time and date—for use by the system's BIOS. (2) in digital imaging, as an image sensor in place of traditional CCDs. Because they are more easily manufactured than CCDs, and require less sophisticated steps, their manufacturing failure percentage is much smaller, leading to a significant reduction in unit cost. *Related Terms:* BIOS; CCD.

CMYK (cyan, magenta, yellow, black) — the subtractive primaries or process colors, used in full color printing. Theoretical result when combined equally at 100% of value would be black. The black (K) is not a color, but is used to enhance the CYM colors and overall image contrast. *Related Term:* RGB.

CNET (The Computer Network) — a network of Web sites, television programs, and a radio station about the Internet; cnet.com, shareware.com, download.com, news.com, search.com, activex.com, browsers.com, are all part of this group. Their TV shows are CNET Central and TV.com plus several others. CNET was the first company to integrate interactive technology with television.

coarse papers — the papers made for linings, wrappings, and other industrial uses, as compared to fine papers for writing and printing or sanitary papers such as napkins and tissues. *Related Term:* industrial papers.

coarse screen — an infrequent and somewhat arbitrary term generally applied to screens with ruling of less than 120 lines per inch. *Related Terms:* fine screen; medium screen.

coated — *Related Term:* coated paper.

coated free sheet — glossy paper which contains less than 25% groundwood pulp.

coated paper — generally, paper manufactured primarily for printing fine screen halftones. Differs from uncoated because of the addition of casein, starch or glue and certain clay pigments that, when processed with heat and smooth pressure rollers, give the finished paper a harder, smoother and less absorbent surface. Improves reflectivity and ink holdout. A coated finish can vary from dull to very glossy and provides an excellent printing surface. *Related Terms:* art paper; chromo paper; enamel paper; and slick paper; uncoated; China clay.

coated paper, cast — the smoothest and most expensive paper available.

coated paper, dull — a paper which is slightly less shiny than gloss.

coated paper, gloss — a shiny paper surface with several quality levels. Next best after cast coated.

coated paper, matte — a paper which is less shiny than dull coated.

coating — (1) in papermaking, pigments mixed with water, casein and other adhesives, usually to form an opaque substance. When applied to paper they impart a smoother and often brighter surface. The process can be done either on the paper machine as part of the manufacturing process or off the machine as a post- manufacture operation. (2) any emulsion, varnish, or lacquer applied to a printed surface to give it added protection or to produce a dramatic special effect. *Related Terms:* coated paper; China clay; UV coating; lamination.

coaxial cable — typically used to connect a television to cable TV services, coaxial cable usually consists of a copper wire surrounded by an insulating material and another conductor with a larger diameter, usually in the form of a tube of copper braid. This cable is then encased in a rubberized protective material. It is designed to carry broadband signals by guiding high-frequency electromagnetic radiation.

COBOL (common business-oriented language) — developed in 1960 by a team led by the National Bureau of Standards (now the National Institute of Standards and Technology, or NIST), COBOL was the first standardized business computer-programming language. COBOL was intended for business use. Many functions, such as payroll and accounting, are still executed using programs written in the language.

cockle finish — a slightly puckered surface on bond paper created by quickly passing the wet paper through an air drier during manufacture.

cockling — the unwanted wavy edges on paper caused by absorption of moisture.

code 39 — (3 of 9 code) a full alphanumeric bar code consisting of nine modules, three of which are wide. See AIM X5-2 USS-39 for specifications.

code — a set of unambiguous rules specifying the way in which data may be represented. Numbers and letters used to represent information. *Related Term:* number system.

code reader — a device that examines a printed spatial pattern and decodes the encoded data.

codec — (1) in broadcasting, a coder-decoder that is used to convert analog signals such as video or voice into digital form for transmission over a digital medium and, upon reception, reconverts the signals to the original analog form; may also perform other signal processing functions; coder-decoder, or compressor-decompressor. (2) in computing, short for COmpressor and DECompressor. Hardware or software that compresses and decompresses digitized audio/video. Codecs encode and decode data types that would otherwise use up inordinate amounts of disk space, such as sound and video files. *Related Terms:* streaming; PCM; TrueSpeech; WAV.

coding — in composition, the inserting of special information in copy before typesetting to control changes in the standard instructions, such insertion of special heads or subheads, changing point size, or to italic or boldface type, etc.

coil bind — *Related Term:* spiral bind.

COLD (computer output laser disk) — a system that captures print information and usually stores it on optical disk for on-demand retrieval and printing as needed.

cold colors — generally considered to be blues, greens and grays. Each suggests cool places and environments.

cold fusion — a Windows NT or SPARC server application used for developing large and complex Web sites. The cold fusion application server provides an extensive system for building and deploying dynamic Web applications.

cold type composition — (1) originally, assembling alphanumeric symbols by strike-on or hand mechanical methods. (2) more conventionally, photographic and digital typesetting. Generally, any reproduction type produced without the use of characters cast from molten metal, such as that created on a workstation or computer monitor. Distinguished from "hot type" composition which utilizes molten metal. *Related Term:* hot type.

collaborative workflow — a workflow that has somewhat defined tasks and uses a defined set of information to arrive at an outcome achieved by consensus. The completion of an effort is determined when all parties, with approval authority, agree no further work is needed. This environment is found in larger organizations with a number of participants each performing important, but nonstandard, tasks.

collage — an assemblage of different kinds of elements such as photos, pieces of cloth, bits of hardware, etc.

collate — (1) in commercial work, to assemble sheets or signatures of paper into proper sequence. (2) in book publishing, to gather separate sections or leaves of a book together in the correct order for binding. *Related Term:* gathering.

collateral — ad agency term for newsletters and other printed materials not directly related to advertising.

collodion — an obsolete term used in early photography to describe a light-sensitive viscous solution of cellulose nitrates, ether, and alcohol that was used to coat glass photographic plates.

colloid — a fine-particle solution which fails to settle out and deflects a beam of light. The particles are usually too small to be resolved via a conventional light microscope.

collotype — a photomechanical printing process of the planographic classification in which a bichromated gelatin-coated plate is exposed through continuous tone negatives. This process is used most often for short runs of fine art prints. It is the only generally accepted printing process which can reproduce full tone original artwork or photographs without the use of any halftone or other screening process.

co-location — a term which refers to a server belonging to one person or group but physically removed from the site and accessed via an Internet network connection to a remote site. This is often done to provide high speed Internet connections and avoid security problems and risks associated with in-house networks.

colophon — (1) generally, a logotype that identifies a printer and/or publisher. (2) traditionally, production information such as typefaces, paper and mechanical techniques, i.e., computers, software, etc. Usually printed at the back of a book.

color — (1) physiologically, a visual sensation produced in the brain when the eye views various wavelengths of the electromagnetic spectrum. Color viewing is a highly subjective experience that varies from individual to individual and instance to instance. In the graphic arts industry, lighting standards and color charts help ensure the accuracy of color reproduction. (2) in photography, a designation for negative chromatic film. Sometimes a contraction for brand names such as Kodacolor,

Fujicolor, Ektacolor or Agfacolor. *Related Terms:* chrome; typographic color.

color analysis — the determination and recording of exact color information from or in a set of color printers.

color attribute — a three-dimensional characteristic of the appearance of an object. One dimension usually defines the lightness, the other two together define the chromaticity, saturation and lightness.

color balance — (1) in video, referring to the relative strength of the signal from each of the three electron guns (red, green, and blue) to excite the phosphors on the inside face of the CRT in such a way as to make each RGB triad be perceived as true white light. If one gun is stronger the other two, a tint of that color will be seen in the screen image. Most monitors provide color adjustment controls, which allow you to adjust the relative strength of the three electron guns to correct the problem. (2) in printing, the combination of yellow, magenta, and cyan needed to produce a neutral gray. Determined by a gray balance analysis, process colors are in balance when perceived as true to the original, with no undesirable/incorrect casts or hues. *Related Terms:* color purity; electron gun; pixel; refresh rate.

color bars — (1) in prepress, a series of color patches carried on the edge of four-color process proofs to show the colors and amount of ink used, the trapping, and the color registration between colors. Used mainly as a guide for the platemaker and printer. (2) in presswork, small patches of color solids, patterns, overprints, tints, and resolution targets printed on the tail edge of press sheets for the purpose of monitoring printing press performance. Usually they help the press operator control ink coverage or density, color balance and registration, trapping, print density, dot gain, and slur by allowing densitometer readings for various sections of the sheet. *Related Terms:* abnormal color vision; calibration bars; color blindness; color control strip.

color blindness — (1) physiologically, a deficiency in vision that permits a person to see only two hues in the spectrum, usually yellow and blue. Very few persons are truly "color blind," and see only in black and white. (2) in graphic communications, a term sometimes used to describe an emulsion that is only sensitive to blue, violet and ultraviolet light.

color break — (1) in art preparation the indication, on tissue or acetate overlays attached to the mechanical artwork, of what image areas print in what colors (usually line and screen tints). This is done so that a different plate can be prepared for each color in a multicolor job. (2) in multicolor printing, the point or line at which one ink color stops and another begins.

color calibration data — information which specifies characteristics of the paper, press, and ink to be used in color printing.

color cast — the modification of a hue by the addition of a trace of another hue, such as yellowish green, pinkish blue, etc. *Related Term:* cast.

color chart — a printed chart containing overlapping halftone tint areas in combinations of the process colors. The chart is used as an aid to color communication and the production of color separation films. The charts should be produced by individual printers using their own production conditions. *Related Term:* Foss Color Order System.

color circle — a Graphic Arts Technical Foundation color diagram used for plotting points as determined by the Preucil Ink Evaluation System. The dimensions are hue error (circumferentially) and grayness (radially). *Related Terms:* Preucil Ink Evaluation System; color triangle; hue error; grayness.

color compensating or correction filter — a high transmittance filter used over a camera or enlarger lens to correct the color balance of transparencies or to change the relationships among colors in a subject and how they are recorded. They are available in six colors and several strengths. Often called "CC" filters.

color control patches — a guide used in color separation and color correction that consists usually of nine blocks that show a record of the color densities.

color control strip — a series of color bars and patterns printed on press sheets designed to help press operators detect problems with color balance, registration, and other printing-related problems. *Related Terms:* color bars; color control patches.

color conversion — a color transparency made from a color reflection original to allow rigid or oversized reflection copy to be color separated using a drum-type scanner.

color correct — to change the color values in a set of film separations or using a software application to correct or compensate for errors in photography, scanning, separation, output, and so on. *Related Term:* color correction.

color correction — (1) in printing, photographic, electronic, or manual adjustments made in the color separation process to bring the printed result as close as possible to the original photograph, despite the limitations of printing inks. (2) in photography, adjustments using filtration to reestablish the proper colors in photographic prints when the original color negative may have deteriorated. (3) any color alteration requested by a customer.

color cycling — a means of simulating motion in a video by changing colors.

color difference — the magnitude and character of the difference between two object colors under specified conditions.

color duplicating — (1) in photography, the process of making a duplicate transparency from an original transparency for purposes of retouching, color cast adjustment, density range normalization, image assembly, or reproduction scale adjustment.

color filter —dyed glass, gelatin, or plastic placed between plates or in a lens' optical path to absorb certain colors and produce a better rendition of other colors. The filters used in color separation are red, green, and blue.

color gamut — the range of colors that can be formed by all possible combinations of the colorants of a particular color reproduction system. *Related Term:* gamut.

color harmony — a pleasing combination of two or more colors used together in the same printed product.

color hexagon — a bilinear plotting system for printed ink films. Adapted for the printing industry by GATF, the method was originally developed by Eastman Kodak. A color is located by moving in three directions, at 120 degree angles, on the diagram by amounts corresponding to the densities of the printed ink film. The diagram is generally used as a color control chart, particularly for detecting changes in the hue of two-color overprints.

Color Key — (1) Imation Corporation trademark for their color proofing system that generates a set of PMS or cyan, magenta, yellow and black tinted film positives from separation negatives so that registration and screen tint combination on process color reproductions can be checked before press proofs are made or the actual press run begins. (2) the term incorrectly applied to any multilayered or overlay proofing system.

color matching system — method of specifying flat color by means of numbered color samples. *Related Term:* Pantone Matching System (PMS).

color measurement scale — a system of specifying numerically the perceived attributes of color.

color pagination system — a production system for performing color page makeup by assembling text and pictures automatically using computer programs under the control of an operator.

color preview equipment — machines which allow control of the quality of digitized separations. In some, the operator sees digital data before they are recorded as film output, while others allow the operator to look at film output before plates are made.

color primaries, additive — the three basic colors which, when properly selected and mixed, produce any hue. The visual spectrum primary colors are green, red, and blue. When combined equally, these colors form white light. *Related Term:* subtractive colors.

color printers — (1) the individual negatives or positives resulting from the color separation process. (2) a generic term applied by the public to printing operations which specialize in spot and/or process color.

color process printing — *Related Terms:* 4-color process printing; 3-color process printing; process color printing; color process.

color process, four — a technique of printing that uses transparent process ink colors, magenta, cyan and yellow, to simulate color photographs and to produce a wide variety of colors from mechanical and electronic art. Due to less than perfect color pigments, an artificial black ink is added to increase shadow detail and add overall contrast.

color proof — (1) in prepress operations, a simulated printed image to evaluate color separation films prior to platemaking and printing. The colorants used are selected so that the proof will produce a close visual simulation of the inks to be used for the final reproduction. (2) in press proofing, a proof should be printed on the same press and use the same inks and paper that will be used for the finished job. Needs approval before the production run. *Related Term:* off-press proof.

color purity — the accuracy of the reproduced colors. In the case of computing, monitors use electromagnetism to control their electron guns. Magnetic fields build up within the monitor which cause dis-tortions to appear as uneven colored patches on the screen. A monitor has good color purity if no such discolorations are visible (they are easiest to see on a white background). The magnetic fields that cause problems with color purity can sometimes be eliminated by degaussing. *Related Terms:* degaussing; electron gun.

color quality index — *Related Term:* color rendering index.

color references — a given set of inks printed at specified densities or strengths on a given substrate, used for color control and matching.

color rendering index — a measure of the degree to which an artificial light source under controlled conditions influences the perceived colors of objects illuminated by the source compared to the same objects illuminated by daylight. *Related Term:* color quality index.

color reproduction guide — a printed image consisting of solid primary, secondary, three- and four-color, and tint areas. It is primarily used as a guide for color correction of the defects of the printing ink pigments and the color separation system. Most experts suggest that the guide be produced locally under regular plant production conditions.

color saturation — (1) the relative ability of a monitor to display subtle color changes distinctly so that the human eye perceives them independently, one from the other. If similar colors blend, are not distinctive, or if they appear dark, they are over saturated; the greater the saturation, the more intense the color. Colors that appear washed out and/or faded are under saturated. (2) generally, the amount of hue contained in a color.

color scanner — a device incorporating a digital or analog computer that separates colored originals electronically. It distinguishes and records four subtractive primary negatives (cyan, magenta, yellow + black) by using additive filters of blue, green and red. An interpreted black printer is made to balance with the color separations. *Related Term:* color separation.

color sensitivity — the chemical change that occurs to a particular silver halide emulsion when exposed to a selective area of the visible electromagnetic spectrum. Blue sensitive films are sensitive only to blue light and are blind to green, yellow and red light. Orthochromatic films are sensitive to blues, greens and yellows, but blind to red light, etc. Panchromatic film is sensitive to all visible colors.

color separation — the process of making, by camera or electronic scanner, separate films from a color original which accurately records the red, green, and blue light reflectance using the four subtractive primary process colors (cyan, magenta, yellow + black). The result is four continuous or halftone films (negatives or positives) which are ultimately used to make printing plates. The colors, when printed in register over one another, create a full color (or "process") image. Color separation was initially done by photographing the image three times through different color filters. However, electro-optical methods using lasers and CCDs are now employed.

color separation service — a business making separation negatives for 3- or 4-color process printing, etc.

color sequence — the color order for laying down the yellow, magenta, cyan, and black inks on a printing press. Sometimes called rotation or color rotation. While there are some recommended semi-standards in the industry for the sequence, individual shops have developed their own, independently, based upon the type of equipment, inks, paper and plates being used.

color space — all the colors that can be represented by the three additive primary colors (red, green and blue), or their subtractive counterparts (yellow, magenta and cyan), black and white.

color specification — tristimulus values, chromaticity coordinates and luminance value used to designate a color numerically in a specified color system.

color strength — *Related Term:* chroma.

color subsampling — a method of using reduced resolution for the color difference components of a video signal compared to the luminance component. Typically, the color difference resolution.

color swatch — a small sample of paper or printed ink.

color temperature — the temperature, in degrees Kelvin, to which a black body would have to be heated to produce a certain color radiation; the sum color effect of the visible light emitted by any source. 5,000 K is the graphic arts viewing standard. The degree symbol is not used in the Kelvin scale. Higher color temperature tends to be more blue; lower temperatures tend to be more red.

color terminology — in the printing industry, color is described in terms of hue (chrome), strength (saturation), and gray (value). Hue is the pure color, strength refers to the color's strength, or saturation, and gray refers to how "clean" the color is. These are not terms used by the artist; they have been suggested by the printing industry to help communication between designer, client, and printer.

color tracking — each pixel on a monitor's screen consists of red, green, and blue spots of phosphor, each lit up by a separate electron gun. Color tracking refers to a monitor's ability to keep all three electron guns operating at equal strength when displaying different brightness levels. In practice, one gun tends to overpower the other two, resulting in slight reddish, greenish, or bluish hues that are visible in dark gray areas of an on-screen image. *Related Terms:* color balance; electron gun; pixel; refresh rate.

color transparency — usually a positive photographic image on transparent film used as artwork. 35mm, 2 1/4 in. x 2 1/4 in., 4 in. x 5 in. and 8 in. x 10 in. formats are commonly used.

color triangle — a Graphic Arts Technical Foundation color diagram based on the Maxwell Triangle and the Preucil Ink Evaluation System. The dimensions are hue error (circumferentially) and grayness (radially). Color masking, color gamut, and ink trapping may be determined from the diagram by using simple geometric techniques. *Related Terms:* Preucil Ink Evaluation System; color circle.

color wheel — a circular tool used by graphic designers that shows the primary, secondary and intermediate colors.

color, blanks — preprinted sheets containing artwork or photographs, with no text material. Often made in advance for publications with different versions, or when an existing publication is known to be updated in the future.

color, typographic — the apparent visual lightness or darkness of a page. It is primarily affected by the typeface and its x-height, the interline spacing, alignment and margin and alley spaces provided in the page design. Darker or bolder faces when combined with minimal leading create a relatively darker page than light-faced types with more leading between the lines.

colorimeter — an analysis device specifically designed to measure color perception by approximating human vision.

colorimeter — an optical measuring instrument designed to make direct measurements of color and to respond to color in a manner similar to the human eye. *Related Term:* tristimulus colorimeter.

column — (1) in publishing, the vertical divisions of a typeset page. (2) an article by an author(s) which appears on a regular schedule and deals, usually, with one subject or subject area. (3) in a spreadsheet table, a vertical collection of cells.

column inch — a measure of area used particularly in newspapers and some magazines to calculate the cost of display advertising. A column inch is one column wide by one inch deep. *Related Terms:* agate; agate measure.

column rule — a (usually) light-faced vertical rule used to separate columns of type.

com — a domain name suffix used in Internet addresses that denotes a commercial entity such as dupont.com.

COM (computer output microfilm/fiche) — a process and the device used to convert computer- generated information into a form that can be "printed" to microfilm or fiche. Allows large amounts of data to be stored inexpensively.

COM port — a contraction of the word communications, it describes the serial port on a PC. COM is generally used in conjunction with a number, as in COM1, COM2, COM3, or COM4.

comb binding — a flexible plastic comb whose teeth fit through rectangular holes in a stack of paper. *Related Term:* plastic comb binding.

combination — (1) generally, any file that has more than one element (text, graphics, voice, video) mixed together. (2) in printing, the presence of both halftone and line art on a single printing plate.

combination folder — (1) generally, a folder which may have both buckle and knife fold attributes which can be used in combination. (2) in web publishing, the given name for the complex series of folds and overlays needed to produce a properly sequenced final product from a series of independent paper webs.

combination press — different printing presses, possibly from different processes or reproduction classes used as a single manufacturing entity. Most common for special purpose printing.

combing wheel — a type of sheet separator usually consisting of a wheel with a circumference consisting of rubber roller elements. As the wheel and its independently rolling outer perimeter rollers contact the top sheet of a pile, it tends to separate it from the remaining sheets, making it ready to be fed into the printing unit.

command — the portion of a computer instruction that specifies the operation to be performed. Carriage return, typeface designation, point size selection, line space settings, etc. are all typesetting commands. *Related Term:* command codes.

command codes — keyboard characters or sequence of characters that specify typesetting parameters and initiate, modify or stop particular functions or operations to be performed, i.e., the type face, size, line measure, leading, etc.

command-line interface — before graphical interfaces like Windows, the Mac OS, etc., users interacted with their personal computers by typing (usually DOS or UNIX) text commands. While the drag-and-drop simplicity of a GUI is attractive to many users, particularly newcomers, command-line interfaces often allow greater flexibility and control and are therefore still preferred by many power users. *Related Term:* GUI.

comment — text that you can view in the FrontPage Editor but that will not be displayed by a Web browser. Comment text is displayed in purple and retains the character size and other attributes of the current paragraph style.

commercial artist — artist whose work is planned for reproduction by printing.

commercial printer — a printer who primarily manufactures print runs of 5,000 or more using larger printing presses than those found in a quick-copy shop.

commercial printing — (1) the phase of the graphic arts industry that relies on piecework from a number of paying customers rather than from one customer (such as in-plant print shops) or one readership (as in newspaper, magazine or book publishing). (2) generally, a printer whose business emphasizes good quality, which has medium and large press sizes and offers full service, normally using metal plates.

commercial register — a measure of color printing in which the allowable misregister is plus or minus one row of dots.

commercial signs — symbols in a type font such as $ (US dollars), ® (registered), © (copyright), £ (pounds sterling), ¥ (Japanese yen), % (percentage), or ™ (trade mark).

commodity — a trade term for the quality rating for inexpensive paper or printing produced quickly and in high volumes.

common carrier — a telecommunications company that is regulated by government agency and offers communications relay services to the general public via shared circuits through published and non-discriminatory rates.

common edge method — a method of multiflat registration in which two edges of each flat are positioned in line with each other; if the edges are lined up with corresponding edges of the plate, the images are in correct positions.

common impression cylinder — a drum-like cylinder setup found on some web and sheet fed offset presses. The common impression cylinder is in contact with several blanket cylinders that are, in turn, in contact with plate cylinders. This configuration is used to print multicolor work on one side of the web or sheet. *Related Terms:* central impression cylinder; McCall press.

communication — the exchange of information among and between people and/or machines using signs, digital pulses, analog waves, symbols, gestures, sounds or printed words.

communication paper — *Related Term:* bond paper.

communications satellite — a satellite in earth orbit which receives signals from an earth station and retransmits the signal, video and/or audio to other stations.

comp — shortened version of designer's "comprehensive dummy." The artwork used to present the general color and tryout of a page. A handmade, full-sized simulation of a newsletter or other printed product, complete with type, graphics, and colors. *Related Terms:* proof; thumbnail; dummy.

compact disk-read-only-memory (CD-ROM) — a 4.72 in. polycarbonate disk capable of storing approximately 650 megabytes of information. The laser-mastered disk can store interspersed audio, video and alphanumeric (textural) files. *Related Term:* CD.

comparable stock — *Related Term:* equivalent paper.

compatability — the level of ability in which two different devices, software solutions, personnel, etc., can successfully interact without discord.

compile — a process of translating high level computer programing into simple native language instructions which can be interpreted by a computing device.

complementary — a color harmony that results from using two colors opposite each other on the color wheel.

complementary flats — two or more flats stripped so that each can be exposed singly to a plate but still have each image appear in correct position on the final printed sheet.

component video — signal format in which the luminance and two channels of chrominance remain separate components. Component video signals retain maximum luminance and chrominance bandwidth (image detail).

compose — to set copy into type. *Related Terms:* composition; typography.

composing room — the area of a printing plant where type is set and arranged for platemaking or printing.

composing stick — in metal hand composition, a metal tray-like device used to assemble type when it is being set. It is adjustable so lines can be set to different measures.

composite art — a black and white interpretation of a piece of color finished art.

composite film — graphic arts negative or image file made by combining two or more images and that contains all color information used to generate separations or final printing image carriers. *Related Terms:* composite; composite negative.

composite font — a special font designed to support large character sets such as Japanese or Chinese.

composite image — a new image created from parts of separate originals.

composite proof — proof of color separations in position with graphics and type.

composite signal — a signal transmission system where chrominance and luminance are combined. *Related Term:* NTSC.

composite video — video signal format that combines luminance and chrominance information in a single signal. It is the standard format used for consumer TVs and VCRs. The format restricts bandwidth (image detail) of signal components, though provides efficient and economical video transmission.

compositing — combining multiple layers of video or still images and/or audio into a single image.

composition — process of assembling font symbols, letters, numbers and other elements in preparation for reproduction and in the precise position defined on the rough layout during image design.

composition and makeup terminal (CAM) — sometimes called a video display layout terminal, is used to compose advertising in actual type sizes and to position the copy directly on the CAM screen.

compositor — the person who sets manuscript or handwritten copy into type in one of several typefaces. *Related Terms:* typesetter; typositor.

compound document — a document with multiple data types often created by different applications and embedded into a single document. Generally, a file that has more than one element mixed together. *Related Term:* combination.

comprehensive — more commonly referred to as a comp. An accurate layout showing type and illustrations in position and suitable as a finished presentation. A "tight" comp closely resembles the final product. A "loose" comp does not and is more of a rough sketch for conceptual evaluation. *Related Term:* comp.

comprehensive dummy — a complete simulation of a printed piece. *Related Terms:* comprehensive; comp.

comprehensive layout — a full-sized layout prepared by hand that shows the exact location, colors, shapes, etc. of the image elements of a printed product; a master plan. *Related Term:* comprehensive.

compress — (1) in computing, to reduce the size of a digital file for the purpose of speedier file transfer and archiving. (2) in presswork, the action taken on certain rubber blankets as the impression and plate cylinder come in contact with it during the printing cycle.

compressed audio — digitally encoding and decoding voice-quality audio. Using a buffer to store the audio information, a limited and controlled amount of audio is delivered to accompany specific, still frame images.

compressed video — video images that have been processed to remove redundant information, thereby reducing the amount of bandwidth required to send them over a telecommunications channel; instead of transmitting full-motion video frames, only the changes in moving frames are captured and transmitted.

compression — the shrinking or flattening of computer files so that the same information is stored in less memory. On CD-ROM, image files are routinely compressed, and text files can be compressed if necessary. Compression and decompression schemes are of a mathematical nature. "Lossy" compression discards a few pixels to reduce the data. "Lossless" compression does not discard any pixels, but is less effective at reducing file size.

compression ratio — the ratio of the size of a compressed file when compared to its original.

CompuServe — a commercial online information network, sometimes called CIS (CompuServe Information Service). It has been acquired as an independent entity of America On Line (AOL).

computer — generally a digital electronic device utilizing manipulation, storing, retrieval, and processing of binary information and data. The precise function of the device is controlled by software.

computer conferencing — group communication through computers; the use of shared computer files, remote terminal equipment and telecommunications channels for two-way group communication.

computer graphics — charts, maps and other pictorial materials generated by use of a workstation or computer and software.

computer literacy — the level of comprehension displayed by a computer user. User ability to utilize computing, peripheral systems and other related components in an organized and productive manner.

computer program — a series of instructions prepared for the computer.

computer security — measures taken to protect systems and data from intrusion. Some consider regular scheduled backup of files an integral part of a computer security plan.

computer, exposure — or calculator; a hand-held, nonelectronic device used to assist in determining halftone exposures.

concentric circle screen — a special effect screen that is constructed of closely spaced concentric circles radiating from a center point.

concertina fold — a method of folding in which each fold opens in the direction opposite its neighbor, giving a pleated effect. *Related Terms:* accordion fold; Z-fold.

condensed — (1) in typography, a typeface in which the height is disproportional to the width, producing an elongated appearance. (2) a shortened version of a book-length work. *Related Term:* condensed.

condenser, optical — an enlarging system in which lenses intercept a comparatively large cone of light rays emitted by the source lamp. The rays are concentrated within a specific area. Very popular when the primary form of printmaking was by projection of the negative image onto print paper. Use of this type of system has been reduced, in recent times, by cold light systems having lower heat, good color properties and which minimize defects which may be in the negative.

conditioning — the placement of paper or other substrate in the pressroom for a period of time ranging from hours to days. The purpose is to improve runability by making the temperature and humidity stabilize at that of the pressroom, thus minimizing static, curl and other running problems. *Related Terms:* mature; cure; season; prestabilize.

conductivity — the ability of a conductor to pass current. It is used in measuring pH of fountain solutions as well as image assistance in platemaking.

conference — a live discussion online in which the topic has been predetermined, often featuring a celebrity guest. *Related Term:* "CO" or "live CO."

conference call — an operator-assisted telephone call connecting more than two individuals.

configuration — the term that, generally, refers to the way your computer's operating system is set up or to the total combination of hardware components: central processing unit, video display device, keyboard and peripheral devices, etc. It also describes the software settings which allow various hardware components of a computer system to communicate with one another. A "vanilla" configuration means a "clean" and "no frills" version of a computer's configuration without device drivers or extra settings. A technician might set up a system as "vanilla" when trying to isolate a problem with a computer's hardware.

configuration variable — information about a FrontPage Web or page that can be displayed when the page is browsed. FrontPage includes standard Web and page configuration variables. New configuration variables can be set by editing the Parameters tab of the FrontPage Explorer Web Settings dialog box.

configure — the process of changing software or hardware actions by changing their settings. The process can be set or reset in software and/or manipulated by changing hardware jumpers, switches or other elements.

confirmation page — a page that is displayed in the browser after a form has been submitted and usually echoes the user's name and other data from the form.

coniferous — cone-bearing, softwood trees that provide wood fibers used in the papermaking process. *Related Term:* deciduous.

connect — to make the connection with an Internet service provider (ISP).

connect time — the amount of time you spend connected to an on-line service provider or ISP.

connectivity — the state of being connected to the Internet or some other type of computer network. On the Internet, if you lose your connectivity then you are no longer online and must redial into your ISP. When someone asks "what is your connectivity?", it usually means "what kind of speed does your Internet connection support?" (like 28.8 or T1).

consecutive numbering — unique serial numbering on individual invoices, tickets, etc., by hand or machine.

conservative typefaces — typefaces having novel or decorative appearance but containing only minor alterations of a standard typeface.

consignment memo — *Related Term:* delivery memo.

consolidate — *Related Term:* amalgamate.

consortium — a voluntary organization loosely affiliated for a specific purpose.

constant aperture exposure method — an infrequently used set-up method for process camera reduction or enlargement. The lens aperture (f/stop) remains the same, while the exposure time is varied.

constant time exposure method — a generally used method of setting a process camera for reduction and enlargement, where every exposure is made for the same exposure time, while the lens aperture (f/stop) is varied on the basis of the reproduction size.

constants — the typographic and graphic elements in a publication that don't change from issue to issue. *Related Term:* standing heads.

consumer unit — in the U.P.C. barcode standard, a specific package quantity of a specific product offered by a specific manufacturer.

contact film — a type of photographic film used to make contact exposures from other negatives or positives.

contact form (template) — an electronic form used to register a new domain name contact and/or agent or to modify information for an existing contact; was once called the contact template. *Related Terms:* modification; contact/agent.

contact paper — a chloride-coated photographic paper which is printed by keeping it in direct contact with the negative during exposure in a printing frame, enlarger baseboard or separate vacuum frame held to an external light source.

contact print — a same size copy made with photographic materials in direct contact with a film negative or positive without the use of lens or optics.

contact printer — equipment used to make photographic exposures with materials in contact. Same size reproductions result. Light passes through one material to strike another material that is in contact with it. *Related Terms:* vacuum frame; vacuum printing frame.

contact printing — *Related Term:* contact print.

contact screen — a screen through which exposures are made. It is used to produce halftone dots by being placed in direct contact with high contrast photographic film during exposure. *Related Terms:* halftone; contact sheet.

contemporary typefaces — a subset of the novelty typeface classification. They have a novel or decorative appearance and are designed to echo as much as possible the thought or word or feeling that they convey.

content edit — an overall evaluation and critique of a manuscript for organization, style, and continuity as well as actual content.

content provider — any site which provides dynamic and updated information in the form of news, entertainment, games, employment listings or dictionary terms. Most major content providers on the Internet, such as CNET and MSNBC, provide their service to users for free.

contents — the list of a book's chapters or a magazine's features and departments that appears as part of the front matter. *Related Term:* table of contents.

contextual selector — a contextual selector is a combination of several simple selectors. *Related Terms:* HTML; cascading style sheets; selector.

continued line — a line of text indicating the page on which an article continues or the carryover line on the subsequent page that identifies the story being continued. *Related Term:* jumpline.

continuity — (1) in typography, the look of belonging together; a design feature of typeface characters. (2) in design, the consistent integration of design elements to define the overall image of a publication.

continuous code — a bar code symbol where all spaces within the symbol are parts of characters. There are no intercharacter gaps in the continuous code.

continuous forms — forms printed as a repeated pat-

tern on long sheets. Often used in dot matrix invoice stations, etc.

continuous ink jet — a nonimpact printing technology in which a steady stream of ink is forced at high pressure through a small nozzle and dispersed as small droplets through a charging field. The stream of charged droplets then passes between high-voltage deflection plates. Because the plate voltage varies, only selectively charged droplets form the desired shape or pattern on the substrate. Excess droplets are diverted and recirculated.

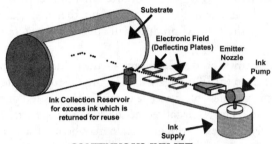

CONTINUOUS INK JET
Ink is continuously pumped from the ink supply and sprayed toward the substrate. Unwanted droplets are diverted by an electrical charge which causes them to be placed in the collection reservoir for recycling.

continuous paper — the paper for business forms or general computer printing with each sheet attached to other sheets as part of an accordion-folded supply. *Related Terms:* fan fold paper; tractor feed paper.

continuous quality improvement (CQI) — in management, the process of making each manufacturing or service step responsive to the customer's needs or expectations.

continuous sheet feeding system — a press feeder design that permits restocking of sheets into the feeding system without interrupting the press run.

continuous tone — an image that has an assortment of tone values ranging from dark to light that does not contain halftone dots. In photography, the process of recording images of differing density on paper or film coated with a light-sensitive emulsion which, when developed, will show varying shades of gray or different hues of color without being broken up into dots. Continuous tones cannot be reproduced by the printing process but must be converted into halftone screened images which translate the continuous tones into variable sized dots. *Related Terms:* halftone screen; contone; continuous tone photography.

continuous tone color separations — color separations that have not been halftone screened. Often in the form of film positives. *Related Term:* direct color separation.

contone — a shortened version of continuous-tone.

contouring — the tendency of materials to lose reproductive details in output as they progress from generation to generation, i.e., positive to negative to negat2ive, etc.

contract artist — a freelance artist.

contract proof — a color proof that printer provides and the client signs in agreement that it is exactly how the printed product will appear. *Related Terms:* contact print; proof sheet.

contraction development — the reduction of contrast to the film negative, bringing the tonal value of an extremely bright scene to within the desired printable range as visualized at the time of exposure. Selected tones of the gray scale may be contracted as necessary to change the overall concept. Contraction should not be confused with underdevelopment. Reduced development indicates the need for one-half to two stops, depending on the degree of contraction.

contrast — (1) in photography operations, the differences between adjacent light and dark tones, including the visual relationship of the tonal values, within the picture in highlight, middle tones, and/or shadow tones. (2) in OCR, the difference in reflectance between the black bars and white spaces of a symbol.

contrast grades — the range of tones available in a photographic emulsion. The grades range from 0 to 6 where the lower numbered grades are higher contrast. Most normal prints are made on a number 2 or 3 paper.

contrast ratio — *Related Term:* opacity.

control character — a delimiter in a computer file such as <, >, ^, /, \, etc.

control chart — a visual tool used to examine variation in a repeating process.

control key — a key on a terminal or computer terminal which expands function. By depressing it simultaneously with another key, an action command is performed. *Related Terms:* alternate key; alt key.

control strip — piece of film used in automatic film processing to gauge the activity level of solutions in the machine; several times a day the photographer sends a pre-exposed control strip through the machine and then examines the processed piece to determine whether machine adjustments need to be made.

controller — a hardware or software device that manages the flow of data between a computer and a specified peripheral. Specific devices have their own unique controllers. Generally, personal computers come with controllers for standard built-in devices such as hard drives, keyboards, monitors, etc., but other add-on peripherals may require expansion cards with new controllers. *Related Term:* jumper.

conventional gravure — a method of preparing gravure cylinders that delivers well openings of the same opening size. *Related Term:* photogravure.

convergence — a computer monitor's ability to produce images without halos across all parts of the screen; the perfect alignment of the three separate electron guns in a color monitor. Convergence is necessary in order to generate crisp white lines without colored halos. Misconvergence creates a blurry picture near the corners and edges of the screen, where the electron beams must bend the most to hit their intended paths. *Related Term:* electron gun.

converter — (1) in computing, a tool that converts a file, or a portion of a file, from one format to another. For example, PageMaker includes a PageMaker to HTML converter and a PageMaker to PDF converter. (2) in printing, a business that manufactures products such as envelopes, display boards, paper bags, etc. from flat raw material which may be made printed or unprinted sheets. Printed sheets are customarily converted into business envelopes, packaging, etc., for a specific application. Blank sheets are converted to produce boxes and blank commercial envelopes for consumer purchase.

cookie — a general mechanism used by Web servers to both store and retrieve information about the client connection. Basically, they are small data files written to your hard drive by Web servers when you open a page in your browser. These files contain activity information used to track, as passwords, pages you've visited, the date when you last looked at a certain page and sometimes merchandise purchase history.

cooking — part of the papermaking process where wood chips are steam-heated, mixed with chemicals and water, and cooked for several hours under high pressure in a kind of large pressure cooker called a digester. The process dissolves the lignin which binds the wood fibers together. The resulting brown stock (pulp) is then washed to remove the chemicals and impure materials prior to bleaching. Depending upon the type of paper being made, it may be further processed through chemical bleaching to make it brighter.

cool color — blue or any color that is made up of a majority of blue (usually a color produced by a shorter wavelength of light).

CoolTalk — a realtime audio and data collaboration tool specifically designed for the Internet where users enhance interpersonal communications and avoid long distance phone charges. It has been on the cutting edge of redefining how people communicate, and provides full-duplex audio conferencing, allowing both users to speak and be heard simultaneously.

co-op advertising — where two organizations, such as an appliance manufacturer and a retailer, share the cost of advertising. The manufacturer usually pays the larger percentage of the advertising costs.

co-op publishing — a situation in which two organizations produce and publish a book together. *Related Terms:* co-publishing; cooperative advertising.

cooperative advertising — a major advertising source, it is shared billing between a manufacturer and its agents. *Related Terms:* co-op publishing; co-publishing; co-op advertising.

copper — (1) metal used in making halftone photoengravings for relief printing. (2) the base coating for many gravure cylinders. (3) a limited application component in bimetal and trimetal lithographic plates.

copper plating — an electrolytic process used to affix a copper jacket onto a gravure cylinder prior to imaging and engraving.

coprocessor — an auxiliary processor that relieves the demand on the main microprocessor by performing a few tasks. Term most commonly used in relation to computing math solutions such as floating point calculations and is often applied to high density data processes such as video.

co-publishing — a situation in which two organizations produce and publish a book together. *Related Term:* co-op publishing.

copy — any material furnished for reproduction (1) for an editor or typesetter, all written material. (2) for a graphic designer or printer, everything that will be printed: art, photographs, graphics and words. (3) in reprographics, the term applied to a single xerographic or electrostatic image made from an original. *Related Terms:* continuous tone copy; line copy; original copy.

copy block — a segment of body type or reading matter in a layout.

copy brushing — an infrequently used term which pertains to electronically repairing damaged areas in scanned originals. The more frequent term is cloning.

copy cylinder — the cylinder of an electronic scanning or engraving machine on which the copy is placed. *Related Terms:* drum; copy drum.

copy density range (CDR) — difference between the lightest highlight and the darkest shadow of a photograph or continuous-tone copy. *Related Term:* contrast.

copy edit — the next level of editing after content editing, where a manuscript is checked for spelling, grammar, punctuation, and consistency.

copy elements — parts of a piece of copy or mechanical, e.g., illustration, typeset or hand-lettered material, photographs. Common elements are line and continuous-tone copy.

copy fitting — determining the area a certain amount of copy will occupy when set in type. *Related Term:* character count.

copy holder — (1) in proofreading, a person who reads the original copy (manuscript) out loud to the proofreader. (2) in machine composition, the device which holds the manuscript from which the typesetting entries are made.

copy paper — the smooth, white paper made for everyday use in photocopy machines and laser printers. *Related Term:* xerocopy paper; copier bond.

copy preparation — (1) in typesetting, marking up a manuscript and specifying type. (2) in paste-up and printing, making the mechanical and writing instructions that ensure proper placement, printing and finishing.

copy protection — any technique designed to prevent unauthorized duplication. It can be an elaborate electronic trap as is used in expensive software or it may be as simple as a movable tab such as on floppy diskettes.

copy range — the densitometric difference between the brightest highlight and the deepest shadow in an original.

copy support — material (paper or board) on which copy is pasted up. *Related Terms:* copy base; paste-up board.

copyboard — part of a process camera that opens to hold material to be photographed (the copy) during exposure. *Related Terms:* front case; back case.

copydot technique — photographing previously printed halftone illustrations and associated line copy without rescreening the illustration. The halftone dots of the original are copied as line material. *Related Terms:* Velox; PMT; diffusion transfer.

copyedit — to check and correct a manuscript for spelling, grammar, punctuation, inconsistencies, inaccuracies and conformity to style requirements.

copyfitting — the procedure used to calculate the amount of space that a given quantity of text copy from a word processor or typewriter will occupy when set in a given type face and size. Copyfitting serves to determine, in advance, the correct type face, size and line length needed to fit copy to layout or layout to copy. Also, the process of adjusting type, either by altering its point size or other type specifications or by eliminating actual words and sentences to make the copy fit a given amount of space. *Related Term:* character count.

copying — any method used to make duplicates directly from an original via photography, scanning, or other imaging systems.

copyright — (1) the right of an author, artist, photographer, composer or other creator of an original work to protect that work against unauthorized copying and/or profiting by someone else. (2) the legal protection and method of recourse given to the originator of material to prevent use without express permission or acknowledgment.

copyright infringement — reproducing without permission, in whole or in part, a printed product protected by a registered or nonregistered copyright.

copyright notice (©) — the symbol or visual image that indicates that a publication has been copyrighted. No longer mandatory to ensure copyright protection.

copyright office — the facility located in the Library of Congress, Washington, D.C., where all copyright business is handled.

copyright page — the page of a publication showing copyright data, ISBN number, Library of Congress Catalog number, etc. In the case of books, verso of the title page.

copywriter — a term which usually means a person who writes copy for advertising.

CORBA (common object request broker architecture) — a set of common interfaces through which object-oriented software can communicate, regardless of computer platform. *Related Term:* OMG.

core curl — *Related Terms:* roll set; white waste.

core logic — a chipset that functions as the computer's traffic control, including bus interface logic, memory control; a cache for instructions, and data path functions. *Related Term:* CPU.

core waste — any paper near the center of a roll that is unusable because of roll set or splicing techniques. *Related Terms:* roll set; white waste; butt end.

corner marks — lines on a mechanical, negative, plate or press sheet showing the trim and corners of a page or finished piece.

corner stitch — affixing flat sheets into a packet by use of a single staple through the corner. Customarily done in the upper left corner unless the design specified otherwise.

corporate identity — the image of an organization conveyed through the tone of its advertising and promotion efforts, its logotype, specific graphic designs, colors and preferred media outlets.

correspondence paper — *Related Term:* bond paper.

corrugated — characteristic of board for boxes made by sandwiching fluted Kraft paper between sheets of paper or cardboard.

cotton content paper — the papers made totally or in part from cotton or linen fibers instead of, or in addition to, tree fibers. *Related Terms:* rag paper; rag content.

cotton paper — *Related Term:* cotton content paper.

counter — (1) in relief printing, sunken area just below the printing surface of foundry type. The central open area of letters where there is no image, such as the bowl of the b, d, p, etc. (2) the device on a printing press or other machine that registers the number of cycles or sheets passing through it. (3) in gravure printing, angle between the doctor blade and the cylinder. (4) the alternate name for the female die used in stamping and embossing.

counter display — a point of purchase merchandising display, as in a retail store.

counter stack — to stack catalogs, magazines or other published works in turns, alternating the spine and front edges to achieve even bundle or package piling.

counterfeiting — the process of imitating a sought-after product, such as paper money, designer clothes or brand-name goods, with the intent to deceive the customer.

counting keyboard — operator-controlled keyboard with a graphic indicator or scale that requires end-of-line decisions by the operator as copy is being typed.

coupon — a certificate entitling the user to discounts, privileges, special offers, payment, etc.

courtesy reply envelope — a preaddressed envelope provided as a convenience to customers, but for which the sender pays postage.

cover paper — a category of printing paper generally used for printing booklet and manual covers, programs, announcements, etc. *Related Term:* cover stock.

cover stock — A variety of heavier papers used for the covers of catalogs, brochures, booklets, and similar publications. *Related Term:* cover paper.

CPI (characters per inch) — one component of copyfitting. *Related Term:* bar code density.

cps (characters per second) — once commonly used to describe the relative speed of printers, cps now sometimes refers to the data transfer rates of modems. For practical purposes, on the personal computer, one character is equal to eight bits or one byte. *Related Terms:* bit; BPS; byte.

CPU (central processing unit) — (1) usually, the most powerful microprocessor chip in your computer. The Intel Pentium, RISC and G3 technology chips, for example, handle the central management functions of a high-powered PC or Mac. (2) also, the inaccurate term sometimes used to describe the whole box that contains the chip including the mother board, expansion cards, disk drives, power supply, etc. *Related Terms:* core logic; FPU.

cracker — an individual who attempts to access computer systems without authorization, often maliciously. Not to be confused with true hackers, who are reputedly benign. Both are equally illegal. *Related Terms:* hacker; Trojan horse; virus; worm.

crash — (1) in bindery, the gauzelike material that is sometimes embedded in the perfect binding adhesive to increase strength. (2) in computing, a general computer failure which necessitates the rebooting of the system. Often caused by memory or interrupt conflicts between devices or poorly written software. (3) the general failure of an economic or management strategy.

crash finish — a textured finish on text paper similar to crash cloth used in bookbinding. *Related Term:* homespun finish.

crash printing — the use of relief letterpress to print premade carbonless form numbers. Because of the pressure employed, the number printed on the top sheet transfers as a carbonless image to all successive sheets in the set.

crawl — the beading of ink, after printing, and failure to adhere to a substrate due to incompatability between the substrate surface and the ink.

crawler — this term is practically synonymous with spider; however, attempts are being made to protect the word as a trademark. *Related Terms:* bot; spider.

crazing — minute cracks or other imperfections which occur in heavy coatings such as inks and varnishes.

CRC (cyclical redundancy check) — CRC is a mathematical technique used to check for errors when sending data by modem. A necessary precaution because many phone lines are notoriously unclean and cause breaks in transmission. If the CRC check sum fails to add up, the receiving end of a data transmission sends an NAK (negative acknowledgment or "say that again") signal until it does add up. CRCs are also used in tape backups and other streaming communications. *Related Terms:* NAK; parity; streaming.

creasing — the process of crushing paper grain with a hardened steel strip to make an indention in paper in a narrow line and weaken the fibers so the paper will fold easily. *Related Terms:* perforating; scoring.

creep — (1) the phenomenon of middle pages of a folded signature or saddle stitched booklet to extend slightly beyond outside pages. (2) the process of compensating for the shifting position of the pages in a saddle-stitched bind.

Cromalin™ — a color, prepress, integral proofing system developed by DuPont that uses powdered pigments instead of ink.

crop — (1) mechanically, to eliminate portions of an image so the remainder is clearer, more interesting or correctly sized to fit the layout. (2) editorially, selecting the part of a photograph to be printed. In conventional photographs, the process is usually done by using crop marks on the original copy to indicate to the printer where to trim the image. In computer prepress, the process usually is accomplished by a cropping tool manipulated by mouse or pen input.

Graphic Communications Dictionary　　　　　　　　　**59**

crop marks — in design, the lines drawn on an overlay or in the margins of a photograph, printed forms, etc., to indicate to the printer where the image should be trimmed. *Related Terms:* crop marks; edge marks.

cropping — *Related Term:* crop.

cross direction — the direction of the fibers in a sheet of paper across the grain, as opposed to with the grain. *Related Term:* against the grain.

cross grain — a term usually relating to folding direction which is perpendicular to the direction of the paper grain. *Related Term:* against the grain.

cross head — a heading set in the body of the text used to break it into easily readable sections.

cross stroke — that part of a typographic character which cuts across the stem, for example, the cross stroke in the lower case "t."

CROSSLINE SCREEN

crossline screen — a glass halftone screen with opaque lines crossing each other at right angles, forming transparent squares or screen apertures. One of the original halftone screens, it is used primarily now as a mastering device for contact halftone screens. *Related Terms:* contact screen; gray screen.

crossover — type continuing across the gutter margin from one page to the next. *Related Term:* DT; double truck; double spread; 2-page spread; across the gutter.

crow's feet — in bindery operations, a folded or wrinkled sheet in the final product. A defect.

CRT (cathode ray tube) — the active video component of monitors and television. An electronic vacuum tube with a neck containing one or more emitters that project an electron beam onto a phosphor-coated screen to form an image. In television, called a picture tube. *Related Terms:* cathode ray tube; monitor; electron gun; VDT.

CSC (computer support collaboration) — the ability of networked personal computers to enhance and expand workgroup collaboration by eliminating time and distance barriers to all forms of electronic communication and the exchange of natural data types. *Related Term:* networking.

CSS (cascading style sheets) — a technique in Web design which allow developers to control the style and layout appearance of many Web pages all at once without individual page adjustments. They work similar to a template or master page system, allowing developers to define an HTML style for element and then apply it to as many pages simultaneously. To effect a change, you simply adjust the style, and the element is updated automatically regardless of how often, or where, it appears. Navigator 4.0 and IE 4.0 support CSS. *Related Terms:* DHTML, HTML.

CT (continuous tone) – (1) in graphics, a file format used to describe high-resolution scan information. (2) in photography, an image which has not been halftone screened.

ctn. weight (carton weight) — weight in pounds of one carton (ctn.) of paper. Varies by paper type and size.

CTP (computer to plate) — technology which permits a PostScript file to directly image a metal, photopolymer or paper image carrier without the use of paste-ups, film negatives, stripping or other prepress functions. All setup parameters are generated on the work station or computer and associated with the image content to control pagination, imposition, etc.

CTRL+ALT+DEL — in the Intel world of DOS and Windows, the key combinations used, together, to cause a system reboot.

CTS (clear to send) — one of the nine wires in a serial port modem communication, CTS carries a signal from the modem to the computer saying, "I'm ready to start when you are." *Related Term:* RTS.

cultural development — (1) the process of improving the education and behavior of human beings; changes in the behavior, customs and technology of human societies caused by education and communication of ideas. (2) the knowledge, customs, arts and tools of a human society transmitted to succeeding generations by means of education and permanent forms of communication. *Related Term:* culture.

cultural papers — *Related Term:* fine papers.

culture — *Related Term:* cultural development.

cure — (1) a term used interchangeably with conditioning. (2) the accelerated drying of coatings such as varnish, ink, etc. to minimize setoff during production. Often done under special heat or UV sources in line with the press. (3) the natural action which takes place as printed materials stand after imaging. *Related Terms:* condition; acclimate.

curl — the bending of paper parallel to its grain due to uneven moisture levels or coating differences between top and bottom surfaces within the substrate.

currency paper — premium 100% rag content paper made to be strong and durable for products such as money and bonds. Typically sub 24 (90 gsm). *Related Terms:* bank note paper; rag content paper.

cursive — The individual unconnected script letter shapes such as in handwriting. *Related Term:* cursive script; incursive.

cursive script — letter forms that do not touch but still look like handwritten or script letters. *Related Terms:* incursive script; cursive.

cursor — a spot of light or icon symbol on a video screen that the user manipulates to indicate the next point of data entry and where action is to be taken or changes are to be made.

curved plate — in letterpress printing, a plate that is precurved to fit the cylinder of a rotary press. Often a duplicate stereotype or cast plate made from a flat relief form from which a mold has been made.

curvilinear type generation — a computer image generation technique that produces letters and symbols by complex mathematics expressions rather than by a series of points. *Related Term:* vector.

custom color — any ink beyond CMYK in process color reproduction, i.e., special mix inks, metallics, fluorescents, etc.

custom dictionary — a dictionary of words that are not in the standard dictionary supplied with an application but that should be accepted by the spelling checker as correct. The custom dictionary is built by the software owner and usually can be edited with any text editor.

cut — (1) in printing, to dilute or thin an ink, lacquer or varnish with one or more solvents or clear base. (2) in engraving and etching, a term originally referring to a woodcut but now generally used to denote a zinc etching, halftone engraving or other illustrative matter or mounted engraving used in relief printing. (3) in bookbinding, to trim book edges during the binding process. (4) in publication work, any incision or shearing

of paper stock with a knife blade or automated machine. (5) in video editing, a single frame transition between two video clips. (6) in desktop publishing, an operating system option which eliminates material from its present position and transfers it to the clipboard where it can be discarded with the next clipboard function or reinserted as desired by the operator. *Related Term:* copy; paste.

cut ahead — paper trimmed without regard for where the watermark appears on the sheet. Not preferred or recommended usage. *Related Term:* cut to register.

cut flush — a method of trimming a book after the cover has been attached to the pages. A typical paperback book would be one example.

cut sizes — (1) all smaller paper sizes used with office machines and small presses. (2) paper distributor term for paper 11 x 17 or smaller.

cut to register — the best quality of trimming paper so the watermark appears close to the same location on every sheet. *Related Term:* cut ahead.

cutline — the descriptive or identifying information printed with art. *Related Term:* caption.

cutoff — the circumference of the impression cylinder of a web press, therefore also the length of the sheet the press will cut from the roll of paper. In a conventional broadsheet newspaper, the length of the sheet from top to bottom.

cutoff rule — a dividing rule between elements, often used in newspaper formats to separate news and editorial matter from boiler plate or filler.

cutout — a halftone where the background has been removed to produce a silhouette. *Related Term:* outline halftone.

cutscore — a knife blade or cutting wheel used to cut only slightly into the paper for folding purposes.

cutter dust — loose particles of paper or paper coating that collect between paper sheets. The dust is created by a dull cutter knife during the trimming process and is a source of ink contamination and related problems in production.

cutting layout — drawing that shows how a pile of paper is to be cut by the paper cutter.

cutting stick — a wood or synthetic material used in a paper cutter as a drop point for the knife to strike at the end of its stroke.

CVS (computer vision syndrome) — a computer age malady. An ergonomic term used to describe the discomfort sometimes felt by operators when they stare too long at a computer monitor. Symptomatically, clues to the ailment include blurry vision or headache, sometimes both. Most physicians believe that the problem is caused by the eyes not blinking as frequently as normal, leading to eye irritation. The remedy is frequent breaks from the constant viewing of the monitor.

cwt (hundredweight) – paper distributor abbreviation for 100 pounds.

cyan — a subtractive color created from a combination of light from the additive primary light colors of blue and green. One of the ink colors used in three- and four-color process printing. Sometimes referred to as process blue.

cyber — a nonword created by the science fiction writer William Gibson, who coined the term cyberspace in his novel Neuromancer. Cyber is most often used to make whatever word it's attached to seem connected in some loose way to the world of computers or the Internet. Overused prefix for anything which is to be perceived as high tech, cool, etc. *Related Term:* cyberspace.

cyberspace — an ill-defined term that is a novelist's word to describe a virtual world of computer networks that cyberpunk heroes "jacked into." Most people overuse the word loosely to refer to virtual reality, the Internet, the World Wide Web, and any other kind of computer system that users become immersed in. *Related Term:* information superhighway.

cylinder cleaner — special chemically formulated paste used to clean unwanted ink and residual material from printing press cylinders.

cylinder gap — the space in the cylinder of a offset or letterpress printing press where the mechanism for the plate and/or blanket clamps are located and where the grippers in a sheet-fed press are housed.

cylinder line — area of a printing plate that marks the portion of masking sheet used to clamp the plate to the plate cylinder; always marked on the flat before stripping pieces of film.

cylinder packing gauge — a calibrated measuring device used to determine appropriate cylinder underlayment to ensure proper image transfer.

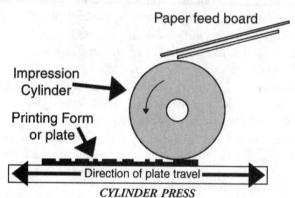

CYLINDER PRESS
The printing plate moves back and forth as cylinder turns.

cylinder press — a press configuration where substrate passes between a cylindrical impression cylinder and a flat printing plate. The impression is made on paper clamped on a cylinder that is rolled over the flat form. Commonly used in letterpress (relief) and screen printing (porous). This style of press is very popular as a device for die cutting operations.

cylinder system — a portion of a lithographic printing press that includes the plate, blanket and impression cylinders.

Dd

D channel (data channel) — a wire used to perform call signaling and setup in establishing a connection over ISDN. In some cases, the D channel is used to carry user data. *Related Term:* delta channel.

d/b/a (doing business as) — a sole proprietorship operating under a name other than that of the proprietor, such as Acme Printing or Ace Photography.

daemon — pronounced "demon," a harmless background UNIX program that waits and runs when a request is made on the port it is monitoring, out of sight of the user. A print daemon could handle print requests from multiple users and/or applications without delaying or slowing operating speed and freeing them to perform other tasks while printing was executed. On the Internet, it is most likely encountered only when e-mail is not delivered to the recipient. You'll receive your original message plus a message from a "mailer daemon."

dagger, double (‡) — a foot note used after the asterisk and single dagger.

dagger, single (†) — a footnote reference mark that is used after the asterisk has been used.

Daguerreotype — the first successful photographic system. A sensitized silver plate is exposed and developed with mercury vapors. A negative image appears as a positive by means of reflected light.

daisy chain — a computer network where two or more computers are connected in a series in a local network configuration. The most basic form of networking. The configuration which is used in modern USB connectivity.

daisy wheel — a metal or plastic circular print element with typewriter characters on spokes radiating from its center; about 3 inches in diameter. Hard copy printers can be equipped with more than one daisy wheel to mix faces. On typewriters, the wheel can produce type with differential spacing.

daisy wheel printer — an impact printer with a metal or plastic disk sliced into thin strips toward the center. A raised letter or character residing at the outer end of each strip. To print, the printer would spin the wheel to the correct character, and a hammer would strike it, forcing the character through an inked ribbon and onto the paper. Daisy wheel printers were able to produce letter-quality text but were slow, couldn't print graphics, and were often incredibly noisy. *Related Terms:* dot matrix printer; ink jet printer; laser printer.

dampener — cloth, paper, or rubber rollers that distribute the dampening solution to the press plate or ink roller in lithography. *Related Term:* dampener system.

dampener fountain — *Related Term:* dampening system.

dampening sleeves — thin cloth or fiber tubes that are slightly larger in diameter than the water form roller when dry; when moistened, they shrink to form a seamless cover.

dampening solution — *Related Term:* fountain solution.

dampening system — the dampening (water) rollers, fountain and controls that regulate and apply dampening solution to the plate on a lithographic duplicator or press. *Related Terms:* dampening solution; dampening fountain.

dandy roll — the wire cylinder device used to produce a weave and watermarks in paper during its manufacture.

dark printer — one of the halftones used to make a duotone; usually printed in black or another dark color; often contains the lower middle tones to shadow detail of the continuous-tone original.

dark slide — the protective slide which covers the film in the film holder when not in the camera or just prior to exposure, if in the camera. The slide is kept in place at all times when the film holder is not in the camera.

darkroom — a room that can be darkened so light-sensitive materials can be processed.

darkroom camera – *Related Term:* camera, darkroom.

DASD (direct access storage device) — a hard disk.

dash — a horizontal typographic mark indicating a break or lengthened pause between thoughts or the separation of thoughts. Several types are available to the compositor and/or editor: an em (—) is longer than an en dash (–) and much longer than a hyphen (-). Generally the longer the dash, the more severe the break. *Related Terms:* em-dash; long dash.

DAT (digital audio tape) — (1) originally, a consumer recording and playback medium for high quality audio. (2) in service bureau imaging, an efficient system for storing large computer files for service bureaus and high volume production shops.

data — information input, output, stored, or manipulated by a computer system.

data bank — the mass storage of information which may be selectively retrieved from a computer. *Related Term:* data base.

data base — *Related Terms:* data bank; database.

data bits — the bits in data transmission that carry the actual information to be recorded and stored. Parity, stop and start bits are not data bits.

data collection terminals — an integrated scanner / terminal where the scanner is built into the terminal with an input port, capable of accepting data from laser and/or CCD scanners. Data collected from scan is routinely stored and transmitted to a host. *Related Terms:* POS; point of sale.

data compression — the use of mathematical algorithms and manipulations to reduce the volume of information in digitized images.

data content codes — *Related Term:* data identifier.

data conversion — to change digital information from its original code so that it can be recorded by a disk or other electronic memory using a different code. *Related Terms:* media conversion.

data identifier — a specified character (or string of characters) that defines the general category or specific use of the data that follows.

data integrity (2D symbols) — in bar coding, a unique encoding scheme which includes start and stop codes, self-checking parity within each character, and check digits which apply to the total message. In 2D symbols, the height of the bars may be expanded to provide for redundant scan paths and allowance for diagonal scanning. Symbols damaged in a small area may retain their integrity because of this redundancy. Among the prominent 2-D symbologies, the data characters are composed of square or near-square elements which do not provide for redundant or diagonal scan paths. This fundamental difference provides for the enormous gain in data density but dictates that an error detection and an error correction system have to be instituted in addition to the character and message orientation and parity checking schemes.

data packet — modems generally send packets of about 64 characters plus some extras for CRC error checking. When downloading files using a protocol like Xmodem, however, the packets are larger. And when using Internet protocols such as TCP/IP, the packets are larger still—around 1,500 characters. *Related Terms:* asynchronous communication; TCP/IP; Xmodem; CRC.

data processing — a generic term for all operations carried out on data according to precise rules of procedure. The manipulation of data by a computer. *Related Terms:* data; data bank; data base.

data rate — data transfer rate; amount of data over time. For example, the rate at which digitized video can be moved from a hard disk or through a computer system. A high data rate usually translates to better picture quality. *Related Term:* data traffic.

data traffic — the number of TCP/IP packets traversing a network over time.

data warehousing — a generic term for a system for storing, retrieving and managing large amounts of any type of data. Data warehouse software often includes sophisticated compression and filtering techniques for fast searches. A database, often remote, containing recent snapshots of corporate data. Planners and researchers can use this database freely without worrying about slowing down day-to-day operations of the production database. *Related Term:* database.

database — a database can be as simple as a shopping list or as complex as a collection of thousands of sounds, graphics, and related text files. Software designed to help organize information. The first databases were limited to simple, searchable rows and columns. Relational databases access and reorganize a variety of data in a variety of ways. Advanced databases let users store and retrieve all kinds of nonstandard data, such as art, sound and video. *Related Terms:* data warehousing; data bank.

database front end — in the context of the Internet, this is an interface which integrates Web applications with sophisticated database programs.

datagram — in networking and the Internet, a packet of information, within network frames, consisting of data and a header. Datagrams are unique to the particular protocol being applied, but in all instances, the header shows the source, destination, and type of data, as well as its relation to any other datagrams being sent. This information enables the packet to be transported from router to router to its final destination.

dated material — written materials that have a limited life span for accuracy, for instance, quotations, price lists, annual projections, etc. Each one will be replaced, after a period of time, by more current data.

dateline — (1) primarily in periodical publishing, a line which is ahead of an article, giving the location from which the story was written or filed. (2) in all publications, the line giving the date of issue of the publication.

Dayglo — a series of specially formulated pigment inks which produce unusually bright reflectance. The colors do not occur naturally in nature and therefore are used for attention-getting promotions and similar.

daylight developing tank — a light-tight container that, once loaded in a darkroom, can be used to process film in a lighted room.

db (decibel) — a comparative measurement between two objects. A standard used in audio and video to determine signal quality.

DBS — the acronym for Direct Broadcasting Satellite service.

DCC (direct client to client) — a protocol used on Internet Relay Chat (IRC) to allow chatting directly without having to go through a server, thereby keeping conversations private.

DCS (desktop color separation) — an enhancement of EPS which produces a digital color format that produces five PostScript files, one for each color separation (CMYK) and one data file. DCS can greatly reduce the cost of four-color publishing.

DCT (discrete cosign transform) — a form of data coding used in many current image compression systems for bit rate reduction.

DDE (dynamic data exchange) — an effort to better integrate applications introduced by Microsoft in Windows 3.0. DDE expanded upon the Windows clipboard by allowing users and programmers to copy data and give commands between applications while maintaining a "live link." Data changed in the source file will also change in the target file. *Related Term:* OLE.

DDN (Defense Data Network) — created in the early 1980s, the Defense Data Network consists of the U.S. military network (MilNet) and other pieces of the Internet under control of the Department of Defense. The DDN provides worldwide electronic communications between military installations and has often been the target of hackers.

DDS — Digital Data Service.

deactivation — the process of deactivating a domain name from the top level domains; the domain name system (DNS) will no longer have the information needed to find its corresponding Internet protocol (IP) number(s), effectively disabling the domain name as a tool for locating the related computers or organizations. *Related Terms:* zone; zone files; domain name system (DNS); hold.

dead form — any form that has been printed and is waiting to be remolded or distributed (if hot type).

deadline — time beyond which copy cannot be accepted, or when production stages are to commence.

debarking — the bark-removing step in the papermaking process.

deboss — to press an image into paper so it lies below the surface. *Related Term:* embossing.

debugging — the process of locating and eliminating defects in a computer program.

decal paper — the smooth, flexible paper used as backing for decals.

decalcomania — process of transferring designs to another surface; products are sometimes called "decals."

deciduous — hardwood trees that provide wood fibers used in the papermaking process. *Related Terms:* coniferous.

decimal — a base 10 number system consisting of digits 0-9.

deck — (1) in publishing, one unit of a headline set in a single type size and style. (2) in the digital world, a shorthand term used to refer to an audio or videotape recorder. (3) in press configuations, a printing unit situated around or on the circumference of a central impression cylinder shared by other similar units. *Related Terms:* deck copy; tagline.

deck copy — the text found underneath the headline of an article or story that provides slightly more detail than the headline and is set in a smaller point size than the headline but larger than the body text. *Related Term:* deck.

deckle — (1) in hand papermaking, the removable, rectangular wooden frame that forms a raised edge against the wire cloth of the paper-forming mold and holds the stock suspension on the wire when handmade paper is produced. The resulting untrimmed, irregular edge on a sheet of paper forms where the pulp flows and sets against the frame. This is referred to as deckle edged paper. (2) in mechanized papermaking, on a Fourdrinier papermaking machine, the arrangement on the side of the wire that keeps the stock suspension from flowing over the edges of the wire. (3) also, the canvas webbing wound around the ends of cylinders on papermaking equipment to control the width of the sheet. (4) a term indicating the width of the web of paper formed on a papermaking machine. *Related Term:* deckle edge.

deckle edge — The untrimmed feathery edges of paper formed where the pulp flows against the deckle.

decoder — the electronic package which receives the signals from the scanning function, performs the algorithm to interpret the signals into meaningful data and provides the interface to other devices.

decompression — to reverse the procedure conducted by compression software and thereby return compressed data to its original size and condition. *Related Terms:* ZIP; WinZIP; ARC; stuffit.

decorative type — an alternate name for display or novelty type. One of the six classifications of type, it has no specific definitive features except that is generally unsuitable for body copy; should be used in large sizes and minimally. An attention-getting type, almost always of unusual design.

dedicated line — communications circuit used for one specific purpose, i.e., for interactive portion of a teleconference.

dedication — the part of the front matter of a book where the author dedicates the work to an individual or group of individuals.

deep etch plate — a lithography positive working plate with the image slightly below the base material and often with a bonded emulsion. The plates are used to print large numbers of copies. Opposite of surface plate. *Related Terms:* bimetal plates; trimetal plates.

defamation — libelous or slanderous statements that cause injury to another person.

default — action that a computer program automatically performs unless manually instructed otherwise.

default drive — the computer drive first accessed by the operating system upon bootup.

default hyperlink — in an image map, the hyperlink to follow when the user clicks outside of any hot spots on the image. You set the default hyperlink by editing the Default Link field in the Image Properties dialog box.

default parameters — process instructions supplied by the computer or software manufacturer if alternatives are not provided by the operator.

definition — (1) in photography, the degree of sharpness in a negative or print. (2) in composition, the style of the second of a pair of paragraphs composing a definition list entry. The first paragraph in the pair is the term. (3) explanation of meaning.

definition list — a list of alternate term and definition paragraphs. Definition lists are often used to implement dictionaries in FrontPage Webs. *Related Terms:* term; definition.

deflected ink jet — *Related Terms:* drop on demand ink jet; continuous ink jet.

degaussing — neutralizing the earth's natural magnetism buildup inside a monitor's CRT which can cause a loss of color purity. Degaussing removes this excess magnetism. Many monitors automatically perform the operation when they are turned on; others include a manual degaussing button. *Related Term:* color purity.

degreasing fabric — removing the oily substance on a fabric prior to use for screen printing.

de-inking — the removal of ink from paper for recycling.

Deja News Research Service — search engine for the newsgroups. It is similar to WebCrawler, but it is a tool exclusively for searching Usenet, the largest information utility in existence.

delayed indent — an indent which does not take effect until after a specified number of lines have been set.

delete — a proofreader's mark meaning "take out." Looks like this: This mark is placed in the margin of the copy, and a line is drawn through the editorial material to be removed.

deletion — the process of removing a domain name and its corresponding record from the domain name system (DNS) domain name database. A deleted domain name cannot be used to locate computers on the Internet. A domain name may be deleted at the request of the domain name registrant, as a result of nonpayment of fees, or due to circumstances particular to individual cases. *Related Terms:* domain name system (DNS); registrant; registration fee; re-registration fee; deactivation.

deletion fluid — a solution used to remove unwanted areas from presensitized lithographic plates or porous printing stencils. *Related Term:* correction fluid.

delimiter — a symbol found at the beginning and end of a character or command string. *Related Terms:* precedence code; control character.

delivery board — a wood board mounted at the front-center of a platen press and used to hold the printed sheets of paper.

delivery cylinder — the cylinder on a press that carries the gripper-equipped chains that control the printed sheet as it leaves the impression cylinder on its way to the delivery end of the press. *Related Term:* transfer cylinder.

delivery memo — a form sent by printers, photographers and stock photo services to clients for signature to verify receipt of merchandise or produce and agreement to contract terms.

delivery system — a portion of a printing press designed to remove paper (substrate) from the press after imaging, and which places products on a pile or into a roll.

delivery unit — the portion of the production machinery that moves the processed sheet from the final production unit to the delivery pile or roll.

Delta (D) — a symbol used to indicate deviation or difference.

demand publishing — *Related Terms:* printing on demand.

demibold — an intermediate weight face between normal roman and true bold.

demo software — a time- or function-restricted version of a software program designed to give the flavor of the real full-blown application. Limits may include features that are disabled, the inability to save anything you create, or expiration after a certain number of days or uses. *Related Terms:* beta software; shareware; freeware.

demodulation — the process of recoding an analog signal into digital data. When data is transferred over phone lines, a modem modulates the digital data into audible tone frequencies. When the data reaches its intended destination, another modem demodulates the signal back into digital data. *Related Terms:* modulation; modem.

demographic printing — personalized printing; print on demand; direct digital printing. These three features are basic to demographic printing where output can be tied directly to a customer or client data base and each individual product can be modified to fit a predefined profile.

demon letters — in foundry type composition, lowercase p, d, q, and b. So named because of the difficulty of distinguishing them apart when composing lines in hand composition where the letters are both upside down and backwards.

denizen — one admitted to residence in a foreign country or an alien given rights of citizenship; a new citizen on the Internet.

densitometer — an electronic or photomechanical device for measuring the optical density of photographic film, paper or printed images. There are two types, reflection and transmission, used by printers and photographers to ensure proper exposure for halftones, density of ink on substrate and tonal density in photographic prints and transparencies. The densitometer measures subtle differences that the human eye cannot discern. It is used to check the accuracy, quality and consistency of output.

density — (1) in production, the relative darkness of copy, ink on paper or emulsion on film, as measured by a densitometer. (2) in photography, the degree of opacity of an image on paper or film. Generally, the ability of a photographic image to absorb or transmit light. Expressed as the logarithm (base 10) of the opacity, which is the reciprocal of the transmission or reflection of a stone. (4) in screen printing, the relative tightness or looseness of fibers. (5) in barcode and shop floor tracking, a term used to describe the relative amount of memory contained in a radio frequency identification tag. (6) in the paper industry, the relative weight of a particular grade of paper. *Related Terms:* bar code density; **D**logE.

density range — difference between darkest and lightest areas of copy, photograph or printed reproduction. *Related Term:* contrast.

depreciation — financial technique of spreading the cost of capital purchases, such as equipment or buildings, over time.

depth — the vertical measurement of a page, figure, table or other block of material.

depth of field — (1) a photographic term for relative sharpness of features in an image regardless of their distance from the camera when photographed. The distance between the nearest and farthest point in a image which is in critical focus. (2) in optical reading, the distance between the maximum and minimum plane in which a code reader is capable of reading symbols.

DES (data encryption standard) — a data encryption method originally developed by IBM and certified by the U.S. government for transmission of any data that is not classified top secret. It uses algorithms for doing private-key encryption. Although fairly weak, with only one iteration, repeating it using slightly different keys can provide excellent security. *Related Terms:* encryption; private key encryption.

desaturated color — a color that appears faded, printed with too little ink, or as though white had been mixed with the colorant. *Related Term:* pastel.

descender — the portion of a lowercase type character

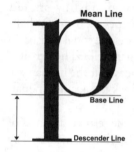

that extends below the common base line (x-height) of a typeface design, such as in "j," "p," "q," and "y". *Related Terms:* ascender; x-height; mean line; ascender height.

desensitize — the process of making non-image areas of a printing plate non-receptive to ink through chemical treatment of the metal, usually a gum arabic solution.

design — (1) process of creating images and page layouts for printing production. (2) the totality of the creative work.

design and layout — the production phase in which planning is done for a printed product.

design brief — written description of how a printed piece is intended to look and the requirements for reproducing it; designed to prevent major breakdowns and/or unscheduled downtime during production.

design principles — the basic guiding suggestions for successful design concepts: contrast, rhythm, proportion, balance and unity.

de-skew — to straighten a slanted item.

desktop — the opening screen when booting up a Windows-based computer. It is the area where software and utility icons are maintained for easy access. It is intended to mimic a true desktop with a variety of objects.

desktop color separation — *Related Term:* DCS.

desktop publishing (DTP) — a term used to describe a process of preparing full page images on a microcomputer; generally intended to circumvent traditional composition and page makeup processes.

desktop video (DTV) — a merging of the camcorder and the home computer, DTV is at the forefront of video-making technology. Only possible because of advances in home computing power and affordability, it opens up a wealth of new possibilities in editing and creativity for both the video maker and the PC enthusiast. *Related Term:* CUSeeMe.

despeckle — to remove extraneous dots or marks generally introduced during the scanning process.

developer — a chemical bath that changes exposed silver halide crystals on photographic film or paper to black metallic silver to make the image on a light-sensitive emulsion visible. Some developers are used in the darkroom to process graphic arts film; others are used in photographic printing, proofing, plate-making, stencil preparation in screen printing, and some areas of masking in gravure. *Related Terms:* reducer; stop bath; fixer; hypo.

developer adjacency effect — underdevelopment caused by chemical exhaustion of small, lightly exposed areas when surrounded by more dense areas.

development — (1) the process of converting a latent photographic image on film or paper to a visible image. (2) in lithographic platemaking, removing the unhardened coating from the plate surface.

device driver — a program that lets peripheral devices communicate with computers. Some device drivers for standard components, such as keyboards or monitors, come with computers. Devices that are added later require the user to install the corresponding drivers. *Related Terms:* real mode; protected mode.

device independence — when an application lets the operating system and its device drivers figure out how to make the hardware handle it. Device independence depends on hardware standards such as VGA and well designed application drivers. APIs and hardware abstraction layer standards are a more robust approach to device independence. *Related Terms:* API; hardware abstraction layer.

device independent color — any device which exhibits the capability of outputting colors which correspond to those produced on the final imagesetter films.

DHCP (dynamic host configuration protocol) — a method of assigning IP addresses to computers connected to a Windows-NT-based local area network (LAN). When a computer logs on to the network, the DHCP server selects an IP address from a master list and assigns it to the system. *Related Terms:* IP address; LAN.

DHTML (dynamic HTML) — an advanced HTML standard, dynamic HTML combines HTML, style sheets, and scripts to make Web pages more interactive. Both Internet Explorer 4.0 and Communicator 4.0 support dynamic HTML features, but do so in different and incompatible ways. Both companies promise compatibility in future versions of their browsers. *Related Terms:* CSS; HTML.

diacritic — accent marks about the character face such as the accent (`) or tilde (~) used to explain the proper pronunciation of words.

diagonal line method – a technique to determine size changes of copy. *Related Term:* proportion; proportion scale.

dialer — a program which establishes a connection to the Internet and provides Winsock support. Most service providers have a built-in dialer. Third party dialers include Trumpet Winsock and the Windows '95/98 Dial Up Networking.

dialog box — a window or menu which appears in graphic user interfaced machines and provides a checklist of options to accept, reject, move, close, etc., pertaining to the presently undertaken operation.

dial-up account — an Internet account that allows you to dial up an Internet service provider's computer with a modem. These types of accounts usually have a UNIX or other command-line interface, although it may be buried by a GUI menu for inserting data. *Related Term:* dial-up connection.

dial-up connection — the most popular form of Net connection for the home user, this is a connection from your computer to a host computer over standard telephone lines. *Related Term:* ISP; dial-up account.

dial-up teleconferencing — using a public phone line to connect with a teleconference, either with or without operator assistance.

diaphragm — a perforated plate or adjustable opening mounted behind or between the elements of a camera lens. It is used to control the amount of light that reaches the film over time. Openings are usually calibrated in f/-stop numbers. *Related Terms:* iris; aperture; f/stop.

diaphragm control – the scales or other calibrated device on a process camera which displays various selectable aperture ratios to accommodate all optical factors of line and halftone photography. *Related Term:* aperture control.

diarylide yellow — a strong organic pigment, fre-

quently used in yellow process inks. Formerly called Benzedrine yellow.

diazo — (1) in printing, a photosensitive non-silver based emulsion usually coated on presensitized lithographic plates. (2) in prepress, a film-coated paper for making contact prints of technical drawings, paste-ups, etc. (3) in engineering, the basis for blueprint systems. Most often developed by exposing to ammonia fumes. *Related Term:* diazo process.

diazo process — a photographic procedure in which a group of light-sensitive compounds are applied to paper, plastic, or metal sheets. The substrate is then exposed to intense blue and/ or ultraviolet light and developed with ammonia vapors or an alkaline solution to form an image. The Diazo process is used to produce proofs from positive film flats and color-pack natural color proofs. It is also used to coat presensitized press plates. *Related Terms:* dyeline; diazoprint; direct-positive; ammonia print; white print.

Didot — the point system used in many European countries. One Didot point is equal to 1.07 American points: .0148 of an inch or .375 mm.

die — (1) a matrix pattern of sharp knives or metal tools used to stamp, cut, or emboss specific shapes, designs, and letters in a substrate. (2) a metal plate, etched in relief or intaglio to provide a raised or depressed impression on paper.

die cut — the technique of using sharp steel rules to make cuts in printed sheets for boxes, folders, pop-up brochures, and other specialized printing jobs.

die press — a relief class press into which dies are placed to perform a die cutting or stamping operation.

die sinking — *Related Term:* die cutting.

die stamp — an intaglio process for creating designs engraved into copper or steel, usually used for producing letterhead, business cards, and other specialized printing jobs.

die, embossing — a brass or steel die that impresses a design in relief into a paper substrate. Unlike a cutting die, the edges are not sharp and will not cut through the substrate but place a distress pattern in the paper, creating an image above or below the normal paper surface.

die-cutting — the use of sharp steel rules in a relief operation to slice paper or board to unusual shapes using dies that look and perform much in the manner of a cookie cutter. Simple die-cutting is inexpensive, usually pricing similar to a single impression of ink. Die-cutting imparts enormous appeal to printed material by altering the square or angular shape normally imparted by traditional trimming. *Related Terms:* die sinking; die.

dielectric printing process — a nonimpact printing technique in which paper with a conductive base layer is coated with a nonconductive thermoplastic material. A set of electrodes applies an electric charge to areas of the substrate, creating a latent image of the original. The paper is imaged by a toner system similar to that used in electrostatic copying devices. *Related Term:* electrographic printing.

dieresis (ö) — two small dots which most often appear over the lower case "o" such as the second "o" in coöperate.

diffuse reflection — the component of reflected light which emanates in all directions from the reflecting surface.

diffused highlight — the lightest area of a full tone original that carries important detail. In printing, these areas are normally reproduced with the smallest dot or printed tone value.

diffusion — the addition of screens or filters over the lens of enlargers or cameras to produce a softer definition. Sometimes used in printing portraits to eliminate minor facial lines or wrinkles. Extreme diffusion is often used for special effects.

diffusion etch process — a process used in gravure printing to transfer an image to a gravure cylinder; involves the preparation of a photomechanical mask that is applied to the clean and polished cylinder; acid is used to eat through the varying depths of the mask creating ink holding cells in the cylinder metal surface.

diffusion transfer — a method for producing positive screened or line prints on paper, film, acetate, or lithographic plates by transferring the latent image from a negative material to nonsensitive receiver. During the single chemical process, the receiver is mated with the negative material in an activator bath. After immersion of the two pieces, they are transferred between two pressure rollers and then allowed to stand to complete the activation. The chemically altered latent image from the negative material migrates to the surface of the nonsensitive receiver which may be a sheet of paper, film, aluminum press plate or acetate. The result, after proper rinsing, is a photographic quality line or halftone image which is ar-

chival permanent. *Related Terms:* photo mechanical transfer; PMT.

diffusion transfer plate — a type of lithographic plate made using the diffusion transfer process.

digerati — a play on the word "literati" that describes the hip, knowledgeable people at the cutting edge of all things digital or otherwise in-the-know regarding the digital revolution.

digest — an abridged version of a full-length article, story, or book; an editorial condensation of content.

digital — a method of storing and/or transmitting data, wherein each code is given a unique combination of electronic bits, each one generally indicating either the presence or absence of a condition (such as on-off, yes-no, true-false, open-closed, on-off, 1-0, etc.) Discrete signals, as opposed to continuously variable analog-type signals.

digital audio — audio tones represented by machine readable binary numbers; in contrast with analog recording techniques. Analog audio is converted into digital by sampling techniques whereby a periodic sample is taken which records the amplitude in a numeric form, which is stored. More frequent sampling results in more accurate digital representations. *Related Terms:* DAT; digital audio tape.

digital camera — *Related Term:* CDD camera.

digital certificate — an attempt by software manufacturers to ensure appropriate and secure identifications for digital data. Basically it is a password-protected file that includes a variety of specific information about the certificate holder, an encryption key that can be used for authentication, the name of the company issuing the certificate, and the term during which the certificate will remain valid. The certificates can be used as on-line identification similar to a driver's license that can verify your identity in the physical world. E-mail messages with an attached digital certificate make the recipient more confident that the document is genuine. There are a number of incompatible systems by competing companies. No universal standards have been established for this technology. *Related Terms:* encryption; digital signature.

digital composition — a composition method in which the images, type, artwork and halftone photographs are created with pixel images. *Related Terms:* PEL; pixel; raster image processor; digital typesetting.

digital computer — a computer that represents and

processes information consisting of clearly defined, or discrete, data by performing arithmetic and logical processes on these data using the binary number system (the number system based on powers of 2 rather than powers of 10). *Related Term:* analog computer.

digital controls — digital controls are mechanisms that alter settings in discrete steps, in contrast to infinitely variable analog dial controls. The advantage of digital controls is that the settings can be set (often automatically) in memory and recalled without running through all incremental steps at a later time. *Related Term:* analog controls.

digital halftone — a halftone image, negative or positive, where the halftone dots are created by varying numbers of laser spots. *Related Term:* dithering.

digital photography — the production or reproduction of images using electronics through the use of photo multiplier tubes (PMTs), charge-coupled devices (CCDs), or complementary metal oxide semiconductors (CMOS). The devices record varying intensities of electronic signals based upon the light and color reflectivity of the subject. The data is then digitized and represented by a series of numbers which can be manipulated by computer and then reconstructed as a photographic representation of the original image. *Related Terms:* area array; linear array; CCD; CMOS; PMT.

digital proof — proofs produced on specialized equipment which produces direct output from the original digital file which resides on the host hard drive or server.

digital signature — *Related Terms:* digital certificate; encryption; authorization; authentication.

digital typesetting — third-generation typesetting systems that generate typeface images from master font characters stored in the typesetter's computer as digital information. The master image digital information is translated into type face images by strokes or points of light from an electron beam pointed onto the face of a cathode ray tube (CRT) or, alternately, a laser light directly onto film. Density of strokes or scan lines per inch determine the quality of typeface characters and imaging speed.

digital video — a video signal represented by sampling of the analog signal and having the samples represented by a series of binary numbers that describe a finite set of colors and luminance levels. *Related Term:* digital audio.

digitize — the process of converting and/or storing analog information as a series of binary codes that can be processed by computer. Digitized data can be stored in solid state memory, magnetic or optical media. *Related Term:* digital.

digitized audio/video — the representation of video images and audio as binary values, 1s and 0s. *Related Term:* digital.

digitized type — *Related Term:* digital typesetting; imagesetting; fourth generation photographic composition system.

digitizer — a computer peripheral device that convert analog signals (images or sound) into a digital signal values. With an image, the digitizer sends position information to the computer, either on command from the user (point digitizing) or at regular intervals (continuous digitizing). Also devices that capture real-life three dimensional images by video. *Related Terms:* CCD; CMOS.

digraph — not a ligature, but a group of two successive letters forming a singular phonetic sound and often joined with a small line in typography, for example "æ" or "œ" or the "oa" in float and the "sh" in shall. *Related Term:* diphthong.

dilutent — a solvent or thinner used to dilute another liquid or solid such as ink, varnish or other coatings.

dimension marks — small L-shaped points or marks on mechanicals or camera copy outside the area of the image to be reproduced, indicating the size limits of reduction or enlargement.

dimensional stability — the tendency of paper and other substrates such as film or plates to retain their exact size despite the influence of external influences such as temperature, pressure, moisture, or stretching, dryness, etc.

DIMM (dual in-line memory module) — a small circuit board filled with RAM chips. A DIMM has a lot more bandwidth than a single in-line memory module (SIMM) and its data path is 128 bits wide, making it up to 10 percent faster than a SIMM. DIMMs are used on both the PC and Macintosh platforms. *Related Terms:* SIMM, RAM.

dingbat — a typographic ornament usually associated with keystroke commands to an imager, employing special non alphabet fonts. There are numerous varieties using names such as wingdings, sorts, webthings, etc.

Representative examples are ⇨ ▼ ✎ ✚ ✭ ✉ and usually used for design emphasis within text.

diode — semiconducting electronic components which allow current to flow through them in only one direction. Light-emitting diodes (LEDs) produce light when current is applied, and liquid-crystal diodes (LCDs) lighten or darken a transparent material when current is applied. LCDs are often used to create flat-panel displays for portable computers. *Related Terms:* flat-panel display; LCD; semiconductor.

DIP switch — *Related Term:* dual in-line package switch.

diploma paper — *Related Term:* parchment paper.

dipthong — a phonetically based combination of two characters into a single graphic and single sound, such as "œ" and "Œ" in words of Greek origin. Not to be confused with a ligature, which is not phonetically based. *Related Term:* digraph.

direct broadcast satellite — satellite designed with sufficient power so that inexpensive earth stations can be used for direct residential reception.

direct color separation — *Related Term:* direct photo screen.

direct connection — a permanent leased line connection between your computer system and the Internet. Sometimes referred to simply as a leased-line because the line is leased from the telephone company.

direct contact print – *Related Term:* contact print.

direct digital proofs — proofs made directly from electronic continuous-tone art and not from film negatives.

direct electrostatic copier — a copier on which the electrostatic charge is placed on a special coated paper and then receives a toner material for imaging onto substrate. *Related Term:* dielectric printing process.

direct emulsion — a photosensitive liquid emulsion that is coated on both sides of a stretched screen in screen printing. When dry, the emulsion coating is exposed through a film positive and serves as the printing stencil. *Related Term:* direct photo screen.

direct entry typesetters — a somewhat, although not totally, obsolete typesetter configuration that contains a complete phototypesetting system and an entry keyboard in a single unit. Typically they include a CRT or other video output device, a memory buffer, a minicomputer and a second generation phototype character generating system. All input and component setups are directed from the keyboard without the use of floppy disks, magnetic tape, etc. Used primarily for small newspapers, agencies and shops to create conventional paste-ups.

direct halftone — the creation of a halftone directly from a photograph or other original using a contact or glass crossline screen. The most popular method of creating analog halftone images.

direct image nonphoto surface plate — a short-run surface plate designed to accept images placed on its surface with a special crayon, pencil, pen, typewriter ribbon, etc.

direct image photographic plate — short- to medium-run surface plates which are exposed and processed in a special camera and processor unit; used mainly in quick print area of printing industry.

direct image plates — paper or plastic plates on which ink-receptive (oleophilic) images are typed or drawn. *Related Term:* direct image nonphoto surface plate.

direct mail — a form of advertising materials mailed directly to the potential customer that are designed to stimulate direct readers' response to purchase, donation, enroll, subscribe, etc.

direct photo screen — a method of preparing a screen printing stencil by coating a light-sensitive emulsion directly onto the screen fabric. *Related Terms:* direct emulsion; direct process; photographic stencils; direct/indirect process.

direct process — a method of photographic screen printing stencil generation in which emulsion is applied wet to the screen fabric, dried and then exposed and developed. *Related Terms:* photographic stencils; direct emulsion.

direct screen — a single-stage color correction process where the processed photographic masks are placed over unexposed separation film in the camera back before producing the separation negatives. *Related Term:* direct screen color separation.

direct screen color separation — a system of producing color separation negatives or positives directly from the original. Each film is color separated and screened in one step. *Related Term:* direct screen.

direct screen halftone — a halftone negative made by direct exposure from the original through a halftone screen.

direct thermal — *Related Term:* thermal printing.

direct to press — *Related Terms:* direct digital printing; digital printing; digital imaging.

direct transfer — a process used in gravure printing to transfer an image to a gravure cylinder by applying a light-sensitive mask to the clean cylinder. Light is passed through a halftone positive as it moves in contact with the rotating cylinder; after exposure, acid is used to eat through the varying depths of the mask into the cylinder metal after the mask has been developed. *Related Term:* diffusion etch process.

direct/indirect process — the photographic screen printing stencil preparation in which stencil emulsion is applied to clean screen from precoated base sheet; when dry, the base is removed and the stencil is exposed and developed directly on fabric. *Related Terms:* photographic stencils; direct process; direct photoscreen.

Direct3D — a part of the larger DirectX standard that simplifies 3D software development by addressing display hardware more efficiently. It is a type of application programming interface (API) hardware abstraction layer that is placed between the application and the display hardware. Developers write instructions to Direct3D, which then translates them to the graphics card. Both the application and the graphics card must support the standard. *Related Terms:* API; DirectX; hardware abstraction layer.

directory — the older name for a sort of table of contents of all files contained on a specific section of a computer disk. Directories generally show file name, ize, date and time created, type, and author. More generally called a folder on newer operating systems.

directory paper — *Related Term:* Bible paper.

DirectX — an application programming interface hardware abstraction layer that acts for Windows 95/98 and specific varieties of hardware. The DirectX standard includes Direct3D to speed up texture mapping and other 3D graphics processes, DirectSound for audio, DirectDraw for vector graphics, DirectVideo for AVI files and other motion pictures, and the DirectPlay and DirectInput team which simultaneously supports sound, drawing, video, networked game play, and joystick standards. *Related Terms:* API; AVI; Direct3D.

dirty copy — copy that has been heavily marked up by editors or proofreaders.

discrete code — a bar code or symbol where the spaces between characters (intercharacter gap) are not part of the code, e.g., USS-39.

discretionary hyphen — a hyphen that is keyboarded with the copy, between the appropriate syllables, but used by the typesetting system only if necessary for line spacing and justification.

discussion group — a Web that supports interactive discussions by users who have Microsoft's FrontPage. Topics are entered as text in a form, and users can search the group using the search form or access articles through a table of contents.

discussion group directory — a directory in a FrontPage Web containing all of the articles in a specific discussion group. The name of a discussion group directory is created automatically by FrontPage and always begins with an underscore (_). Discussion group directories, although not necessarily visible from the FrontPage Explorer, can be searched by a WebBot Search component.

dish — a parabolic antenna that is the primary element of a satellite earth station.

disk — *Related Terms:* disk; hard disk; floppy diskette; magnetic tape; CD-RW; diskette.

disk mirroring — writing data to two disks at the same time. This is used to ensure that the data is written accurately and not lost if one device fails.

disk operating system (DOS) — software for computer systems with disk drives which supervises and controls the running of programs. The operating system is 'booted' into the computer from disk by a small program which permanently resides in the memory. Common operating systems include MS-DOS, PC-DOS (IBM's version of MS-DOS), CP/M (an operating system for older, 8-bit computers), UNIX and BOS.

disk refiner — machinery that rubs, rolls, disperses, and cuts pulp fibers during the papermaking process.

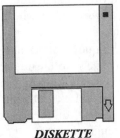

DISKETTE
Most standard diskettes use a circular oxide-covered Mylar disk which is used as the recording medium.

diskette — the 2 in., 3½ in., 5¼ in. or 8 in. flexible (floppy) or rigid magnetic, optical or M/O recording medium used in most microcomputers. *Related Term:* disk.

display ad — *Related Term:* display advertising.

display advertising — area space advertisements in newspapers, magazines and other publications which utilize art, photos, display or decorative types, etc., in their design. It contrasts with classified advertising, which generally utilizes small type and no illustrations and is confined to single column widths.

display file — text as it appears on the screen or in printout, with field tags and file markers invisible. The exceptions are Hypertext links, which are visible because they are used to "jump" between related URLs or sections of the text.

display type — any size type larger than the predominant body size, used for headings or to otherwise attract attention, etc. The term is sometimes applied in error to type of the decorative or novelty classification.

dissolve — transition in which one video clip simultaneously fades into the next. *Related Term:* fade.

distortion — accidental or intentionally induced output image variations from the original. Accidental can be caused by improperly aligned camera planes or bad optics or improperly selecting image handles in graphics and photo manipulation software. Intentional deviations are made by, electronically, selecting a side or other specified image handles and moving their position(s). *Related Terms:* proportional; anamorphic.

distributing rollers — in offset and letterpress press inking systems, the series of rubber-covered rollers that moves the ink from the ductor roller and works it into a thin, uniform layer before transferring it to the form rollers which contact the image carrier. Not a part of most flexographic or gravure presses.

distribution — a method of limiting where Usenet postings go. Useful for things such as "for sale" messages and discussions of regional politics.

distributor — (1) in commerce, a business that buys paper from mills and sells it to printers and printing buyers. (2) in production, a roller which is part of a water or inking system between the fountain ductor and the form rollers. *Related Terms:* jobber; wholesaler; merchant; paper merchant.

dithered dots — a bitmap technique for alternating the values of adjacent dots (or pixels) to create the effect of intermediate values or colors. Dithering can give the effect of shades of gray on a black and white display or additional colors on a color display. This technique is used when a full range of colors is not displayable or available.

dithering — a digital pixel averaging technique used to add detail or minimize the difference between pixels. This is accomplished by filling in a gap between two pixels with another pixel that is the average of the other two. Generally used by software programs

when a computer's graphics card doesn't support a particular color or resolution is low. The result is less ragged edges. *Related Term:* antialiasing.

Ditto — a brand name for a planographic class duplicator using the spirit process of reproduction.

divider sheet — a sheet of paper, usually made from card stock, that segments a publication into various sections.

divider tab — the portion of a divider sheet that extends beyond the trim size of the rest of the publication, often folded to fit under the book's trim size and left for the customer to unfold to reveal the tabs.

DMA (direct memory access) — one of two ways hardware devices attached to PCs can be designed to send their instructions to and from main memory. Computer default settings have the CPU to do the work, but DMA channels send instructions directly to memory, leaving the CPU free to do other tasks. DMA channels are limited in number, however, and one channel cannot be allocated to more than one keyboard, sound card, etc. *Related Term:* plug and play.

D-max — the maximum optical density that can be achieved in a given photographic or photomechanical system. *Related Term:* D-min.

DMI (desktop management interface) — an efficient means of reporting system problems. Compliant computers report status information to a central management system by the network. With DMI, if a hard drive's error rate exceeds a specified limit, the computer reports the problem to a central management system, which notifies systems personnel. The standard is maintained by an industry group of Microsoft, IBM, Compaq, and Hewlett-Packard.

D-min — the minimum recordable density that can be archived in a given photographic or photomechanical system. *Related Term:* D-max.

DMS — Document Management System.

DN (directory number) — like an ISDN telephone number which is used primarily to track billing on the ISDN line. It corresponds to the contact number that gets called in establishing a connection.

DNS (domain name system) — (1) the conventions for naming hosts on the Internet. (2) the way the names are handled and processed. E-mail or point a browser to an Internet domain such as cnet.com causes the domain name system to translate the names into Internet addresses (a series of numbers looking something like this: 123.123.23.2).

dock — the process or equipment which enables a laptop computer to be semipermanently installed on a desktop and render full access to desktop peripherals. The process, called docking, is usually done "hot" without the need to shut down the computer.

doctor blade — (1) in all intaglio printing and some flexographic printing, a steel blade that wipes the excess (surface) ink from a gravure cylinder before printing. (2) in papermaking, a steel or wooden blade used to keep cylinder surfaces clean and free from paper, pulp, size, or other material during papermaking. (3) also, a blade is used on some Flexographic presses to remove ink from the surface of the Anilox roll. *Related Term:* (in screen printing) flood bar.

doctor roll — the fountain cylinder on a Flexographic press.

document — on the Web, any file containing text, media or hyperlinks that can be transferred from an HTTP server to a client program.

document info — a mode feature of some browsers which can give you information about the Web page you are currently viewing, including the document's structure, composition, and security status. Structure information contains URLs of images in the document. Composition includes location, MIME type, source, local cache file, modification and expiration dates, content length and character set. Security information informs you about encryption and certification. You can usually select a menu item to select this mode.

document source — a mode feature in most browsers that will open a window and display the source code (HTML, JavaScript, etc.) of the Web page you're currently viewing. These contents can be copied in Windows by pressing CTRL-A to select the contents, followed by CTRL-C to copy, and CTRL-V to paste into another document. On Macintosh systems, the source code is automatically saved into a text file on the desktop each time you view a document source. You can select this menu item by pulling down the VIEW menu.

dodge — to block light from selected areas while making a photographic print to reduce intensity or density in selected areas. The procedure can be accomplished with special tools or by manipulation of the hands and fingers between the enlarger projection light and the print paper. *Related Term:* burn in.

dodging — the procedure of holding back light from

the enlarger on portions of the projected image to reduce density in the final print. *Related Terms:* burning; burning in.

dogeared — the term applied to damaged edges and corners of paper, usually caused by improper handling.

DOM (document object model) — one of the core technologies of HTML. It is a specification under development by the World Wide Web consortium. Ultimately, it will allow Web page elements such as graphics, text, headlines, styles, etc., to be individually manipulated and acted on by programs and scripting languages and to define each individual object and then to assign separate qualities such as color, size, and style, etc. Closely integrated with DIM are cascading style sheets (CSS) and traditional HTML. *Related Terms:* cascading style sheets; DHTML; HTML; JavaScript; VBScript.

domain — a subset of the total domain name space. A domain represents a level of the hierarchy in the domain name space, and is represented by a domain name. For example, the domain name prof-lyons.com represents the second level domain prof-lyons which is a sub-domain of the top level domain com, which is in turn a larger subset of the total domain name space. *Related Terms:* domain name space; second level domain; top level domain; third level domain; domain name.

domain name — the address to the right of the @ sign in an e-mail address, or about ten characters into a URL. The author's domain name is prof-lyons.com. Domain names come with different extensions based on whether the domain belongs to a commercial enterprise (.com), an educational establishment (.edu), a government body (.gov), the military (.mil), a network (.net), or a nonprofit organization (.org). Some domains use a geographical notation too (such as the Las Vegas, Nevada-based xbca.lv.nv.us). *Related Term:* domain name system (DNS).

domain name disputes — conflicts that arise over who has the right to register a specific domain name. Most occur when a domain name that is the same, or similar to, a valid registered trademark is registered by a party that is not the owner of the registered trademark. A central factor in domain name disputes is the fact that domain names at the same level of the hierarchy must be unique. *Related Terms:* trademark; domain name.

domain name registration agreement — the legally binding contract between the registrar for .com, .net,

.org, and .edu domains, and the individual or organization applying for a domain name. The registration agreement must be completed and submitted electronically to register a new domain name. The agreement is also the form used to update information in a domain name record. *Related Term:* modification.

domain name space — *Related Terms:* domain name; domain name system (DNS).

domain name system (DNS) — a distributed database of information that is used to translate domain names, which are easy for humans to remember and use, into Internet protocol (IP) numbers, which are what computers need to find each other on the Internet. People working on computers around the globe maintain their specific portion of this database, and the data held in each portion of the database is made available to all computers and users on the Internet. The DNS comprises computers, data files, software, and people working together. Some important domains are *.com* (commercial), *.net* (network), *.edu* (educational), *.gov* (government) and *.mil* (military). Most countries also have a domain, for example, *.us* (United States), *.uk* (United Kingdom) and *.au* (Australia).

domestic satellite — a satellite that provides communication services primarily to one nation.

dominance — the design characteristic that describes the most visually striking portion of a design.

dongle — (1) in security, a device that prevents the unauthorized use of hardware or software and usually consists of a small cord attached to a device or key that secures the hardware. (2) in general computing, a loose term used to signify a generic adapter for peripherals.

Doom — a popular fast-moving 3D virtual reality game. To escape alive, you must out fight legions of grisly fiends and solve lethal puzzles. Doom is popular on the Internet because of its ability to allow two players to compete via a modem, enabling worldwide play.

DOS — refers specifically to operating systems developed by Digital Research Corporation and sold to Bill Gates of Microsoft, for IBM and compatible computers. Pronounced "doss." *Related Term:* disk operating system.

dot — the individual, most basic element of a halftone or a mechanical screen tint. It may be square, elliptical, or a variety of other shapes. *Related Term:* dot shape.

dot address — the common notation for Internet

protocol addresses such as 177.236.30.02. Each section between the dots represents, in decimal, one byte of a four-byte IP address. *Related Term:* dotted decimal notation.

dot area — the proportion of a given area which is occupied by halftone dots. Usually expressed as a percentage.

dot area meter — a transmission or reflection densitometer designed to display actual dot sizes and determine the percentage of light that does or does not pass through the film area being measured.

dot etching — a manual technique for chemically changing the dot size on halftone films, for purposes of color correction or adjustment of individual areas. Similar techniques can be used for adjusting continuous tone images on film or areas on metal relief or intaglio image carriers. Can be localized or general.

dot file — a file type on UNIX public-access systems that alters the way messages interact with that system. As an example, if your .login file contains parameters for things such as the text editor you use when you send a message, it will make appropriate accommodations for it.

dot for dot registration — the process of passing a sheet through the press twice and fitting halftone dots over each other on the second pass, often without the benefit of external registration markings.

dot formation — (1) the arrangement, size and shape of dots on halftone negatives, positives and printing plates to give as accurate a translation of detail and tone values as possible. (2) in dot etching, the photographic construction of individual dots to provide a proper density gradient suitable for dot size reduction.

dot gain — (1) in presswork, a printing artifact in which dots print larger than desired, causing changes in colors or tones. (2) the term used to describe the increase in size of the imaged dot on a substrate when compared to the exact same dot on the image carrier. It is usually expressed as an additive percentage. For example, an increase in dot size from 50% to 60% is called a 10% gain. Dot gain has a physical component, the gain in the dot area, and an optical component, the darkening of the white paper around the dot caused by light scatter within the substrate. Compensating for press dot gain is a key element in calibrating a digital prepress system. *Related Terms:* dot growth; dot spread.

dot leaders — a series of evenly spaced dots on the base line between typographic elements on a line, as between the chapter title and page number in a table of contents. Sometimes other typographic characters are used such as leaders, hyphens (hyphen leaders) or dashes (dash leaders).

dot loss — the natural decrease in dot size which occurs during the various prepress activities.

dot matrix printer — a printer in which each character is formed from a matrix of dots caused by impacting a series of mechanically driven pins fired at a ribbon in contact with the substrate. The wires leave a controlled series of inked dots on the page to form a character. Alternatively, thermal and electro-erosion systems are also used which rely on chemical changes caused by heating the wires and making the contact specialized receiver materials which turn dark where touched. *Related Terms:* thermal printing; daisy wheel printer; ink jet printer; laser printer.

dot mosaic — the imaging system used in third-generation phototypesetters where a cathode ray tube (CRT) converts each type character into a pattern of dots for output.

dot or "."" — the top of the hierarchy in the domain name system (DNS). *Related Term:* root.

dot pattern — the design formed by the dots in a halftone screen or a screen printing stencil. The light and dark tones produced by the dots, which vary in size, compose the image.

dot pitch — in computer monitors, the dot pitch is the distance (measured in millimeters) between the holes in the shadow mask: the smaller the number, the sharper the image. Generally, the smaller the number, the higher the resolution of a given monitor size; a .26mm dot pitch is preferable to a .31mm, etc. A dot pitch larger than .31mm will generally produce text which is difficult to read. *Related Terms:* CRT; focus; shadow mask.

dot range — the difference between the smallest printable halftone dot and the largest nonsolid-printing dot.

dot shape — the shape of the dots that make up a halftone. Dot shapes can be round, square, elliptical, linear, etc.

dot size — (1) in scanning, the diameter of the beam of light used to scan a bar code symbol — ideally, the beam width should be the same as the width of the narrow bar. (2) in a desktop printer, the size of the printed dot laid down on a substrate in a matrix or line to form characters.(3) in code and symbol scanning, the diameter of the beam of light used to scan a

bar code symbol — ideally, the beam width should be the same as the width of the narrow bar.

dot structure — the physical makeup of halftone dots that cause the eye to see values or percentages of gray or color. In a checkered pattern the percentage of dot area is 50%.

dot gain or spread — *Related Term:* dot gain.

dot, elliptical — one type of halftone screen dot that has an oval, rather than circular, shape. Developed by Kodak as one means of improving better tonal gradations. *Related Term:* elliptical dot screen.

dots per inch (dpi) — the measurement of resolution for page printers, phototypesetting machines and graphics screens. Currently most page printers work at 600 dpi and a photo imagesetter can print 2,540 dpi or more. Conventional graphics screens are generally between 10 and 300 dpi. *Related Term:* dpi.

double buffering — a programming technique using two buffers to speed up computer tasks. Double buffering is often in video cards to store the upcoming frame of a video clip in an off-screen frame buffer while it is displaying the current frame. Consequently, when the present frame is finished, the next frame is ready and waiting to write to the screen. *Related Term:* frame buffer.

double bump — printing a single image twice so it has two layers of ink. *Related Term:* double kiss.

double bump/double kiss — the printing of a single image two or more times to produce multiple layers of an ink. It is often used whenever light inks must be printed on dark substrates to provide sufficient opacity for coverage of the darker color.

double burn — exposing several different line and halftone negatives in register and sequentially on the same photosensitive surface to create a composite.

double byte — a software method that enables the system level to access more than the usual 256 single byte characters or normal character sets. Most helpful in some foreign composition such as Japanese kanji or Chinese.

double click — in graphic user interface (GUI) systems, a command initiation technique using a double depression of a mouse button while holding the mouse steady.

double density — a method of recording on floppy disks using a modified frequency modulation process that allows more data to be stored on a disk.

double dot black duotone — (1) the reproduction

of a continuous tone original made by printing two halftones, both with black ink; its purpose, when compared with a halftone, is to improve tonal range and the general quality of tone reproduction. (2) the single halftone plate produced by the separately made halftone negatives when one is exposed to accentuate the highlights and midtones, and the other for the shadows. *Related Term:* duotone.

double gate fold — a gate-folded product which is then folded, face in, at the point of juncture between the outside panels. *Related Term:* closed gate fold.

double hit — *Related Term:* double kiss.

double kiss — to print a single image multiple times in the same position to achieve a greater ink film thickness and increase opacity.

double overlay masking — *Related Term:* 2-stage masking.

double overlay method — the Australian and British term for color correction using 2-stage masking. *Related Terms:* double overlay masking; 2-stage masking.

double page spread — two facing pages of newspaper or magazine where the textual material on the left- hand side continues across to the right-hand side. Abbreviated as DPS.

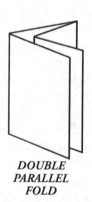

DOUBLE PARALLEL FOLD

double parallel fold — two parallel folds, the first made singly and the second a folding of the first two overlaying reveals.

double pyramid — specified placement of advertisements on a page or facing pages to form a center "well" for editorial material.

double sheet detector — usually an adjustable micro-meter device which measures sheet thickness as it is fed from the feed pile of a press. Excessive thickness, such as caused by multiple sheets causes a diverting or shutdown action to prevent the multiple sheets from being imaged.

double tone ink — a printing ink that produces the illusion of two-color printing with a single impression. The special inks contain a soluble toner that bleeds out to produce a secondary color.

double truck — a single advertisement that occupies two facing pages. *Related Terms:* double truck; double page spread.

double wire bind — the wire binding process typified by the Wire-O brand of machine, although several similar and competitive products exist. *Related Term:* Wire-O bind.

double-faced paper — *Related Terms:* duplex paper; duplex.

double-fold — a type of brochure fold in which one sheet of paper is parallel folded over twice, creating four panels.

double-sided — printing on both sides of a sheet of paper or other substrate.

doubling — a printing defect, particularly in halftone imaging, that appears as a faint second image slightly out of register with the primary image. Can be caused by improper make-ready, faulty plate mounting, etc. *Related Term:* slur.

doughnut — an incomplete or defective halftone dot, usually with a missing center area.

down — (1) the sequencing of colors in multicolor work, e.g., yellow is the first color down. (2) mechanically, broken; nonworking. A term often used to refer to equipment in that condition, e.g., "The press is down." (3) on the Internet, nonoperational time due to technical troubles on a public-access site. When this occurs and you can no longer gain access to it, it is said to be "down."

downlink — satellite receiving antenna.

download — the transfer of data from one electronic device to another. Information can be downloaded from one computer to another with a modem, for instance, or you could download information from a hard disk to a floppy disk. The most common downloading technique on the Internet is FTP. *Related Terms:* Internet; network.

downloadable fonts — type faces which can be stored on a disk and then downloaded to the printer when required for printing. These are, by definition, bitmapped fonts and, therefore, fixed in size and style.

downstyle — a headline publishing style in which only the first word and proper nouns are capitalized.

downtime — (1) in production, a usually unchargeable time span during which a production machine is malfunctioning or not operating. (2) the time spent waiting for instructions, O.K.'s, etc., during which work is held up. (3) in inventory, the time spent waiting for materials.

DP bond paper —abbreviation for dual purpose bond paper.

dpi (dots per inch) — an abbreviation for "dots per inch." Refers to the resolution at which a device, such s a monitor or printer, can display text and graphics. Monitors are usually 100 dpi or less, and laser printers are 300 dpi or higher, while imagesetters and film recorders can be 2600 – 5000 lpi. An image printed on a laser printer looks sharper than the same image on a monitor because higher DPI results in clearer images and printouts. *Related Terms:* resolution; lines per inch.

DQPSK (differential quadrature phase shift keying) — a digital modulation technique commonly used with cellular phone and cable modem systems.

drafting board — a hard, flat surface with at least one perfectly straight edge that is used to hold drawing paper or layout board during paste-up.

drag — in GUI computing, the act of depressing a mouse button to select an element and holding it down as the mouse is moved to a new location. The effect is, usually, to move the element to the new position.

DRAM (dynamic RAM) — an inexpensive, slightly inferior readable/writable memory compared to speedier RAM types. DRAM data resides in a cell made of a capacitor and a transistor. The capacitor tends to lose data unless it's recharged every couple of milliseconds and this recharging tends to slow down performance. *Related Terms:* EDO RAM; SRAM; RAM.

draw down — (1) in presswork, a sample of a specified ink and paper, used to evaluate color. (2) in other prepress operations, the process of evacuating air from a vacuum printing frame in preparation for exposure. (3) an ink testing technique used to roughly determine color shade in which the chemist places a small amount of ink on paper and draws it down with the edge of a spatula.

draw program — a type of graphics program that create illustrative matter or images, as opposed to text, using vectors (line and curve segments) rather than a mass of individual dots. *Related Term:* paint program.

draw sheet — a term applied to the tympan sheet or tympan paper of platen or cylinder letterpresses. In platen presses, the sheet is "drawn" taught between the tympan bales to provide a smooth surface for the gauge pins and subsequent substrate insertion. On cylinder machines the sheet is typically held by clamps in the cylinder gap which draw the sheet taut about the circumference.

drawn on — a method of binding a paper cover to a book by drawing the cover on and gluing to the back of the book.

dressing the press — the process of placing standard packing on the platen of a relief printing press. *Related Term:* make-ready.

drier — material added to ink to make it dry more quickly.

drill/drilling — (1) in printing, the boring of holes into paper, such as that done for 3-ring notebooks, etc. (2) on the Web, the term for negotiating the hierarchy of an Internet site, e.g., drilling down to the page with the required data. (3) in software programs, working through the hierarchy of menus, submenus, etc. to be able to select a specific action, feature or subject.

driography — printing plates that consist of metal for image areas and rubber or polymer for non-image areas for printing without water. *Related Term:* waterless plates.

drive side — the side of a press on which the main bearing train(s) are located. The opposite of the operator side. *Related Term:* gear side.

driver — a computer program interface which interprets data and allows peripheral devices to perform a function, i.e, printer drivers, mouse drivers, etc.

drop — (1) in telephony, the portion of outside telephone plant which extends from the telephone distribution cable to the subscriber's premises. (2) in publishing and advertising, to stop running an advertisement, a story line, etc.

drop cap — a large initial letter, often decorative, at the start of the text that drops into one or more indented lines of text below. *Related Terms:* drop initial; initial letter.

drop folio — a page number, or folio, that has been placed at the bottom of a page.

drop initial — a upsized display letter that is inset into the text which surrounds it. Usually placed at the beginning of chapters or sections of a book or article. *Related Term:* drop cap.

drop list — a list of options that drops down when you click on a task bar icon or button. *Related Term:* drop-down menu field.

drop on demand ink jet — asynchronous ink jet. A

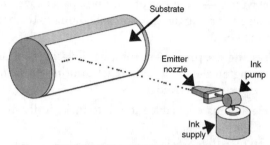

DROP ON DEMAND INK JET
Common to most desktop systems, the nozzle emits ink only as required. There is no recovery or recycling. A simpler device than continuous ink jet; also somewhat slower.

nonimpact printing method in which ink droplets are emitted only when required for imaging. *Related Terms:* bubble ink jet; valve jet; continuous ink jet.

drop out — (1) in camera and prepress, the accidental or purposeful elimination of halftone dots or fine lines by overexposure. (2) in camera work and platemaking, the intentional removal of image areas by using masks to reveal substrate in shapes of type or art. (3) in audio, a non-signal gap or silence in an audio reproduction or brief pauses in sound in web-phone and other communications connections as data is lost somewhere on the wire. (4) in code and symbol scanning, a description of a color that does not show up when scanned. If the background color on a document matches the same wavelength of the light used for scanning, the color is not present in the scanned image. Some systems use electronic drop out technology to remove form information and improve compression and character recognition. *Related Terms:* nonread color; surprint; highlight print; facsimile halftone; high key; dropout halftone.

drop shadow — a graphic effect in which display type is repeated behind itself often in a darker hue, creating a "shadow."

drop-down menu field — a form field that presents a list of selections in drop-down menu style. Both Windows and the Mac operating system utilize drop-down menus. The form field is what appears after making a selection and can be configured to permit the selection of additional fields. *Related Term:* drop list.

dropout halftone — *Related Term:* drop out.

dropout ink — on forms to be optically scanned, those reflective colors producing lines and instructions visible to the human eye, but not to optical scanners. *Related Terms:* nonread color; nonread ink.

dropout type — *Related Terms:* drop out; reverse type.

drum imagesetter — an imagesetter which uses sheet file mounted on a rotating drum. Depending upon manufacturer, the imaging material may be mounted outside or inside. *Related Term:* capstan drive.

drum scanner — a scanner where the original is wrapped around a spinning cylinder with a synchronized moving detector in the form of photo multiplier tubes or CCDs. *Related Terms:* scanner; flat bed scanner; slide scanner.

DRUPA — an acronym for Druck und Papier, an international exhibition of printing equipment and supplies held in Düsseldorf, Germany every fifth year.

dry back — the change in density of a printed ink film from wet to dry caused by the penetration of ink into the substrate.

dry end — in papermaking, the end part of the machine where the raw stock has been reduced from more than 95% water into a roll of paper containing about 5 percent water. The entire process is performed at extraordinary speed, often in excess of 3,000 feet per minute.

dry etching — a technique for creating selective or overall change in dot areas by manipulating contact printing exposures onto photographic material. *Related Terms:* contact printing; dot etching.

dry gum paper — label paper which has been coated with a water-activated glue. *Related Term:* lick and stick; moisture-activated label.

dry offset — printing done from relief plates by transferring the ink image from the plate to a rubber surface and then from the rubber surface to the paper. Printing with this technique eliminates the need to use water. Popular with check manufacturers to lay down safety designs to prevent alterations. *Related Terms:* indirect letterpress; letterset; relief offset; offset printing.

dry transfer (lettering) — cold type characters, drawings etc., that can be transferred to the artwork by rubbing the surface with a stylus or burnisher which transfers the images from the back of the base to the substrate. *Related Terms:* dry transfer type; dry transfer; rub on; burnisher.

dry trapping — a method of trapping in which wet ink is printed over dry ink. *Related Term:* trapping.

dryback — a term referring to ink, dyes or paint which lose density and become dull in appearance as they dry. Wet surfaces appear darker or more dense because they reflect more light than when dry.

dryers — the catalytic materials added to printing ink to accelerate the drying process.

drying oil — petrochemical distillates which become more solid upon exposure to air or particular types of radiation.

drying oven — a heated enclosure containing portable holding racks for screen-printed products to expedite drying.

drying time — the time it takes for something to dry or for liquid ink to harden.

drypoint engraving — the process of producing images from hand-prepared intaglio plastic plates.

DSL (digital subscriber line) — high speed copper telephone wires. With DSL, data can be delivered around 30 times faster than through a 56-kbps modem and users can receive voice and data simultaneously. Offices can leave computers plugged into the Net without interrupting phone connections. The technology is expensive because specialized equipment needs to be installed at the subscriber's location. A more consumer-ready version of DSL that requires no such splitter promises comparable access speeds at a cheaper rate, but is not expected to be generally available before 2001. *Related Term:* ADSL.

DSP (digital signal processor) — a reprogrammable special-purpose microprocessor designed to quickly handle signal-processing. They are used in several types of computer hardware such as sound cards, modems, and audio and video compression boards. *Related Term:* codec.

DSSG — Distribution Symbology Study Group.

DSVD (digital simultaneous voice/data) — DSVD is a communications protocol that allows both voice and data to be transmitted and received at the same time, over a single connection. *Related Term:* modem.

DTP — *Related Term:* desktop publishing.

DTV (desktop video) — video production using low cost video equipment and desktop computers. Once very expensive and rarely done, it is becoming much more common for home use due to the increased speed of processors and the reduced cost of computer memory and video boards.

dual in-line package switch — the small switches found on computer add-on boards and printers which enable the settling of compatibility issues with other installed peripheral boards. *Related Term:* DIP switch.

dual knife — a type of cutting tool with two knives that can be adjusted in width of separation and at-

tached to a pen-shaped handle for use on screen printing film and stripping film.

dual scan — a technique used in flat panel displays where the screen is split into two parts and each is refreshed simultaneously. Dual-scan displays produce an image that is brighter and clearer than single-scan LCD screens. They are not as bright as active-matrix displays. *Related Terms:* active matrix, LCD, passive matrix.

dual-purpose bond paper — a bond paper suitable for printing by either lithography (offset) or xerography (photocopy). Abbreviated DP paper.

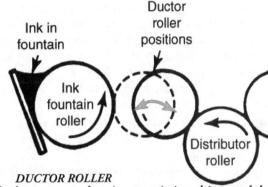

DUCTOR ROLLER
The ductor rotates about its own axis, is undriven, and the axis changes laterally to cause it to contact the fountain roller and distributor roller alternately, delivering ink or water from the respective supply.

ductor roller — a cylinder that alternately transfers ink to the ink distribution rollers or dampening solution from its respective fountain to the corresponding roller chain on a lithographic or letterpress printing press. Not a part of flexographic or gravure presses.

dull finish — the flat (not glossy) finish on coated paper; slightly smoother than matte and reflects relatively less light. *Related Terms:* suede finish; velour finish.

dummy — (1) in design and production, a folded representation of a booklet, pamphlet or other multiple-page printed product, used to plan the location of pages in a signature. (2) in newspapers, a diagram of each newspaper page, prepared by the editorial department, to guide compositors in placing and fitting stories and illustrations. *Related Term:* mock-up.

dump — (1) in retailing and sales, a display for books, often made of cardboard. (2) in computing, to print out the contents of a system's memory when a crash has occurred with the intent of enabling repair or recovery of data files, software, etc.

dumping — removing foundry type from a composing stick. To print out the contents of a system's memory when a crash has occurred. *Related Term:* pi (pronounced "pie").

duotone, black — a duotone reproduction in which the halftones are made to different specifications but both are printed in black ink to increase the density range beyond the normal screen range of a single halftone.

duotone, fake — a standard halftone printed over a solid color block or colored tint block in an attempt to simulate the effect of a true duotone.

duotone, true — a special effects reproduction of a monotone photograph into separate halftone films with different screen angles and different exposures. In the most common type of duotone, the two halftones are printed in different inks, one in a light color (a normal halftone negative) and the other in a darker one. *Related Term:* duograph.

dupe — Short for duplicate. To reproduce a page or an image exactly as it originally appeared. *Related Term:* color duplicating.

duplex — (1) in desktop publishing, a type of printer that can print on both sides of a sheet of paper automatically. (2) in printing, a press which images both sides of the sheet in a single pass. (3) in communications, a communication connection that lets you send and receive data (talk and listen) at the same time. When used in reference to sound cards and Internet phones such as WebPhone and IPhone this means the ability to send and receive audio at the same time like a standard telephone. (4) in micrographics, a method of recording on film, in a single exposure, the images of the front and back of a document. The micro-images appear side by side across the width of the microfilm. The term applied to any scanner capable of performing duplex work as described above.

duplex paper — two-ply paper, with each ply being of a different color and/or texture, projecting different colors on the front and back. *Related Terms:* double-faced paper; two-tone paper.

duplicate — to reproduce a page or an image exactly as it originally appeared.

duplicate forms — type forms that are exactly alike. They are placed next to each other and printed on one large sheet. Later the large sheet is cut into the correct sizes. *Related Terms:* step and repeat; photocomposing.

duplicate negative — a contact or projection film printed from an original negative. *Related Term:* dupe; dupe film.

duplicate plates — generally describes relief plates made from a (master) form that was not intended to be used as a printing surface. *Related Term:* stereotype.

duplicate transparency — a transparency created by photoduplicating.

duplicating film — film designed to produce either duplicate film negatives from original negatives or duplicate film positives from original positives by direct contact exposure.

duplication — a class of imaging often used in departments and small offices. Duplication is usually less expensive than true press printing processes. Common methods are small offset duplication, spirit duplication, mimeograph duplication.

duplicator — (1) generally, any press that is without bearers and smaller than 11 x 17 in. (279 x 432 mm). Duplicators are regularly used to print simple single- or two-color work, but can also be used to print multicolor jobs. (2) sometimes used to refer to press for quick printing. Not to be confused with spirit duplicator or photocopier. Generally it requires less skill to operate and produces products of lower quality than a true heavy-duty printing press.

duplicator paper — the category of reproduction paper having a hard, smooth surface and generally used when producing materials with the spirit duplicating process.

dust cover — The outer paper wrap on a hardcover book. *Related Term:* dust jacket.

dust jacket — The outer paper wrap on a hardcover book. *Related Term:* dust cover.

DV — (1) generically, digital video. (2) both the standard digital videotape format and the type of compression used on DV tape.

DVD (digital versatile disc) — a high-density, multilayered compact disk system which has evolved from and is similar in appearance to a CD-ROM. They have significantly more storage capacity than the 650mb standard CD and can store up to 17GB when recorded on both sides. Introduced in 1997 as a purely read-only medium, it has further evolved into a recordable medium. The technology is different in that rather than a single layer of information, there are two. The data tracks of each layer are much smaller and more closely spaced. The wave length of the laser is much shorter. This combination makes each layer store up to 4.7GB of data. The system can automatically read both layers simultaneously, creating an 8.5 GB capacity per side. Initially directed at home entertainment, it has been embraced by the computer industry, and drives supporting both it and CD-ROM are now fairly common on microcomputers. Although this disk format is incompatible with your old CD player, the new DVD equipment can read your old CD-ROMs and audio CDs. *Related Term:* CD-ROM; DVD-ROM.

DVD-audio — a DVD format that focuses on music and other forms of audio-only content. Readable in DVD computer players.

DVD-R (recordable) — one DVD format which supports write-once, read-many. Designed to be used for archiving, software development and low-volume data distribution.

DVD-RAM — the DVD format which supports write-many, read-many storage. Uses include short-term archiving, software development, and media recording and system backups.

DVD-ROM — a read-only storage intended strictly for PCs. Essentially a much higher capacity CD-ROM, that can also store MPEG2 video, AC3 audio, and traditional PC content. The format works well for games, reference materials and other data-intensive applications. *Related Term:* DVD.

DVD-video — a read-only storage format intended for the playback of video content, such as movies, in consumer DVD players or DVD drives in a PC.

DVI — Intel's underlying component and algorithm technology for bringing video to the PC.

Dvořak keyboard — a typewriter or computer keyboard key arrangement which places the characters in a more convenient location and is claimed to produce faster, more accurate input.

dwell — the time the ductor roller is in contact with the ink and/or dampening pan roller.

dye — a soluble coloring material, used as the colorant in color photographs, flexographic inks, sublimates, etc.

dye sublimation – a process of printing digital continuous tone quality images using thermally sublimated dyes where heat determines the amount of dye transferred from a carrier sheet. Output from dye sublimation machines offers accurate pre-film proofs for commercial, publication or newspaper printing processes. *Related Term:* sublimation.

dye transfer — a method of producing color prints, first involving the making of red, green, and blue filter separation negatives, and then the subsequent transfer of yellow, magenta, and cyan gelatin images from dyed matrices.

Dylux — (1) in prepress, specifically, a DuPont brand name for photographic paper used to make blue line proofs. (2) in printing, a type of photopolymer letterpress plate.

dynamic bandwidth allocation — the technique used by ISDN lines to handle voice and data at the same time. The user could be simultaneously downloading files from the Net and be talking on the phone over the same ISDN line. *Related Terms:* DSL; DSVD; ISDN; bandwidth; MPPP.

dynamic imbalance — a defect in the cylinder balance on a press such that the cylinder differs in density or balance from one end to the other, causing harmonic vibrations when rotating. Both damaging and dangerous without correction, especially on high speed equipment.

dynamic RAM — in computing, a competent but slow and volatile random access memory module which is erased if not continuously powered.

dynamic range — the extent of the range of densities which can be captured by a scanning device, as compared to the total possible range, from the lightest highlight to the darkest shadow.

dynamic voice override — similar to dynamic bandwidth allocation. *Related Term:* dynamic bandwidth allocation.

Ee

ear — (1) in publication work, the small boxes of type which appear at the top and on either side of a newspaper or magazine nameplate or flag. Often contain weather or inside previews. (2) in typography, an appendage to some letter forms such as the extra stroke at the top right of the lower case "g."

earth station — the ground equipment, including a dish and its associated electronics, used to transmit and/or receive satellite communications signals.

easel — a device used to hold photographic paper in place while making projection prints with an enlarger.

Easter egg — Easter eggs are "hidden features" placed by programmers in software applications, operating systems, and even some hardware. Discover the hidden command sequence, and an Easter-egged product will perform an action, such as displaying a secret message (oftentimes the development team members), provide an extra game level or secret area, play a sound or animation. Programmers often "bury" Easter eggs in their programs to add a certain extra depth to their program or Web site and to challenge users to find them.

EBCDIC (extended binary coded decimal interchange code) — (say "E B see-dick") an alphanumeric code set used in telecommunications to represent data.

ECI – *Related Term:* electronic color imaging.

ECP (extended capabilities port) — an industry specification to provide 2-way higher speed parallel ports. ECP uses direct memory access (DMA) and buffering. Windows 95 supported ECP, but a lot of hardware that uses the parallel port such as scanners, CD-ROMs, etc., didn't always respond to the way the spec used DMA. Both ECP and EPP support the IEEE 1284 specification and newer parallel ports often support both. *Related Terms:* EPP; parallel port.

edge enhancement — a pixel based computing operation which makes edges appear less aliased by manipulation of contrast differences and dithering.

edge roughness — the edge irregularities printed bar code elements most noticeably resulting in non-uniform edges, causing reading errors.

edge spectrogram — a visual representation of a film's reaction to light across the visible spectrum.

EDI (electronic data interchange) —a method by which data is electronically transmitted from one point to another.

edit — the modification and correcting of a manuscript to conform to publication standards.

editing — checking copy for fact, spelling, grammar, punctuation, and consistency of style before releasing it to the typesetter.

edition — all printings of a book from the same original materials. Once changes have been made to the original materials, the next printing becomes a new edition.

edition bind — *Related Term:* case bind.

editor — (221) in computing, an interactive program that can create and modify files of a particular type. For example, the FrontPage Editor is an HTML editor and PageMaker has its own editing tools built in. (2) a person who selects words and visual elements such as photographs so they accomplish their communication goals within the space and budget allotted them.

editor in chief — the top editor at a magazine or book publisher responsible for all editorial decisions.

editorial changes — modifications requested by the customer, after production commences, to change a text or graphic materials to differ from that originally specified.

EDO RAM (extended data-out RAM) — a form of dynamic RAM that speeds access to memory locations by assuming that the next time it is accessed, the data will be at a contiguous address. While not always correct, the assumption speeds up memory access times by up to 10 percent over standard DRAM. *Related Terms:* DRAM; IDE; SCSI; RAM.

EDP (Electronic Data Processing) — any system which employs computers, magnetic storage and retrieval, laser disks, scanners and related hardware as the core technology for capture, sorting, refining, analysis, storage and/or dissemination of data.

.edu — a domain name suffix used in Internet addresses that denotes an educational institution.

EEPROM (electrically erasable programmable ROM) — a type of ROM which can be electronically erased and have a brand new BIOS instruction set introduced. A manufacturer can easily distribute BIOS updates via floppy diskette or even the Inter-

EF

ELECTRON GUN

net, eliminating the need for hardware replacement at the computer level. This capability is sometimes called flash BIOS, which is the term often used in add-in devices such as modems, graphics/video cards, etc. *Related Terms:* BIOS; ROM; flash BIOS.

EF — *Related Term:* English finish.

effective aperture — in photography, the diameter of the lens diaphragm as measured through the front lens element; the unobstructed useful area of the lens. *Related Term:* f/number; iris.

EFT (electronic funds transfer) — a system for payments made via an automated clearing house (ACH). The Federal Reserve's and MasterCard's remittance processing system (RPS) are two examples. EFTs may also be made directly to a merchant (point-to-point) by transferring data directly to the merchant (by fax, tape or modem) with a single, aggregate ACH or wire transfer payment. Long touted as the future, it has not yet taken off as fast as was predicted and remains a relatively small component of the flow of cash. It is expected the advent of smart cards, home computer banking and similar systems will accelerate wider implementation.

EGA (enhanced graphics adapter) — an older graphics standard which replaced CGA (color graphics adapter) on the PC. Can be added on or built into a system to give sharper characters and improved color with the correct display device. Standard EGA resolution is 640 by 350 dots in any 16 out of 64 colors, and has been supplanted by VGA and S-VGA, both with more dots of resolution and more color depth.

eggshell finish — a paper finish simulating the surface and color of an egg.

Egyptian — a term for a style of type faces having heavy square serifs and almost uniform thickness of strokes. *Related Term:* square serif.

eight-sheet (8-sheet) — a poster measuring 60 in. x 80 in. and, traditionally, made up of eight individual sheets.

EISA bus (enhanced industry-standard architecture bus) — a 32-bit bus for PCs built around 386, 486, or Pentium chips. EISA was developed as an alternative to IBM's Microchannel bus, and is more compatible with the original ISA bus. EISA computers can generally use ISA cards as well. *Related Terms:* bus; ISA; PCI; USB.

e-journal — an electronic publication, similar to an e-zine or zine. It is typically found in academic circles and is regularly published and made available solely in electronic form.

elasticity — the property of any material to resume original shape after being physically distorted, such as a rubber band.

electric eye mark — (1) a mark added to the gravure cylinder as an image to be read by special electronic eyes that monitor image registration during production. (2) carton and packaging manufacture, a solid bar which can be optically detected to indicate to the packaging machine the proper cutoff and/or folding locations.

Electro Ink – the name which designates the proprietary ink developed for use in the Indigo E-print digital printing system.

electrochemical — the process of using electrical current to produce chemical changes needed when making an electrotype printing plate. *Related Terms:* electrolysis; electroplating.

electrolysis — *Related Term:* electrochemical.

electromechanical — the combining of electronic and mechanical principles and procedures to produce results, such as an imaged gravure cylinder, without using photographic film. *Related Term:* electromechanical engraving.

electromechanical engraving — (1) any electromechanical method of producing a relief plate by cutting an image into a base material with a device that is electrically controlled. (2) in gravure, the electromechanical process — a name given to the specific technique used in gravure printing to place an image on a gravure cylinder by cutting into the metal surface of the cylinder with a diamond stylus.

electron gun — the electronically active portion of a cathode ray tube or CRT. A color CRT is made up of three separate electron guns that each produce a stream of electrons, specifically directed at particles of red, green, or blue phosphor on the inside face of the tube. The electron stream hits the tube face and causes the respective phosphors to glow or emit visible red, blue or green light. With their particle direction controlled by electromagnet fields or yolks, the electron streams sweep across the screen, their strength varying to produce the desired strength in the combined RGB phosphors on the tube face. Nearly any color can be created by the additive color mixing of these red, green, and blue phosphors. *Related Terms:* color balance; color purity; CRT.

Graphic Communications Dictionary

85

electronic air brush — a tool often found in raster art programs which allows the manipulation of pixels to change color values, density and tones by manipulation of a mouse or stylus.

electronic check — an electronic commerce replacement for paper checks. A normal checkbook, pens and signatures are replaced by electronic card functions, digital signatures; stamps and envelopes by electronic mail or the Internet. *Related Terms:* credit card; debit card.

electronic color correction — the process of correcting color output by adjustments on a color scanner or similar electronic imaging system during output.

electronic distribution — sending of data, images, etc., via electronic means such as the Internet, intranet, video, etc.

electronic image assembly — *Related Term:* electronic imaging systems.

electronic imaging systems — computer-controlled equipment used to merge, manipulate, retouch, airbrush, and clone images, create tints and shapes, and adjust and correct individual color areas within an image that has been scanned, stored on magnetic disk, retrieved and displayed on the monitor, and positioned according to a predetermined layout.

electronic mail — communications between individuals or groups via electronic means, e.g., computer mail and facsimile. *Related Term:* e-mail.

electronic manuscript/mechanical — any document, with text and graphics that are merged, that is sent through the production process as a "soft copy" via modem or network rather than as a "hard copy" such as a paper galley or using a paste-up board format.

electronic masking — the simulation of photographic mask color correction by the circuits of a color scanner.

electronic memory — disk, chip, tape or other magnetic or solid state electronic device which holds, stores or shares information in digital form.

electronic page assembly — the assembly and manipulation on a computer screen of type, graphics and other visual elements stored in memory. *Related Terms:* electronic imaging systems; desk top publishing; electronic publishing; area composition.

electronic planimeter — a measuring device used for the visual and mathematical evaluation of dot area. It is comprised of a microscope, a television camera and receiver, and a small computer.

electronic publishing — (1) any system using a computer and related word processing and design and page-makeup software to create paginated text and graphics, which are output to a laser printer with a PostScript interpreter and/or imagesetter at varying degrees of resolution from a minimum of 300 dots per inch to maximum quality levels exceeding 3000 dots per inch. (2) generally, the systematic process of creating, assembling, editing, proofing, storing, retrieving and preparing publications for printing and binding through the use of networking computers, printers, scanners and image setters. (3) a generic term for the distribution of information which is stored, transmitted and reproduced electronically. Teletext and Videotext are two examples of this technology in its purest form, i.e. no paper. *Related Terms:* desk top publishing; electronic page assembly.

electronic retouching — the use of a computer to enhance or correct a scanned photograph or other art.

electronic scanner — a type of photoelectric equipment for scanning black and white or full color hard copy by reading the relative densities of the copy to make negatives, positives, or color separations.

electronically integrated publishing (EIP) — a concept that assumes the ability to have freely transportable bits and bytes from the creative process through to the final printed product.

electro-optical — systems which combine lenses or photographic techniques with computer technology and/or electronic properties of materials to produce their results. Typical examples include digital photocopiers and graphics scanners.

electrophotography — in modern terminology, processes including xerography and laser printing, that produce images by passing toner particles over an intermediate photo conductor drum, which has received an electrical charge that enables it to transfer and fuse the toner particles to plain (untreated) paper, forming the image. *Related Term:* dielectric printing process.

electroplating — a process of transferring very small bits (called ions) of one type of metal to another type of metal. *Related Term:* Electrotyping.

electrostatic — a method of printing utilizing an electrostatically charged image area on a drum, belt or paper, as a vehicle to attract oppositely charged toner and ultimately transfer it to a substrate to create an image.

electrostatic assist — a device using an electronic charge that is licensed by the Gravure Association of America to help pull ink from the cylinder wells in gravure printing and transfer it to the substrate.

electrostatic copier — *Related Term:* electrostatic copying.

electrostatic copying or printing — (1) generally, an image replication process that uses an intermediate photosensitive plate or drum or a coated takeoff sheet that is electrically charged to accept an image-producing agent in certain areas of the sheet. (2) specifically, a printing method in which electrically charged, powdered colorant particles are transferred from the image carrier to a substrate moving in their path. The particles are fused to the substrate to form the permanent image. *Related Terms:* electrostatic copying; xerography.

electrostatic plate — (1) an organic photo conductor that serves as an offset image carrier. (2) a lithographic printing plate on which the oleophilic image is placed with an electrostatic copier. *Related Term:* electrostatic transfer plate.

electrostatic transfer plate — *Related Term:* electrostatic plate.

electro-thermosensitive — the printing processes that rely on electricity to heat thermal print heads which, in turn, are used to print on heat-sensitive paper or heat-sensitive ribbons that transfer the print to plain paper. *Related Term:* thermal printing.

electrotype — high quality duplicate relief plates made mainly of copper and produced from a mold from an original form; the final printing plate is made from silver and copper or nickel is produced in the mold by an electrochemical exchange. *Related Term:* electroplating.

element — (1) generally, a single part of page, layout, publication or image. Elements of a publication may include the index, contents, logotype, individual pages, illustrations, etc. (2) in bar coding, a single bar or space.

elite — the smallest size of two standard typewriter faces, having 12 characters per inch as compared with 10 per inch on the pica typewriter.

ellipsis (...) — a punctuation symbol consisting of three dots separated by thin spaces, used to indicate an omission. Often used when editing and omitting copy from a quote. *Related Term:* ellipse.

elliptical dot screen — one type of halftone screen dot with an elliptical (oval) dot, rather than circular shape. It sometimes produces better tonal gradations and smoother midtone reproduction. *Related Term:* dot, elliptical; halftone.

Elmdorf test — a test performed on papers to determine tear resistance.

Elrod — a hot metal casting machine used to make lead and slug line spacing (material) for hot metal or hand composition.

em — (1) in typography, a measurement of linear space, or output, used by typographers and is a square measurement of the body dimension of a specific foundry type size. The unit of measurement is exactly as wide and as high as the point size being set (usually approximating the measurement of the uppercase M). For instance, an 18-point type "em" would be 18 points square; in 12-point type it is 12 points square, etc. In photographic typesetting, it is called the em space. (2) commonly used shortened term for em quad. *Related Terms:* em quad; em space.

em dash — a dash the width of an em space, and that is approximately centered on the x-height of characters. Used to indicate a pause in the written material. *Related Term:* ellipse; long dash.

em quad — *Related Term:* em.

em space — in photocomposition, a fixed space equal to the width of the type size being used. It is not "stretched" for justification purposes. *Related Term:* em; em quad.

e-mail (electronic mail) — a system whereby a computer user can exchange messages with other computer users across a local area network, via the Internet, or through an on-line service like CompuServe or America Online (or groups of users) via a communications network. Electronic mail is one of the most popular uses of the Internet. *Related Terms:* snail mail; spam.

e-mail address — the domain-based or UUCP address that is used to send electronic mail to a specified recipient destination. For example, djlyons@ip.every.com is the e-mail address for the user djlyons on the machine IP that is part of the every.com domain.

emboss — a technique using metal dies to press or distress an image into paper so it lies above the surface. When done on blank, unprinted paper, it is referred to as "blind." *Related Terms:* embossing; debossing; blind embossing.

embossed finish — the surface pattern pressed into

dry text paper and having a name such as leather, linen, pebble, or canvas. Mills put embossed finishes on paper after it comes off the paper making machine. It is created by passing a web of paper between the nip of an engraved metal roll and a mating soft backing roll. The rolls may be engraved to produce various patterns.

embossing — to impress paper into bas-relief by pressing it between special dies on a heavy duty press. Special female dies are used with a male counter created by making ready with a special compound. The process of producing a printed raised (relief) image in paper (substrate). *Related Term:* blind embossing.

embossing die — a heated or cold brass or steel form

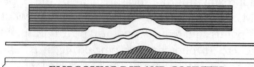

EMBOSSING DIE AND COUNTER
Substrate is placed between the female die and its male counter. Under pressure, the material is slightly deformed. If the image is above the normal plane it is called embossing. If the image is depressed below the plane, it is called debossing.

used to impress a design in relief into paper or other substrates. Usually used in a male/female configuration, with unsharp edges between the fitted elements.

emoticon — ASCII glyphs used to indicate an emotional state, typically used in e-mail or Usenet messages. Emoticons are clusters of punctuation marks used to set the tone for the sentence that precedes them because the lack of verbal and visual cues can otherwise cause what were intended to be humorous, sarcastic, ironic or otherwise nonserious comments to be badly misinterpreted. They can be understood by tilting your head sideways to the left and looking for a facial expression or symbols. A sampling of emoticons includes: — :) a grin; :-) another grin

emphasis text — an HTML character style intended for mild emphasis. Some browsers display emphasized text as italic.

EMS (Expanded Memory Spec) — a memory system used on Intel and compatable machines to increase processing capabilities.

emulsification — the condition which exists when solid material becomes suspended in a liquid. In press work, usually refers to the situation when fountain solution is too acetic and mixes with and dilutes the ink, often tracking back to the ink foun-

tain. The ink becomes paste-like and produces a highly mottled image on the substrate.

emulsion — (1) in film or platemaking, a thin coating that may or may not be light sensitive and that is applied to plastic, polyester, paper or other substrate. Photographic film is a combination of light-sensitive emulsion and clear base material. The emulsion normally always is positioned for exposure with it facing the lens system. (2) in screen printing, the liquid, light-sensitive material applied to a screen fabric to make direct photo screen stencils. (3) in conventional stripping and art preparation, the emulsions on mechanical masking film and hand-cut stencil materials that are not light sensitive. (4) in general photography, an additional screen or filtration used over the lens of cameras or enlargers to produce softer definition of the subject. Often employed in portraiture to flatter by eliminating facial lines or wrinkles.

emulsion coating blade — a tool made of non- oxidizing metal used to apply light-sensitive liquid emulsion to the fabric of a screen printing frame.

emulsion down — emulsion away from viewer. The orientation of film negatives when in contact with negative acting material (film, proof material, printing plate); right reading because in this orientation photos appear "right" and type can be read on a light table. The printer usually decides whether emulsion should be up or down.

emulsion speed — the sensitometric measurement of an emulsion's ability to produce a certain density under controlled exposure conditions. Usually rated in ASA or ISO numbers where the lowest numbers indicate a slower film. The range can be from ASA 3 to ASA 6400 and higher, with most commercial films rated between ASA 125 and 400.

emulsion to emulsion — the orientation of processed film to unexposed photosensitive material (film, paper, plate) in preparation of image transfer by light exposure; also contact.

emulsion up — emulsion towards the viewer. The orientation of film negatives during the stripping phase (assembly on masking material).

en — a unit of measurement exactly one-half as wide as the point size being set, or half of an em. In 18-point type the en is 9 points wide and 18 points high; in 12-point type it is 6 points wide and 12 points high. In photographic typesetting, it is called the "en" space. (2) one half of an em. *Related Terms:* em; em quad.

en dash — (1) a dash sitting approximately centered on the x-height of characters, one "en" long, used in compounded words and numbers. (2) one half (1/2) of an em dash. *Related Term:* em dash.

en quad — also called a nut. *Related Term:* en.

en space — a fixed space equal to one half of the width of the type size being used. One-half of an "em" space. It will not be "stretched" for justification purposes.

enamel — a coating material used on paper.

enamel paper — *Related Term:* gloss paper.

encapsulated PostScript – a standard file format for storing high-resolution graphics which enables device- independent transfers between application programs. EPS files can be resized without sacrificing image quality. Programs capable of displaying, creating and editing EPS graphics include Adobe Illustrator, QuarkXpress®, etc.

encoding — (1) in computer file management, the process of creating a compress version of a file, usually for the purpose of reduced storage and/or faster data transmission. (2) in general computing, particularly e-mail and secure file transmissions, a term sometimes used incorrectly to refer to encryption. *Related Term:* character encoding.

encrypted password — *Related Terms:* encryption; modification.

encryption — a process of scrambling, or encoding, information in an effort to guarantee that only the intended recipient can read only the communication. The receiver must have the proper decryption key. Traditionally, the sender and the receiver use the same key to encrypt and decrypt data. Public-key encryption schemes use two keys: a public key, which anyone may use, and a corresponding private key, which is possessed only by the person who created it. *Related Terms:* PGP; DES.

end book price — the lowest price for paper shown in a price book.

end leaf — the paper used to bind a book's cover to its interior pages.

end of line decisions (EOL) — generally concerned with hyphenation and justification (H&J). Decisions can be either by the keyboard operator or by the computer.

end papers — the four page leaves at the front and end of a book which are pasted to the insides of the front and back covers (boards) of a case bound book. *Related Term:* end sheets.

end user — the ultimate consumer of a service.

English finish — a grade of book paper with a smooth finish. It is not considered to be as rough as uncoated machine finish paper. *Related Term:* EF.

engrave — to produce a raised printed surface by printing with a cut-away plate.

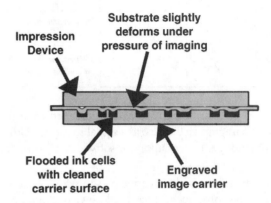

engraver — (1) in intaglio or relief printing, a person who makes hand or photoengraved plates for printing. (2) often a somewhat incorrect term used to refer to trade camera service.

engraver's proof — a proof copy pulled from the original engraving as a check, prior to final finishing or duplication.

engraving — the removal of material, such as in making the ink cells on a gravure cylinder or etching away of the nonprinting areas of a relief printing plate. In the modern world, this usually means doing so by some sort of chemical erosion process using acids. *Related Terms:* intaglio; relief.

engraving, electronic — engraving accomplished by use of a scanning mechanism guiding a stylus.

engrossing — the descriptive term for the formalized calligraphy used on diplomas, certificates and other socially important documents.

enlargement — more than 100% of the original copy; increasing the comparative size of the reproduction. If X and Y size ratios are preserved it is called proportional; if they are independently changed, anamorphic. *(See illustration on following page).*

enlarger — a projection device used in conjunction with small, medium or large format camera negatives, usually to produce photographic prints of a larger size than the original negative image.

enter key — the initiating key for a computer ac-

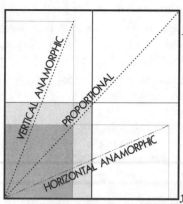

ENLARGEMENT
(See previous page)
Note that a diagonal will produce a proportional enlargement. Any point on it, when projected to the horizontal and vertical edges, will retain the original ratio of height to width. Any deviation from the diagonal line will yield an anamorphic size change.

tion. Most commonly used as a carriage return at the end of paragraphs, but also as a signal to execute other commands dependent upon the software environment in which it is used.

entrepreneur — the creative person who initiates a business or product and has the knowledge and ability to be successful in the endeavor.

envelope — an enclosure used for mailing. It can be made of paper of artificial substrates, and usually includes some type of closure device such as moisture-activated glue, string and eyelet, etc.

enzyme — a product that causes a biochemical reaction to indirect photo screen film, thus permitting it to be removed from screen printing fabric.

EP — the abbreviation for envelope.

EPA (Environmental Protection Agency) — the government department responsible for monitoring and enforcing environmental laws. *Related Term:* OSHA.

epilogue — a closing section of a book or article which is intended to update the reader and provide the author an opportunity to make conclusions, final arguments, comments, etc., as opposed to a prologue, which appears at the beginning of a story and offers introductory comments.

EPP (enhanced parallel port) — a specification that builds on the design of the standard parallel port, providing faster data exchange and two-way communications. It is popular because of the increased demand for peripherals, especially in laptop computers. Both EPP and ECP are supported by IEEE specifications and most new parallel ports can run in either fashion. *Related Terms:* ECP, parallel port

EPROM (erasable programmable ROM) — chips that do not require power to retain their data, unlike

dynamic and static RAM. EPROM chips are used to store BIOS information and basic software for modems, video cards, and other peripherals. EPROMs are erased by ultraviolet light and can then be reprogrammed using a device called a PROM burner. *Related Terms:* BIOS; DRAM; EEPROM; RAM; ROM; SRAM.

EPS (encapsulated PostScript) — an extension of the PostScript graphics file format. EPS is often used for high-resolution images that will be added to another document. A file format used to transfer PostScript data within or between compatible applications. Nearly all graphic editing programs support the importing of EPS files. *Related Term:* encapsulated PostScript.

EPSF (encapsulated PostScript file) — *Related Term:* EPS file.

Epson emulation — the industry standard control codes for dot matrix printers that were developed by Epson and adopted by virtually all software packages. Most other dot matrix printers followed but were generally unable to improve on these basic codes.

equal spacing — the typographic system of allocating the same amount of space to each character regardless of its width "i" gets the same space as "p" and "w" or a punctuation mark. *Related Terms:* fixed spacing; non-proportional spacing; equal spacing.

equalization — a method of enhancing perceived color by enhancing shades within the reproduction.

equivalent neutral density — a balanced value for process colors to achieve a neutral gray when overprinted.

equivalent weights — (1) the differences in basis weights that describe different papers which seem different but which are identical, because each was computed using different basic sizes. For example, 16# bond and 40# book are equivalent weights. (2) a paper that is not the brand specified, but looks, prints and may cost the same. *Related Terms:* comparable stock; equivalent paper.

erasable bond — a paper with surface treated for easy erasure.

erasable optical disk (rewritable optical disk) — a type of optical CD that can be rewritten many times at the user's computer or workstation.

errata — a loose insert sheet in which the errors found in a printed book are listed.

error codes — standard codes that are used to indicate common errors in the domain name registration agreement, the contact form, or the host form that prevents the processing of the domain name registration agreement or forms. *Related Terms:* domain name registration agreement; registration forms.

escape key — a keyboard key which performs a variety of tasks and is software dependent. Its precise function depends upon the program in which it is used. Most generally used to revert to the previous menu or cancel a specific action which was previously initiated and is in the process of being performed.

escapement — the side by side movement distance for placement of type characters sequentially to form a line.

esparto — a coarse grass. A recent development in papermaking, a coarse grass whose fibers yield strong, economical paper. *Related Term:* Spanish grass.

estimate — a price that states what a total job or its associated parts will probably cost based on initial specifications from the customer, not reflecting any additional production changes made after work has commenced. Different from a quote, which is a legally binding, signed agreement between a printer and a publisher in which the cost is guaranteed not to fluctuate for a specified period of time.

estimating — the process of creating an estimate. *Related Term:* estimate.

estimator — an individual who performs calculations and produces estimates or cost in advance of production. *Related Terms:* estimate; quotation; bid.

etch — (1) in platemaking, using chemicals or tools to carve away metal leaving a relief or intaglio image in a metal image carrier. (2) in lithographic presswork, a casual term for the fountain solution.

etchant — a substance, like an acid, used in etching or in fountain solution.

etching — (1) in relief platemaking, the eroding of a metal or polymer material which has been chemically treated with a photo-resist to make a printable relief surface. The term is also sometimes applied to platemaking when using deep etch offset lithographic plates. (2) in general photography, the scraping away of small unwanted spots on a print. The removal or lightening of a portion of a negative to reduce density using chemical etchants. *Related Terms:* etch; engraving.

e-text — an electronic text document. Often included with program installation files, e-text may take the form of a short pamphlet, a README file or a note.

Ethernet — the most commonly used local area network (LAN) technology that uses a collision avoidance protocol for network information. It is a standard for connecting computers into local area networks. Of several varieties, the most common form is called 10BaseT, which denotes a peak transmission speed of 10 MBPS using copper twisted-pair cable. *Related Terms:* fast Ethernet (100BaseT); hub.

Euclidean dot shapes — round, elliptical, square, or linear halftone dots that invert with their cell after 50% intensity. This strategy helps reduce dot gain problems sometimes experienced with elliptical, square, and linear dots facing pages.

evaporation — the conversion that occurs when a liquid combines with oxygen in the air and passes from the solution as a vapor.

event handler — an action that occurs on a Web page and may be performed by the user or forced by a programming script. In JavaScript, it is a mechanism that causes a script to react to an event, such as clicking a link or changing the value of a text area. The code will be executed when the user clicks the link.

exception dictionary — in computer-assisted typography, that portion of the computer's memory in which exceptional words (do not hyphenate in accordance with the logical rules of H&J) are stored. Some programs, PageMaker for example, only use an exception dictionary and some have no routine of their own.

excerpt — a portion taken from a larger work, such as when portions of a book appear as a magazine article.

excess density — the density difference after the basic density range (BDR) of a halftone screen is subtracted from the copy density range (CDR).

exclusive — (1) in publishing, the granting (or selling) of sole distribution or publishing rights to only one distributor or publisher in a particular geographic area. (2) in news publishing, a news or feature article published by a publication before any of its competitors.

execute — to initiate an action and carry it to completion. For example, clicking on "save" in the file menu of a program causes the file to be written to disk and preserved for future use.

exhibition — the display of a body of photographic or other graphic materials in a gallery, museum or exhibition hall for public or private viewing.

exhibition mount — the mounting board on which a print is affixed before matting or framing. All photographic prints should be mounted on 100% rag board to assure permanency.

expanded — the attribute pertaining to a typeface that has characters whose width is greater than their height would generally dictate. Expanded typefaces look "stretched" horizontally. *Related Terms:* expanded type; expanded; condensed.

expanded type — *Related Term:* expanded.

expansion cards — circuit boards which can be inserted into expansion slots to enhance a computer's performance or implement specialized functions. The adoption of USB, Firewire and similar technologies has reduced the need for some cards as their electronics can now be a part of the operating device. *Related Term:* add-in boards.

expansion development — increasing development of a low value, low luminance scene to render it printable on the photographic paper as visualized at time of exposure. *Related Term:* contracted development.

expansion envelope — an envelope with gussets to allow for expansion as its contents grow. Sometimes called a gusset envelope, an accordion envelope, a concertina envelope or a portfolio. *Related Term:* gusseted envelope.

expert reading — a review of a manuscript by an expert in the field. *Related Term:* technical edit.

exposure — the process during which light produces a latent, invisible image on light-sensitive photographic paper or film.

exposure factor — the adjustment to exposure of films due to the use of filters which reduce light intensity or necessitate changing of shutter speed or f/stop.

exposure meter — a not perfect name for a luminance meter. Sometimes called a light meter.

exposure time — (1) in process photography and platemaking, the period of time during which a light-sensitive surface (photographic film, paper, or printing plate) is subjected to the action of actinic light.

(2) generally, the time required for light to record an image while striking a light-sensitive emulsion.

exposure, basic line — the exposure time required to produce an accurate reproduction of the original line or two-tone black and white copy.

exposure, halftone main — the camera exposure made through the halftone screen to produce highlight detail in a halftone negative. It is made through the lens with the copy in the copy board of the process camera.

exPRESS — a proprietary printer control language developed by OASYS.

extended attribute — a Microsoft term for an HTML attribute not directly supported in FrontPage. In it, extended attributes are assigned using the Extended button in the object's properties dialog box.

extended type — *Related Term:* expanded; wide; extra wide.

extenders — any material added to an ink to reduce its color strength and/or viscosity.

extension — (1) in process photography, the distance between the lens and the photosensitive material or between the lens and the copyholder in a camera. (2) in computer makeup, the derivative name given to add-on all third party enhancement software for QuarkXpress®. (3) in computing, the 3-letter combination following the period in a file name. The naming system is often called 8.3, meaning that the name can have an 8-letter name followed by a period and then a 3-letter extension, usually to indicate the type of file which was created. Generally seen extensions include ".exe," ".com," ".doc," etc. *Related Terms:* MS-DOS; Quark extension.

external hyperlink — a hyperlink which addresses any file that is outside the current Web page, often on a remote server.

extra leading — increments of white space added between lines or blocks of type for visual purposes and to facilitate vertical justification. *Related Terms:* interline spacing; leading.

extranet — the connection between two or more intranets which allows users inside one company to communicate and exchange information with other companies' intranets and gains the ability for them to share resources and communicate over the Internet in their own private space. They are, basically, a corporate nonpublic information networking system whose access is limited to a select group of people. Users have to enter a password or use digital encryption to access it. Once access has been granted, it operates much like any Internet site. An excellent example is Federal Express's extranet that lets customers track packages by simply entering a tracking number. Many banks and credit unions use extranets to let users transfer funds or look up account balances on-line. Extranets help companies save money by allowing customers to find relevant information without adding personnel. *Related Terms:* Internet; intranet.

extrusion — (1) in hot metal composing, a method of making leads and slugs for hot metal composition; the metal is forced from the machine in strips. A typical example of the machine is the Elrod strip caster. (2) the production of a continuous sheet of film by forcing hot thermoplastic material through a die or orifice. *Related Term:* Elrod.

eye – (1) in typography, the enclosed part of the lowercase e. (2) in video, a slang term often applied to the lens of a video recording device.

eye markers — small squares of process or other colors printed outside of the image area of packaging materials, particularly in flexography. They are used for automated control of register and density and as a guide to slitting and trimming and packaging machines. Usually they are seen by the consumer on the inside flaps of cartons and boxes.

eye span — the maximum distance that a person can comfortably follow from left to right while reading. An important consideration in establishing line lengths, column widths, etc. If the length is too long it causes visual stress, reduces comprehension and reduces inclination to complete a document.

eyelet — a metal or plastic reinforcement for holes in printed matter, banners, etc. A small grommet.

e-zine (electronic magazine) — sometimes just "zine." An online magazine, it also can be a regular magazine published in electronic form. The technology is new and major publishers are struggling with the best formats and the most cost-effective methods to gain readership and advertisers to support the ventures. *Related Term:* e-journal.

Ff

f keys — special keys on a computer keygoard. *Related Term:* function keys.

F&G — an abbreviation for folded and gathered pages which form the unbound pages of a book.

f/number — the numerical value of the effective aperture of a lens. The numbers are derivatives of the lens focal length, with smaller numbers representing larger aperture openings. Each numbered value will halve or double its adjacent number, i.e. f/8 is twice as large as f/11 and half as large as f/5.6, etc.

f/stops — (1) fixed sizes at which the aperture of the lens can be set. The values of the f/stops are determined by the ratio of the aperture to the focal length of the lens. (2) the measure of aperture setting on a camera. Large numbered f/stops (f/16 and f/22) signify small aperture openings; small numbered f/stops (f/3.5 and f/1.9) indicate large aperture openings. Changing from one f/stop to the next adjacent one doubles or halves the amount of light that will reach the film. *Related Term:* f/number.

fabric mesh — the fibers making up the screen fabric in screen printing; the fabric holds the stencil and permits ink to pass through.

face — (1) in typography, an abbreviation for typeface referring to a family in a given style. (2) in relief composition, the part of metal type that prints.

face out — the displaying of books, videos, etc., on a shelf so that their front covers are showing, as opposed to spine out.

facilitator — the individual responsible for the local component at a teleconference site. May or may not be an expert in the subject matter.

facing material — the top substrate in pressure-sensitive labels. The part that is printed, as opposed to the carrier or release sheet.

facing identification marks — the vertical bars near the top of business reply mail that identifies the category of mail to automated sorting machines. Not generated locally, they are supplied by the postal service as film negatives.

facing page — a companion, opposite facing page that forms a spread.

facsimile — the transmission of graphic matter (letters, documents, charts, spreadsheets, pictures, etc.) by wire or radio. Some fax machines are capable of transmitting full-color images, while others have been integrated with personal computers in order to transmit data. *Related Term:* fax.

FACT — Federation of Automated Coding Technology. A bureau of AIM consisting of organizations that use and promote automatic identification among their members.

factor number — in typography, an average of the width of characters, by typeface and size. It is used in the calculations for copyfitting type to a layout.

factory floor data collection — the utilization of software applications that use data collection terminals at job stations to help manage inventory and keep track of work flow within a production facility. Data collected is routed directly to a central data collection system where it is processed to create management reports on inventory, wage and hour costs, efficiency levels, etc.

fade in — in video, a term applied to the transition from a static scene, often black, to a moving one.

fade-O-meter — a measuring device designed to determine the amount of fading which occurs to a pigment or ink upon exposure to known amounts of light.

fair use — a copyright law concept which grants limited use, without the permission of the copyright holder, of short quotations, excerpted from copyrighted works for the purpose of review, education, etc.

fake color — a one-color reproduction printed on a colored sheet. *Related Term:* fake duotone.

fake duotone — a normal halftone in one ink color printed over screen tint or color block of a second ink color.

family — one major grouping of specific typefaces having similar characteristics but specific differences, e.g., Times Roman, Times Roman Italic, Times Roman Bold, etc. All the type variants of a particular type face (roman, italic, bold, condensed, expanded, etc.) *Related Terms:* type series; series; type font; font; family of type; font family.

fan guide — a sample booklet of colors used for choosing and specifying color in which the pages fan out so that various colors can be compared.

fanfold paper — *Related Term:* continuous paper; continuous form paper.

fanout — (1) the distortion of paper on press due to absorption of moisture at the edges of the paper. (2) the distortion of a straight stack of loose paper to facilitate manual sheet counting.

FAQ (frequently asked questions) — A common type of document on the Internet that is supposed to answer all the questions a newcomer to an on-line site might have. On the World Wide Web, questions are often hyperlinks to the answers. Pronounced "facks" or sounded out by each letter.

farm out — *Related Term:* buy out.

Farmer's reducer — named for the inventor, Howard Farmer, a chemical solution (potassium ferricyanide and sodium thiosulfate) that is used to reduce the density and increase the contrast of developed film negatives. A strong solution of Farmer's reducer is used to slowly dissolve the developed silver image on a photographic film or paper. It clears fog in the transparent areas of negatives or positives and reduces halftone dot size during tone or color correction.

fast color inks — the term used to describe fade-resistant inks, particularly those used in fabrics, etc. They are resistant to degradation from washing, sunlight, and other external influences.

fast Ethernet — an upgraded standard for connecting computers, it works just like regular Ethernet except that it has a higher peak transfer rate (100 mbps). Referred to as 100BaseT, it is more expensive and less common than regular 10BaseT Ethernet. *Related Terms:* Ethernet; (10BaseT).

fast film — any film that requires relatively little light (short exposure or small aperture opening) to record an image. A film with a higher ISO or ASA rating number.

fast lens — a lens or lens system capable of being opened to a relatively large aperture, such as f/1.9.

fast scan direction — the raster direction designating the line element along which successive pixels are arrayed, perpendicular to the slow-scan direction. Most often the array is the moving part and the copy is stationary.

FAT (file allocation table) — the filing system used by most PCs to store and retrieve files on hard disks.

The disk is divided into "clusters" of bytes and then it files data into these clusters. When a program calls for a file, the FAT looks up the locations of all the clusters where the data is stored. The cluster size depends on the size of the hard disk, and a single cluster can only store data from a single application or file; if the data doesn't fill the whole cluster, then the rest of that hard disk drive space is wasted. Let's say a cluster is 16 bit; a 19K word processing document would take up a full cluster, plus 3K of a second cluster. The 13K remaining in the second cluster will remain unused, no matter how full the hard disk appears to be. Newer 32-bit FATs have been developed which can increase the available storage space on larger hard drives by as much as 40 percent by eliminating some of the unused space.

fax — *Related Term:* facsimile.

FCC (Federal Communications Commission) — the national agency responsible for administration and allocation of airwaves in the broadcast spectrum.

feather edge — *Related Term:* deckle edge.

feathering — (1) in typography, a ragged or feathered edge on printed type. (2) in printing, the tendency of ink on a rough, porous surface to spread. (3) increasing leading to lengthen a column of type, often to balance with an adjacent column (s) on the same page. (4) in digital imaging, blurring both inside and outside the selection outline. *Related Terms:* antique; vignette.

feed board — a flat surface, generally of wood, mounted on the front-right of a platen press, used to hold the unprinted sheets of paper. Opposite of delivery board.

feedback — (1) in broadcasting video, a distortion of the picture caused when a video-signal re-enters the switcher and becomes overamplified; (2) in broadcasting audio, an unpleasant howl from a loudspeaker, caused when the sound inadvertently is fed into the microphone and is overamplified. (3) in advertising, response or reaction to test advertisements, product surveys, etc.

feedback control chart — (1) in print production, a printed press image that supplies actual production environment information to the color separation department for calculations in producing color separation films. Dot gain and gray balance charts, color charts, and color reproduction guides are examples of feedback control charts. (2) in advertising and promotion, an organized chart or other presentation identifying marketing parameters, favorable/unfavorable responses, comments, etc.

feeder — (1) the portion of a production machine which removes substrate from the supply pile and places it into proper position for processing. (2) the person who manually performs the feeding operation.

feeding edge — *Related Term:* gripper edge.

feeding system — the part of a printing press designed to feed paper (substrate) from the supply pile into the press.

feeding unit — *Related Term:* feeder.

feet — the base or bottom of the body of a piece of metal type.

feet per minute (FPM) — the rate at which paper passes through a web press or at which a Fourdrinier machine makes paper.

felt — a continuous cloth pickup belt on a paper-making machine. It pulls paper from the Fourdrinier wire and carries it into the dryers.

felt finish — a soft woven pattern in text paper.

felt side — the top side of the paper that was not in contact with the Fourdrinier wire during papermaking, as compared to the wire side that was in contact. It is the preferred side for printing. *Related Term:* wire side.

ferrotype plate — a shiny, smooth and usually heated sheet of metal on which photographic prints are dried.

Fetch — a Macintosh program that uses FTP (file transfer protocol) for transferring files to and from a server, or vice versa. It enables you to write files on your local computer in whatever editor you like, and then place the files on the server when finished.

fiber — wood or other fibrous particles used in the papermaking process.

fiber optics — a method of transmitting light beams along glass fibers. Fairly secure medium, usually buried due to its frail nature, it is very hard to access or bug. It usually uses a light beam, such as that produced by a laser, that is modulated to carry information. Each fiber is capable of carrying from 90 to 150 megabits of digital information per second or 1,000 voice channels. *Related Terms:* HFC network; fiber optics; fiber-optic cable.

fiber puff — in paper, fiber swelling which results in a rougher textured and less smooth surface.

fiber-based paper — photographic paper consisting only of an emulsion coating on a high quality paper base.

fiberboard — *Related Terms:* chipboard; cardboard.

fiber-optic cable — transmission cables that consist of thin filaments, usually of glass, which can carry beams of coherent light. A modulated laser transmitters emits pulses of light that are sent down the optical fiber to a receiver. It translates the light signals. Fiber optics are much less susceptible to noise and interference than other kinds of cables, and can transmit data over great distances without amplification. The glass filaments are fragile, however, and optical fiber is usually run underground or in protected channels. It is rarely exposed in the manner common to copper wires. *Related Terms:* HFC network; fiber optics.

fiche — an abbreviation for microfiche. *Related Term:* microfilm.

field — (1) in video, every other line of an interlaced video image. In the United States, NTSC TV signals are transmitted at 60 fields/sec, each field equaling half a frame. The combined total makes up 30 separate and complete pictures per second. (2) in computer data bases, a single item of information, such as a name or zip code, that is part of a record in a data base. *Related Term:* cell.

field of view — the amount of a scene as revealed through the view finder of a camera.

FIFO (first in, first out) — a method for managing data or inventory storage queues so that the first information, materials, etc., placed in the queue is also the first out of the queue.

file — (1) in general computing, a named collection of information that is stored on a computer disk, for example, a document, resource, or application. (2) on the Internet, a protocol that allows the operator to create file hyperlinks (file://) in the FrontPage Editor using the Insert Hyperlink command.

file compression — a method of making computer files smaller so less time and space is needed to represent the information. More sophisticated data such as those to produce audio, video, JAVA, VRML, etc., and other multimedia are usually compressed extensively before being put on the Internet. Most shareware and freeware programs found on the Internet come in compressed formats like .ZIP, .HQX, .BIN or in a self-extracting EXE form.

file extension — the group of letters after a period or "dot" in a file name is called the file extension. This extension refers to the type of file; for example, if the filename is readme.doc, the extension doc denotes this is a document file and can be viewed using a document editor such as Wordpad, MSWord, etc. Operating systems such as MAC OS or Windows 95/98 will refer to a file's extension when choosing which application to launch when a user clicks on a particular file name.

file protection — any method of preventing media files from being altered, erased, accessed, etc., either accidently or on purpose. Usually switchable in magnetic media but used with some optical disk systems which are not rewriteable as a means of making unalterable and permanent data records.

file server — a large storage capacity central computer attached to a network and used to store, route and retrieve files for persons who are connected to that network. It facilitates the sharing of files by multiple users in the network.

file transfer — any process for copying a file from one computer to another over a network. *Related Terms:* file transfer protocol; Kermit; FTP.

file transfer protocol (FTP) — (1) on the Internet, one protocol which allows a host user to access and transfer files to and from another host over a network. (2) the name of the program the user invokes to execute the protocol. *Related Terms:* anonymous FTP; file transfer protocol; Kermit.

file type — the format of a file is usually indicated by its file name extension. Computer editing programs usually work on a limited set of file types. For example, Photoshop works primarily in graphic file types, i.e., PSD, JPG, EPS, GIF, TIFF, CGM, WMF, PIC, BMP, etc., while Microsoft Word works primarily in word processing file types such as .txt, .doc, .rtf, Unicode, etc.

fill flash — a flash or strobe light used to supplement available light, usually in shadow areas of scenes. *Related Term:* fill-in flash.

fill pattern — an alternate term for screen tint and used with reference to operations within the software for graphics and desktop publishing.

filler — (1) in newspaper publishing, extra material used to complete a column or page, usually general in nature and of little real importance. (2) in papermaking, the clay, titanium dioxide, calcium carbonate and other materials added to pulp to increase opacity, brightness, and printing surface.

fillet — an internal curve that is part of a character in alphabet design.

filling in — a condition in which the ink fills the area between halftone dots or plugs up the type, such as in the letter "e." *Related Term:* filling up.

filling up — a condition in which the ink fills the area between halftone dots or plugs up the type, such as in the letter "e." *Related Term:* filling in.

film — (1) in photography, sheets or rolls of flexible translucent or transparent acetate, vinyl, paper or other base materials that are coated with a photographic or light-sensitive emulsion. (2) in reproduction, any thin,

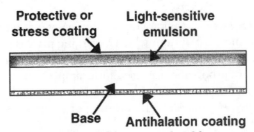

FILM LAYERS
The basic structural components and the major chemical coatings on light-sensitive film or paper. Often this number is limited to four: a protective upper coat, the emulsion, the base or substrate, and an antihalation backing. Certain adhesive layers may be present to bind the others together or to the base, but they are not counted.

organic, nonfibrous flexible material (usually not more than 0.010 in. thick) that is used as a substrate, particularly in flexography. Some examples include cellophane, polyethylene, Saran, acetate, and Mylar. *Related Term:* film base.

film assembly — the manual assembly of films made by process cameras and scanners. *Related Term:* stripping.

Film at 11 — a somewhat sarcastic reaction to an overwrought argument: "The sky is falling. Film at 11."

film base — (1) generally, any transparent or opaque material on which an emulsion is placed. (2) in papermaking, a paper with a very thin surface coating. *Related Term:* wash coating.

film board — that portion of a process camera that opens to firmly hold the film during exposure, usually by means of a vacuum.

film coating — a very lightweight mineral layer added to paper; sometimes by the paper machine size press. *Related Terms:* wash coating; film base.

film holder — (1) in general photography, a frame with removable light-tight panels, used to hold camera film flat and in correct position during exposure. (2) in process cameras, it was originally a coating of sticky "stay flat" solution. Modern cameras more commonly have a vacuum system to restrain the film during exposure. *Related Terms:* film back; film plane; back case; focal plane; dark slide.

film laminating —the affixing or bonding of a thin sheet of plastic, acetate or polyester to a printed product to provide scuff protection, waterproofing and/or increased gloss. Common laminated items are menus and record jackets.

film master — a photographic film representation of a specific symbol from which a printing plate is produced. Typically applied to film supplied by controlling or authorizing agencies, for example, when the U.S. Postal Service provides accurately sized and positioned negatives for the printing of business reply envelopes.

film negative — a piece of film with a tonally reversed image, in which dark areas of the original copy appear white (or clear) and vice versa.

film pack — a film package which contains numerous individual sheets for exposure. As a piece of film is exposed it is transferred to the back of the pack, revealing a new sheet in front. A dark slide mechanism protects film from exposure and is replaced immediately after.

film plane — *Related Term:* focal plane.

film sensitivity — a reference to such factors as contrast, color sensitivity and relative ASA/ISO rating or speed.

film speed — the number assigned to light-sensitive materials that indicates sensitivity to light; large numbers (like 800) indicate high sensitivity; low numbers (like 6 or 12) indicate low sensitivity. Fast film is highly sensitive, requiring less light to register an image, but with greater apparent grain or coarseness in the image. Slow film is less sensitive but tends to carry fine detail without artifacts; therefore, it is better suited for enlargement. *Related Terms:* fast film; ASA; ISO.

film-coated paper — *Related Term:* wash-coated paper.

filter — (1) in general black and white and color photography, a colored sheet of transparent material such as gelatin, acetate or glass, that is mounted in front of a camera or enlarger lens to emphasize, eliminate or change the color or density of the entire scene or certain elements in the scene by absorbing specific colors or wavelengths. They may also be used to soften or otherwise alter image characteristics. (2) in typesetting and data processing, a rule or script that can be used to manage incoming and stored mail. Software that supports filters lets you create rules that perform actions, such as automatically routing messages to various folders based on the sender's address, sending prewritten replies to certain people, or deleting messages that include specific subject lines. (3) in prepress camera operations, a transparent material characterized by its selective absorption of certain light wavelengths and used in a variety of applications, for example, to separate the red, green and blue components of an original when making color separation films. (4) in computer graphics, to distort or offset pixels.

filter factor — the number of times exposure must be increased due to the use of filter(s). The available light is reduced when a filter is used, therefore requiring an increase in light intensity or exposure time to render an optimum result. Each emulsion reacts differently to filtration and lights. The exposure adjustment can be from 2X to more than 8X over normal unfiltered exposure calculations.

filter, UV — a specialized filter which absorbs the ultraviolet but allows all visible light to be transmitted to the film. UV filters greatly reduce haze and minimize the bluish cast which can be present under some natural lighting conditions.

filtering — (1) the act of selectively restricting color rendering through the use of filters on a camera. (2) a process used in both analog and digital image processing to reduce bandwidth. Filters can be designed to remove information content such as high or low frequencies, for example, or to average adjacent pixels, creating a new interpolated value from two or more adjacent pixels.

FIM (facing identification mark) — in mail processing, used on reply mail to identify the front of the envelope during presorting or bulk mail.

final count — the number of products delivered and charged for by a printer or other production operation.

final film — after all elements are assembled, the completed films ready for proofing or platemaking.

final layout — *Related Terms:* mechanical; base art.

find — an element in Windows 95 and 98 under the start menu where specific files may be located based upon name, content, extension, etc.

finder — the Mac operating system that manages all the other files; the finder is like an index: it saves, names, renames, and deletes things in a file.

fine etching — the process of etching dots on metal to correct tone values when making printing plates.

fine line copy — any copy with tone lines that are narrower than the dots produced by the screen, for example, topographical map lines. Fine line copy is very difficult to reproduce and often requires the use of special production techniques such as still development.

fine papers — any of a group of papers made specifically for writing or printing, as compared to coarse papers and industrial papers. *Related Terms:* cultural papers; graphic papers.

fine screen — a halftone screen with ruling of more than 150 lines per inch.

fineness of grind — in ink compounding, the relative evenness of pigment dispersal in the vehicle.

finger — (1) an Internet search program that displays particular information about one or all users who are logged onto a local or remote system. It typically shows full name, most recent login time, and other user permissions or it may divulge plan, project and other files created by the user. Not all systems support finger. (2) a verb, meaning to apply the program to a username.

finish — (1) in papermaking, generally the characteristic texture and feel of a paper's surface finish. (2) with regard to printing, any process taking place in the bindery.

finish size — the size of a printed product after all folding, cutting and other production is complete. *Related Term:* trim size.

finishing — an inclusive term used to describe nearly all operations that take place after the printing operation, such as inserting, padding, perforating, diestamping, punching, scoring, coating, varnishing, diecutting, laminating, embossing, and mounting, etc.

fire off — an expression meaning to send e-mail, i.e., "I fired off an e-mail responding to their charges!"

firewall — any of several ways designed to keep unauthorized outsiders from tampering with a computer system and, therefore, increasing server security. Primarily used to protect a network from damage (intentional or otherwise) or an untrusted host or other network. Typical among firewalls are dedicated computers equipped with security measures such as a dialback feature, or software-based protection called defensive coding. *Related Terms:* encryption; hacker; password protection.

FireWire — the IEEE 1394 high bandwidth I/O protocol for high-speed connections between computers, cameras, TVs, stereos and VCRs to directly transfer digital data.

firmware — software instructions stored within a computer as part of its basic instruction set. *Related Term:* ROM.

first down color — the first ink film thickness laid down on a substrate as part of a multicolor laydown.

first edition — the original printing from the original, manuscript or author file, after initial typographic and editorial changes. The original press run and (usually) copyright and ISBN number.

first proofs — the proofs submitted for checking by proofreaders, copy editors, etc.

first read rate — in bar code and other optical scanning, the ratio of the number of successful reads to the number of attempts.

first serial rights — the right to publish a serialized version of a work before the work in its entirety is actually published.

first virtual payment method — a payment system provided by First Virtual, Inc. that allows customers to pay domain name registration and renewal (re-registration) fees via the Web without having to provide their credit card information over the Web. Customers fill out an application for a Virtual PIN™, which is then used over the Internet to represent their credit card. First Virtual contacts the VirtualPIN™ holder every time a transaction takes place and requests verification and approval before charging the specified amount to the VirtualPIN™ holder's credit card. *Related Terms:* registration fee; renewal fee.

first-surface reflection — the reflected light scattering from the surface of a printed ink film.

fit — (1) in composition and layout, the term used to describe the horizontal spacing or relationship between two or more characters. Fit can be altered by kerning or modifying the horizontal width (set width) assigned to characters. Some programs use terms like tight or "loose, but it is generally subjective. (2) on press, the relationship of images both to the paper and, if multicolored, to each other; often confused with the term "registration," which refers to position of the press sheet in the press. *Related Term:* image fit.

five- and six-color printing — (1) normally, a photomechanical variant of the four-color subtractive color reproduction process which adds additional chromatic colors such as light magenta, light cyan, or red or green in addition to the process primaries. Their use expands the available color gamut. (2) in magazine publishing, a term sometimes used to refer to a Pantone color used in addition to process color on a cover. *Related Terms:* hi-fi color; Hexachrome.

five-em space (5-em) — the small fixed space equal in width to the period or comma (usually about 1/5 of an em space).

fix — (1) to stabilize the photographic image on film or paper after development by dissolving or neutralizing the remaining unexposed silver salts in the emulsion. (2) the chemicals used in the third step of normal development to remove unexposed emulsion and to "fix" or stabilize the developed image. *Related Terms:* sodium thiosulfate; fixer; fixing bath; hypo.

fixative — a clear spray applied to mechanical art or drawings as a sealer and protective coat.

fixed costs — nonvariable costs which do not decline with volume. Such costs might include creative art, photography, conceptualizing, etc. Costs which will be the same for a given project regardless of whether the final count is expected to be 100 or 100,000. *Related Term:* variable costs.

fixed space — a particular amount of white space, such as an "em," "en," or thin space, which will not be "stretched" for justification purposes as will a space band. A uniform space assigned to all type characters, regardless of their actual widths.

fixed-beam bar code reader — a scanning device where scanning motion is achieved by moving the object relative to the reader (as opposed to moving beam reader). Grocery store readers are fixed-beam devices.

fixer — a process which makes a photographic material insensitive to further exposure. *Related Terms:* fixer; sodium thiosulfate; fixing bath.

fixing bath — one of the basic developing chemicals, it is used to dissolve the unexposed silver halides and remove residual sensitive silver from a developed film or print. Removal of these residual substances renders the print permanent and prevents staining and yellowing. This chemical, often referred to as hypo, contains sodium sulfite, sodium thiosulfite, acetic acid and a hardening agent to preserve the emulsion coating. *Related Terms:* fix; fixer; hypo.

flag — (1) the designed title of a newspaper as it appears on the first page. (2) alternate name for nameplate or banner.

flame — a strong opinion or criticism of something, usually as a inflammatory statement, often heated and/or abusive, in an electronic mail message, bulletin boards or community areas. Flaming is an impassioned personal attack on another individual and when a flamer responds in kind, the exchange is called a flame war.

flame war — an online discussion that degenerates into a series of personal attacks against the debaters, rather than discussion of their positions. A heated exchange. *Related Term:* flame.

flap copy — the text that appears on the front or back flap of a dust jacket.

flare — an exposure problem caused by uncontrolled reflection of stray light passing through the camera's lens.

flash exposure — (1) the supplementary exposure given in halftone photography to strengthen the dots in the shadows of negatives, thereby lengthening the screen's tonal reproduction range. (2) in general photography, a brief, intense burst of light produced by a bulb or an electronic unit, generally used where the scene lighting is inadequate for available light photography or as a supplemental light to provide shadow detail in the photograph. (3) a phototypesetting function that is keyboard controlled. Allowing flash will produce an exposed character; the normal function. Canceling flash will allow character escapement without the associated image, a feature often useful during typographic input. Virtually all phototypesetters and imagesetters have this function regardless of the light-source technology used. *Related Terms:* exposure, main; exposure, bump.

flash ROM — jargon that refers to ROM (read only memory) chips that can be reprogrammed with new BIOS instructions after the chips have left the factory. Such ROM chips are technically called EEPROMs.

FlashPix — a file format that displays images only at the resolution needed for a specific task, thereby granting users greater speed when working with graphics. It is a fairly sophisticated and promising technology, but is not yet a well established standard.

flat — (1) in prepress, a noun describing completely assembled films ready for platemaking, the result of the stripping operation. A sheet of film or goldenrod paper to which negatives or positives have been attached (stripped) for exposure as a unit onto a printing plate. (2) in photography, an adjective; a nontechnical description of a print or proof lacking contrast, color, or brilliance.

flat bed cylinder press — a type of press, usually relief, designed so that the sheet rolls into contact with the type form as a cylinder holding the sheet moves across it. *Related Term:* cylinder press.

flat bed scanner — (1) a scanner configuration where images to be scanned can be placed directly on a flat

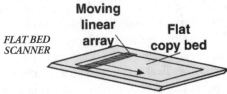

FLAT BED SCANNER — Moving linear array — Flat copy bed

platen or surface which is then read by a moving linear array of sensors. This method is in contrast to drum scanners, where images must be mounted on a cylinder. *Related Terms:* drum scanner; slide scanner.

flat color — a term generally used to refer to solid colors, ink tints, or mixed to order; inks other than process colors.

flat etching — the chemical reduction of the silver deposit in a continuous-tone or halftone plate, produced by placing it in a tray of etching solution.

flat fee — a one-time payment, as opposed to a royalty fee which involves payments at regular intervals. *Related Term:* work for hire.

flat finish — *Related Term:* low finish.

flat proof — often a part of the makeready process. It is a preliminary look at how a project might reproduce and provides a last opportunity to make adjustments prior to a press run.

flat shading — a type of shading in a 3D graphic where the wire frame polygons which make up the image are simply one-color fills. The result lacks realism and is not considered acceptable for games and other applications where a degree of realism is necessary. *Related Terms:* Gouraud shading; Phong shading; shading; wire frame.

flat-back bind —in case binding, leaving the signatures flat prior to application of the cover, without rounding. It is somewhat less expensive than roundback spines.

flatbed platen press — a press where a flat, inked form is pressed directly against a sheet of substrate

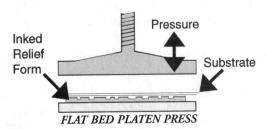

FLAT BED PLATEN PRESS

placed on top of it. The pressure mechanism is reciprocal, moving alternately toward and away from the form, while printed sheets are removed and unprinted sheets are placed in printing position.

flat-panel display — any of a variety of ultrathin displays commonly used for notebook computers. *Related Terms:* active matrix; LCD; passive matrix; TFT; plasma display.

fleurons — the flowered ornaments often used for separating sections of type.

flexography — a method of rotary relief printing characterized by the use of flexible rubber or plastic plates with raised image areas and fluid, rapid drying inks. Traditionally used mainly for packaging but now being adopted by small to medium sized newspapers and other commercial applications. *Related Terms:* aniline printing; letterpress; relief plate; relief printing.

flier — a one-page, unfolded printed promotional piece.

flip — to rotate an image along a horizontal line.

float — in page layout and/or paste-up, the placement of a smaller element in a larger area than it would require.

floating accent — an accent mark which is set separately from the main character and is then placed either over or under it.

floating flag — a publication flag or banner set less than full page width and displayed in a position other than the top of the front or section page.

floating rule — a rule, usually placed between columns, whose ends do not touch other rules.

floating-point math — a branch of mathematics often used in spreadsheet operations, and floating-point operations (called FLOPs) are often used to measure computing performance. *Related Term:* MFLOPS.

flocking — (1) the coating on simulated suede papers. (2) in production, the affixing of cloth fibers to all or sections of a substrate to enhance production values.

flood — total coverage of a printed sheet with ink, varnish or aqueous coating, without regard for individual image elements.

floor sheet — *Related Term:* house sheet.

flop — (1) in photographic printing, the reproduction of a photograph or illustration so that its image faces opposite from the original. (2) in prepress, the placement of a film positive or negative with the emulsion facing up instead of down. *Related Term:* lateral reverse.

floppy disk — a magnetic storage system usually used in micro- and minicomputers consisting of a spinning oxide coated disk inside a protective sleeve or case. Recording heads place a magnetic field onto the oxide layer, which can then be retrieved by similar reader heads when the data is to be accessed. *Related Terms:* diskette; disk.

flow — the ability of an ink to spread over the surface of the rollers on a press.

fluorescence — (1) the emission of visible light from a medium following the absorption of light of a shorter wavelength such as ultraviolet. In addition to the light reflected by the color in the normal way, fluorescence gives an extra brightness. The phenomenon often occurs through the conversion of ultraviolet radiation into visible radiation and can occur in printing inks, papers, or original photographs, artwork, or retouching dyes and pigments. (2) the light from fluorescent lighting tubes as compared to tungsten bulbs.

fluorescent ink — inks with special spectral values that produce a bright glowing appearance, particularly when viewed under ultraviolet light or strong daylight.

flush — to align text or images along one edge of a page layout.

flush cover — a cover trimmed to the same dimensions as the inside pages. Typical of paperbacks, and soft cover literature.

flush left — (1) copy which is aligned on the left margin. (2) (quad left) a code which directs preceding copy to be set flush against the left margin; ragged right; not justified.

flush paragraph — a paragraph set without the (usual) indention of its first line.

flush right — (1) copy which is aligned on the right margin. (2) (quad right) A code which directs preceding copy to be set flush against the right margin. (3) ragged left; not justified.

flute — the wavy paper pleat between the walls of corrugated cardboard.

flyer — an inexpensively produced circular used for promotional distribution; an advertising handbill or circular.

flying ink — *Related Term:* misting.

flying paster — an automatic device on web presses that retrieves a new roll of paper and attaches it to an expiring roll without a press stop. *Related Terms:* festoon; paster; reel-tension paster (RTP).

FM screening — frequency-modulated screening. A type of screening that employs irregular clusters of equally sized CMYK pixels to represent continuous-tone images. The placement of these pixels, although seemingly random, is precisely calculated to produce the desired hue and intensity. This process differs from traditional halftoning in which the distance between CMYK dots remains constant while dot size varies to create the desired hue and intensity. *Related Term:* AM screening.

FM synthesis (frequency modulation synthesis) — one of two main ways to synthesize music. It is not as desirable as wavetable synthesis. *Related Term:* wavetable synthesis.

foam board — plastic sheeting made of foam or corrugated plastic with image-receptive surfaces suitable for printing. Used extensively for exterior signs and posters where durability and strength are major considerations.

FOB (Free On Board) — often used when delivering printed materials, in the context of "Free On Board origin," which means the addressee pays the shipping costs, or "Free On Board destination," which means the shipper pays the shipping costs.

focal length — the distance from the center of a lens to the film board when lens is focused at infinity.

Focoltone Color System — a European system of color specification based on the four-color process.

focus — (1) in computing, a monitor's ability to display sharp images, characterized by easy-to-read text. Good focus can be undesirably reduced by antiglare treatments. (2) in camera work, the process of adjusting the lens so light reflected from the object or copy is sharp and clear on the film. *Related Terms:* antiglare treatment; dot pitch.

focus, long focal length — a lens with a long distance from the front nodal point of the first element of a lens to the same position in the most extreme rear element. A telephoto lens.

focus, short focal length — a lens with a short distance from the front nodal point of the first element of a lens to the same position in the most extreme rear element. A telephoto lens.

focusing cloth — a large piece of material used to shield the photographer from extraneous light while composing and focusing a scene.

focusing magnifier — a magnifying glass often used in conjunction with a ground previewing glass to ensure sharp focus.

fog — (1) in photography, a photographic defect in which the image is either locally or entirely clouded due to uncontrolled exposure. Fogging is often caused by stray light, but can be caused by improperly used or compounded chemical solutions. Fog can supress a photographic image to where the whites become dead and lifeless in a print. Potassium bromide and benzotriazole can be added to reduce the chance of fog and brighten the higher values. (2) in 3D graphics, a setting that adds a graduated opacity to areas of the image to simulate fog.

foil — metallic or pigmented leaf used in cold- or hot-stamping lettering and designs on a surface. *Related Terms:* foil stamping; foil blocking; ribbon; foil emboss.

foil blocking — a process of stamping a design on a book cover without ink by using a colored foil with pressure from a heated die or block. *Related Terms:* hot foil stamping; foil stamping; foil embossing.

foil emboss — a relief process used for the placement of metallic or pigmented colors onto a substrate using specially engraved or cut dies, heat and pressure. Flat foiling transfers only the image. Foil embossing transfers the image and distresses the paper to give the foil additional depth and dimension. *Related Term:* foil stamping.

foil papers — the name given to papers with surfaces that seem metallic and/or shiny in appearance.

foil stamping — *Related Term:* foil embossing.

fold — to double a sheet of paper over itself.

fold lines — the (usually dashed) lines drawn on a paste-up in non-reproducing pencil that represent where the final job is to be folded; small black lines are sometimes drawn at the edge of the sheet to be printed and used as guides for the bindery. *Related Terms:* fold marks; trim marks; crop marks; corner marks.

fold marks — lines drawn on a mechanical, negative, printing plate and/or press sheet indicating where the product is to be folded. *Related Term:* fold lines.

FOLDED SIGNATURE
In this example a single sheet, printed on two sides, is folded at right angles. The resulting reveals are numbered to form an 8-page booklet, after trimming.

folded signature — (1) a single press sheet folded appropriately to create pagination for a section of a book. (2) a folded press sheet which may be a complete product.

folder — (1) in computing, an electronic holder of documents. (2) in bindery and finishing, a machine that creases, scores and/or folds printed sheets of paper to particular specifications during binding and finishing. The process itself is called folding.

Folder View — in Web site development, the view option in FrontPage developer that shows the relationship between folders (directories) of a Web project or site. You can create, delete, copy, and move folders in Folder View.

folding dummy — a dummy signature or compiled product sample made of paper specified for the printing job. Used to illustrate imposition of pages on press sheet. *Related Term:* imposition dummy.

folding to paper — in bindery and finishing, folding paper without regard for the location or alignment of the images on the substrate.

folding to print — in bindery and finishing, folding to assure precise or exact image location according to predetermined markings on the substrate.

foldout — extra page that may be pasted or folded into a book or booklet.

folio — (1) in printing, a page number, often placed at the outside of the running head, at the top (head) of the page. (2) in descriptive bibliography, a leaf of a manuscript or early printed book, the two sides designated as "r" (recto, or front) and "v" (verso, or back). (3) formerly, a book made from standard-size sheets folded once, each sheet forming two halves or four pages. (4) in page numbering, the evens go on the left, the odds on the right. *Related Term:* header.

folio lap — the "lip" of extra paper which is beyond the trim size of a signature. It is often used as a gripper point during the folding and gathering operations.

folio lines — originally page numbers, but usually now the line giving date, volume, and number; or page number, name of publication and publication date, in small type on the inside pages.

folio, blind — a folio counted in numbering pages but not printed (as on the title page).

followup — a Usenet posting response to an earlier message.

font (or fount) — in hot metal composition, a complete assembly of all the characters (upper and lower-

LIST OF CHARACTERS IN TWO-LETTER FONTS
WITH ITALIC AND SMALL CAPS

ABCDEFGHIJKLMNOPQRSTUVWXYZ
ABCDEFGHIJKLMNOPQRSTUVWXYZ
ABCDEFGHIJKLMNOPQRSTUVWXYZ&

12345   abcdefghijklmnopqrstuvwxyz   67890
VBCDE   *abcdefghijklmnopqrstuvwxyz*   FGRTJ

,.:;?!(|)*'`-- Æ Œ ﬂ & £ $ ... ﬁ ﬂ ﬀ ﬃ ﬄ æ
,.s;?!ʌıǫo''-- Æ Œ ﬂ ɴ ʟ P L ﬁ y ﬀ w M K

12345   Z & : ( ) " " QU Qu ﬆ ﬆ ﬂ ﬁ ﬃ ﬄ $ æ œ   67890   ;;
12345   Z & : ( ) " " *QU Qu ﬆ ﬆ ﬂ ﬁ ﬃ ﬄ $ æ œ*   67890   ;;

⅛ ¼ ⅜ ½ ⅝ ¾ ⅞   H X Z & Æ Œ   @ % † ‡ § ¶ - [ ]

SWASH CHARACTERS   LONG DESCENDERS
ABCDEFJKLMNPRTVY&   g j p q y
Made in All Point Sizes and Supplied on Special Order   These characters may be substituted for those regularly furnished
                                             with a font, if so ordered, or they may be added as an extra

MODERNIZED FIGURES
1234567890   *1234567890*
Made only in 6, 8, 9, 10, 11, 12 and 14 point Granjon with Italic and Small Caps. These figures will be substituted
for those regularly furnished with a font, if so ordered, or they may be added as an extra

TWO-LETTER LOGOTYPES
F. P. Ta Te To Tr Tu Tw Ty T. Va Ve Vo V. Wa We Wo Wr W. Ya Ye Yo Y.
F. P. Ta Te To Tr Tu Tw Ty T. Va Ve Vo V. Wa We Wo Wr W. Ya Ye Yo Y.

fa fe fo fr fs ft fu fy ﬀa ﬀe ﬀo ﬀr ﬀs ﬀu ﬀy f, f. f- ﬀ, ﬀ. ﬀ- f ﬀ
fa fe fo fr fs ft fu fy ﬀa ﬀe ﬀo ﬀr ﬀs ﬀu ﬀy f, f. f- ﬀ, ﬀ. ﬀ- f ﬀ

ONE-LETTER ITALIC LOGOTYPES
FA PA TA VA WA YA Th Wh

TRADE **LINOTYPE** MARK

A COMPLETE TYPOGRAPHIC FONT
Traditionally all of the letters, numbers and figures of a single face in a specific size, i.e., 8, 10, 12, etc. In digital and desktop, the term has come to more closely describe a face, as a single loading will produce an infinite variety of sizes.

case letters, figures and punctuation) of one size of one typeface, for example, 10-point Bodoni roman. Font sizes (characters in a font) vary from 96 to 225, depending on the makeup of the font. Special characters (those not in a font) are called "pi" characters. In electronic publishing, the term "font" is used to mean a complete typeface design in any or all point sizes.

font downloader — a device used in computer imaging to send digital font information from a computer to an output device such as a printer.

font metrics — character width of a particular character in a font. *Related Term:* font hints; hints.

font substitution — the replacement of a specified

font by another whenever the specified font is unavailable on a device.

foo/foobar — a sort of online algebraic place holder, for example: If you want to know if another site is run by a for-profit company, you might look for an address in the form of <foo@foobar.com>.

foot — the bottom of a page; the top of the page is referred to as the head. Used to orient pages in signature imposition, as in head-to-head or head-to-foot, etc.

footer — a book title or a chapter title printed at the bottom of a page. A drop folio (page number) may or may not be included. *Related Terms:* running foot; folio, drop.

footprint — (1) in broadcasting, the geographic region on the earth which can easily receive and interpret a signal from a communications satellite. (2) in computing, the physical area(s) occupied by the various components of a desktop system, i.e., CPU, monitor, printer, keyboard, etc.

for position only (FPO) — inexpensive or low-resolution images placed in page layouts to indicate what final art should be placed in its respective position, or to expedite file transfer by creating less complex data files. Often defaced with "FPO" to prevent anyone from utilizing it as the final art. A common occurrence in modern electronic imaging systems where a low resolution is provided for screen viewing but whenever the file is raster image processed (RIPed), a high-resolution image is substituted to enhance reproduction. The process often occurs automatically as long as both resolution files are provided as part of the data transfer process.

for your information (FYI) — on the Internet, a subset of the request for comment (RFC) series. They tend to be more informational than technical in nature. There are FYI documents that describe what the Internet is, how to use anonymous FTP, and how to get your school connected to the Internet. *Related Terms:* request for comments (RFCs).

foreign rights — a subsidiary right that allows the book to be translated and published in countries other than the one in which the book was originally published.

foreword — introductory remarks found in the front matter of a book, often written by someone other than the author.

form — (1) in computing, a set of data entry fields

on a page that are processed on the server. The data is sent to the server when the user submits the form by clicking on a button or, in some cases, by clicking on an image. (2) in imposition and presswork, all the typographic elements on an image carrier for a particular printed piece or sheet, with the individual elements arranged to allow folding and binding after imaging of the substrate. (3) in presswork, one side of a press sheet.

form bond — any lightweight bond made for business forms. *Related Term:* register bond.

form handler — a program on a networked server that executes to process submitted query data from a user.

form letter — used in word processing to describe a generic or repetitive letter where the names and addresses of individuals are automatically generated from a data base or, in the case of preprinted letters, typed individually.

form rollers — any press rollers that contact the image carrier or plate.

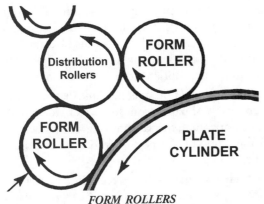

FORM ROLLERS
Only the rollers in the roller chain which contact the image carrier.

formal balance — a design characteristic in which images of identical weight are placed on each side of an invisible center line.

format — (1) in production planning, generally, the shape, size, type area, margins and overall visual design and appearance of a page, book, magazine or other publication. (2) in composition, the arrangement or sequence of typesetting parameters and commands into format codes that can be accessed with a simple keyboard command or keystroke. (3) in elec-

tronic publishing, a verb that describes the process of applying type specifications to text files. (4) in general photography, a descriptive statement relating to the size of the film used in the camera. 35mm, 2¼ in. x 2¼ in., 2¼ in. x 3¼ in., 4 in. x 5 in., 5 in. x 7 in., 8 in. x 10 in., 10 in. x 12 in., and 16 in. x 20 in. are all standard formats.

formation — the distribution of fibers in paper as perceived when the sheet is lighted from behind. Good formation means fibers appear uniformly distributed; in poor formation they appear in concentrated areas and produce a mottled appearance.

formatted text — uniform-spaced style in which all white space (such as tabs and spaces) are displayed by the Web browser.

formatting — (1) in typesetting, the translation of the designer's type specifications into command codes or machine instructions to direct the typesetting equipment. Formatting is gradually replacing markup. *Related Terms:* format; markup. (2) in computing, the division of a hard disk, diskette, etc., into sectors, tracks and blocks to enable the operating system to address data.

formatting tool bar — a selection tool bar in any computer program which contains icons relating to commands that reformat selected paragraphs or text.

formed font impact — in addressing and labeling, a printing method for labels consisting of a rotating drum etched with raised bars and characters. A one-time ribbon and the label move between the drum and a micro-controlled hammer to impact an image on the substrate.

formula pricing — a printing pricing system based on standard papers, formats, ink color and quantities.

fortune cookie — inane/witty/profound comments found on the Net.

forum — a section on the Internet or on an online service devoted to a particular topic where people can exchange information and ideas.

Foss Color Order System — a printed color chart that features color order, equal tone spacing, and a full range of black.

fotomatrix — the original name for the image matrices used on the Intertype Fotosetter, the first-generation phototypesetter. Closely modeled after conventional linecasting matrices, a film negative image was mounted

FOTOMATRIX
The original format for the first photocomposition macmachine. When introduced in 1946, it closely paralleled the hot metal matrices in use in Intertype and Linotype machines of the day.

into the wall of each brass matrix. The thickness of the matrix determined the letter escapement and each one was photographed sequentially, in place of the conventional hot metal casting process. See illustration on next page. *Related Term:* Fotosetter.

foundry type — Gutenberg type. Individual letters cast from a metal alloy and designed for repeated use by hand composition and for assembly, individually,

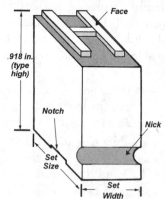

FOUNDRY TYPE
It provides the basis for most modern typography measurement systems. The raised face is the printing area which sits upon the body. It is from the bvody that the point size is derived.

to form words, sentences, etc. *Related Terms:* hot type; cold type; phototype-setting.

fountain — a reservoir metering and control system for the dampening solution or ink that is fed to the plate on a lithographic press. *Related Terms:* water pan; dampening system.

fountain blade — on a printing press, the strip of flexible steel that forms the bottom of the ink fountain. The fountain roller forms the other side of the ink trough. Moving the blade closer to or farther away from the fountain roller controls the thickness of the ink film across the roller and determines how much ink will be conducted into the roller train by the ductor roller. *Related Terms:* fountain keys; fountain roller; doctor blade.

fountain keys — a series of thumb- or motor-driven screws or cams behind the fountain blade that con-

trol the distance between the fountain blade and the fountain roller, thus providing control of the ink flow onto the ductor roller. *Related Terms:* fountain blade; fountain roller.

fountain roller — a metal roller that rotates intermittently or continuously in the ink or dampening fountain and carries the ink or dampening solution on its metal surface onto a ductor roller and then, subsequently, on to the respective portions of a printing press. *Related Terms:* fountain blade; fountain solution; fountain; dampening solution; water roller.

fountain solution — in lithographic printing, usually a combination of water, gum Arabic, and other chemicals used to wet the printing plate and keep the nonimage areas from accepting ink. Some fountain solutions contain alcohol. *Related Terms:* dampening solution; water; fountain etch.

four-color — a publication that is created using four colors, usually, but not always, printed with cyan, magenta, yellow, and black.

four-color process — the photomechanical method of reproducing full color copy (original artwork, transparencies, etc.) in printing by separating the original color image into its three primary subtractive colors — magenta, yellow and cyan—plus black and overprinting the plates with the specified amounts of each color in the appropriate areas of the reproduction.

Fourdrinier — (1) a paper machine that forms a continuous web of paper from furnish or stock metered onto a horizontal, forward-moving, endless wire belt. (2) the name of the person who invented the machine. Pronounced four-drin-ear.

Fourdrinier wire — the screen on a papermaking machine that catches furnish and carries it while it begins to form paper.

fourth-generation imagesetter — imagesetters that use a laser to expose characters, lines, tints or halftone dots onto photographic film, paper or plates, usually by area composition.

four-up — The imposition of four items to be printed on the same sheet in order to take advantage of full press capacity and minimize paper consumption. May be almost any number, depending upon the size of the finished product and the sheet size capacity of the press on which it is being printed.

four-wire circuit — a circuit that has two pairs of conductors (four wires), one pair for the send channel and one pair for the receive channel; allows two parties to talk and be heard simultaneously.

foxing — the characteristic yellow or brown spots on old paper. Foxing is caused by chemical impurities and poor storage conditions.

fpm — the abbreviation for feet per minute. Often used to designate the speed of a web printing press.

FPO (for position only) —a low-resolution or simulated version of a graphic that is used only as a placeholder and not for final reproduction.

FPRR (first pass read rate) — *Related Term:* first read rate.

FPU (floating-point unit) — a computer CPU is designed to work with integer mathematics. When floating-point math is used, the CPU assigns the processing responsibility to the FPU, which is designed to handle floating-point math more efficiently. The FPU was originally called a math coprocessor and it can be either a separate chip like Intel's 80387 or the Motorola 68881, or built into the CPU such as in the Pentium or Cyrix processors. *Related Term:* CPU.

fractal — a word coined in 1975 by Benoit B. Mandlebrot from the Latin fractus ("to break"). *Related Term:* Mandlebrot fractals.

fragmentation — in computing, a process which occurs as a normal part of writing and deleting files. Files are not stored linerally as in tape recording, but are broken up into parts and spread across the disk to any available or unused location. Subsequent deletion of the file may cause small unused segments to not be used or cause newly written files to be broken up into smaller and smaller segments for storage. This greatly increases access times for these files. Periodic defragmentation recovers the small segments and re-assigns data to larger blocks of space, thereby minimizing the number of locations which must be accessed to recover stored data. *Related Term:* defragment.

frame — (1) on the Internet a named element of a frame set that appears in a Web browser. When you create a hyperlink to the page, you assign a page to a frame. A term used to describe a viewing and layout style of a Web site, it refers to the simultaneous loading of two or more pages at the same time on the same screen. Some sites come in two versions: "frames" and "no frames." Frame versions usually take a little longer to load and may contain other "enhanced" features such as Java scripts and animation. (2) in newspaper page makeup, the pattern in which the left and right outside columns are each filled with a single story. (3) in multimedia, full screen of video data where the video is composed of 30 frames/sec. (4) a single picture from a series as a roll of pictures, i.e., motion picture film. (5) in design, a name often used to describe a border or box which bounds art or text.

frame buffer — a storage area on a graphics card that memorizes information not currently being displayed on screen. It "buffers" or stores rendered frames off screen before they are displayed. *Related Term:* double buffering.

frame grabber — in video, a utility that freezes a single image from a motion video source. Popular with home computer enthusiasts for capturing still images from their video camera output.

frame grabbing — seizing a single screen image from a series of frames viewed on a film or video screen or another media so that it can be manipulated and combined with other individual images for reproduction purposes.

frame rate — (1) in video, the speed at which video images are displayed. (2) in conventional motion picture photography, the rate at which individual film frames are viewed. *Related Term:* frames per second.

frame relay — a-standard packet interface protocol for data only that has a few advantages over ISDN. It can be purchased in increments and the protocol has a flat-rate billing structure as opposed to per-hour usage charges. It is used extensively to extend a LAN between corporate branches.

frame source — a term which is just about the same as document source but relates to a particular frame on a Web site.

frames per second — the rate at which film or video switches from one still frame to the next to create the illusion of motion. Film is normally displayed at 24 frames while video uses 30 frames per second that, through persistence of vision, convinces the eye that it is seeing real movement. Higher frames per second tend to make a smoother motion illusion.

Franklin Printing Catalog — a privately published pricing guide that printers use to determine average costs for nearly all operations or products.

free sheet — a paper made from cooked wood fibers mixed with chemicals and washed free of impurities, as compared to groundwood paper. *Related Term:* wood-free paper.

freehand drawing — compositions drawn without

the aid of matrices, templates or other mechanical guides.

freelance — to work on a client-by-client and job-by-job basis, as opposed to being employed full time by one particular company. *Related Term:* work for hire.

freeware — freeware is software you can download, pass around, and distribute without payment. However, unlike public domain programs, it is copyrighted, so you can't decompile it, sell it or use it as part of your own software creation.

French fold — a sheet which has been printed on one side only and then folded with two right angle folds to form a four-page uncut section.

frequency — an analog signal; waves. The frequency of occurrence of these waves passing a specific point in a given amount of time, it is often reported in cycles per second, or Hertz. *Related Terms:* amplitude; phase.

frequency spectrum — a frequency spectrum refers to a range of frequencies. For example, cable television commonly occupies the 5-MHz to 550-MHz frequency spectrum, while the visible spectrum occupies another range of frequencies, etc. *Related Term:* spread spectrum.

FRENCH FOLD
A folded signature, printed on one side. Typical of many greeting cards. Provides color and/or art on all four pages. Inside pages from back side of sheet are left blank.

friction clip binding — the binding of multiple sheets together with a metal or plastic clip which expands to accom-modate the pages and contracts to hold them firmly in place. Inexpensive and usually not permanent, i.e., pages can be removed simply by expanding the clip.

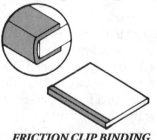

FRICTION CLIP BINDING
Sheets are simply held together with friction supplied by a metal or plastic clamp.

frisket — a paper placed between two grippers fingers to hold a press sheet while an impression is made. Most common in platen letterpress with heavy ink coverages or light stock, the form prints through an opening made in the frisket.

front end phototypesetter — the output is controlled primarily by coded paper or magnetic tape, or other electronic signal. Operators on separate keyboard units produce the tape signals that direct the typesetter to compose the phototype on paper or film. After development, it is ready for paste-up as camera ready copy. *Related Terms:* separate entry typesetters; direct entry typesetters.

front guide — one of a series of stops that halt the forward movement of a press sheet on the feed board. The front guides square the sheet in relation to the printing cylinders and determine the lead edge margin. *Related Terms:* front lays; front stops.

front list — newly released books, as opposed to back list, which are previously published titles still available from the publisher.

front matter — in publishing, materials appearing at the beginning of a book, such as acknowledgements, Foreword, table of contents, etc. Any elements appearing prior to the first text elements. The pages are frequently numbered with miniscule Roman numerals.

front plate — an illustration that faces the title page. *Related Term:* frontispiece.

frontispiece — an illustration that faces the title page. *Related Term:* front plate.

FrontPage server extensions — a set of programs and scripts that support Microsoft FrontPage and extend the functionality of the Web server being used.

FrontPage web — a home page and its associated pages, images, documents, multimedia, and other files that is stored on a World Wide Web server or on a computer's hard drive. A FrontPage web also contains files that support FrontPage functionality such as WebBots and that allow the web to be opened, copied, edited and administered in the FrontPage Explorer.

FTP (file transfer protocol) — the most common procedure used for downloading and uploading files with an Internet browser. FTP can log in to another Internet site and transfer files. Some sites have public file archives that you can access by using FTP with the account name "anonymous" and your e-mail address as password. This type of access is called anonymous FTP. Macintosh users use a program called Fetch and one of the best FTP programs for Windows is WS-FTP. *Related Terms:* anonymous FTP; Archie; Gopher; WAIS.

fugitive color inks — inks which readily fade when exposed to sunlight. The opposite of light-fast inks.

fulfillment — the process of filling orders for a product through order taking, packing, shipping, and collecting payment.

full bleed — printing which extends beyond the trim or bind points on all four sides of a printed piece. A bleed on all sides.

full bound — a case bound book whose entire exterior is of one material, such as leather. *Related Term:* half bound.

full color — 4-color process color printing using the subtractive primary colors of ink: cyan, magenta and yellow, plus black which uses an artificially created black plate image.

full duplex — a communication channel over which both transmission and reception are possible in two directions at the same time; e.g., a four-wire circuit. Another way of saying duplex, so that it's not confused with half duplex. *Related Terms:* duplex; half duplex; four-wire circuit.

full frame — in photography, the making of a print which contains the entire negative image area, without cropping. Relatively rare in reproduction, but utilized in the form of contact prints for frame selection and content editing.

full measure — type set to the full line measure or length.

full motion video — video production at 30 frames per second when using the NTSC standard or 25 frames per second when using the European PAL system.

full page pagination — the display of one full page of text and illustrations on a video display terminal for the purpose of making copy, text or layout changes.

full point — a full stop. A period.

full ringing support — an ISDN device that enables phones, fax machines, and other POTS (plain old telephone service) devices to use ISDN just like a standard analog phone line. ISDN adapters that don't have full ringing support will usually have the computer or the adapter itself alert an incoming call using a blinking light or a pop-up menu. *Related Terms:* POTS; ISDN.

full text search — a search method based on finding any word or phrase within the entire text of the document.

full-coated carbon — a carbon paper sheet which is coated on both sides.

full-scale black — a black printer that will print in all tonal areas of the reproduction from the highlight to the shadow. *Related Terms:* GCR; gray component replacement; UCR; under color removal.

fully saturated — a photographic term indicating rich color.

function code — a computer code that controls the machine's operations other than the output of typographic characters, i.e., advance film or paper, cut off, etc.

function keys — a series of 10 or 12 keyboard keys which are indicated as F1, F2, etc. They are provided as special execution keys for use within software. For example in many software packages, F1 is designated as an access key for help files. Their specific use, however, is dependent upon the specific software under which they are used.

functional typography — a philosophy of design in which every element used does an efficient and necessary job.

furnish — in papermaking, the slurry mixture of fibers, water, dyes, chemicals and other additives which are distributed onto the Fourdrinier wire from the headbox. As approximately 93+ percent of the water is extracted through the wire, furnish is reclassified as true paper. *Related Term:* stock.

furniture — any line spacing material that is thicker or larger than 24 points; used primarily in relief (letterpress) printing. *Related Terms:* reglets.

furniture within furniture technique — a method of arranging furniture around a relief form when the form is of standard furniture dimension. *Related Term:* chaser method.

fuzz — loose fibers on a paper's surface. Undesirable for high quality printing, as they may be picked from the surface by the ink and tracked back into the ink chain. When they are removed, it can cause uninked blemishes in the printed area.

FWIW — a chat room or e-mail shortcut acronym; for what it's worth.

FYI (for your information) — an acronym often used in e-mail, posting, and chat, that rarely adds much to the information that follows it, and is often interpreted as a snide remark.

Gg

G — (1) the abbreviation for giga, or one billion, as in gigabyte. (2) The seventh letter of the English alphabet.

gain — (1) in electronics, the increase in signaling power as an audio signal is boosted by an electronic device. It is measured in decibels. (2) in printing, the increase in size of the printed dot when compared to either the negative or plate dot size. *Related Term:* dot gain.

gain tradeoff — image density variability due to ink film thickness. Thin ink films will produce low dot gain along with light solids. Thick ink films will produce darker solids along with increased dot gain.

galley — (1) the raw output of a phototypesetter, usually in the form of single columns of type on long sheets of photographic paper and used as a preliminary proof. (2) the final typeset (or image set) copy output to photographic paper or directly to film. (3) in metal relief composition, a shallow three-sided metal tray that holds the type prior to page makeup for printing.

galley proof — a rough proof. An impression of raw type taken on a galley proof press and usually not spaced out or fully assembled. It allows the typographer, proofreader or client to see if the job has been properly set and is free of errors.

gamma — a measure of contrast in photographic images. The ratio of the density range of a negative or reproduction to the density range of the original. A gamma of 1.0 means that the tones in the reproduction show the same separation as those in the original.

gamma correction — compressing or expanding the ranges of dark or light shades in an image through photographic or electronic manipulation. In digital imaging, gamma is usually the brightness level of the medium gray (1.0 density).

gamut — the greatest possible reproducible or detectable range of a device or system.

gamut chart — a color chart for comparing the gamut or color limits of any given display device, printer, ink set and substrate combination.

gamut mapping — the process of color matching in which differences in color gamuts between the source

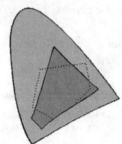

GAMUT MAP
The outer area is the total visible spectrum. The darker gray area represents the colors reproduced by a typical CRT or computer monitor. The dotted line indicates the colors reproduced by CYMK. Note that both systems produce less that the total possible number of colors, and that each produces some that the other can not.

device and the target device are taken into consideration.

gamut simulation — the process by which a device with a wider gamut can simulate the behavior of a device with a narrower gamut.

gang — to combine unrelated jobs on one printing plate in order to save costs and setup charges. *Related Term:* ganging.

gang run — a print run in which two or more print jobs are combined on one printing plate in order to economize.

gang shot — a halftone or line shot where more than one image is photomechanically or electronically created using only one exposure step. Typically used as a method of reducing both time and costs associated with expensive production steps. *Related Term:* ganging.

ganging — (1) in imposition, a grouping of different or identical forms arranged to print together in one impression. (2) in photographic gang printing, multiple images exposed as one unit; a gang shot. (3) in presswork, the production of two or more printed pieces or multiple copies of the same piece simultaneously on one sheet of paper. Also called 2-up, 3-up, etc. *Related Terms:* gang up; ganging; gang printing.

gap — (1) the opening on press cylinders where plates, blankets, stereotypes, etc. are attached. (2) the space between die-cut labels. The waste areas.

garbage — (1) in computing, unwanted information in memory. (2) bad input; corrupted data, etc.

gate — an alternate name for a foldout, particularly a gatefold.

gatefold — an oversize four-page insert or cover

with foldouts on either side, where both outside edges fold toward the center of the sheet, often in overlapping layers. Sometimes used to accommodate maps in books and in special products such as greeting cards, formal presentation pieces, etc.

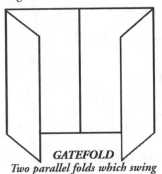

GATEFOLD
Two parallel folds which swing toward one another at a center point, opening like a "gate".

gateway — any computer system used for exchanging in-formation across in-compatible networks that use different pro-tocols. Used when you either log into an Internet site or when passing e-mail between different servers.

gather — (1) arranging signatures or pages in the order in which they will appear when finally bound. (2) the operation of combining the printed pages, sections or signatures of a book in the correct order for binding. *Related Term:* collating.

gathered signatures — multiple signatures which are stacked together in preparation for combining them into a larger pub-lication. Typical of both perfect and case binding.

GATHERED SIGNATURES
Typically used for larger books to combine individual sheet signatures to reach the required page count.

gauge pins — mechanical fingers that hold sheets in position when printing. They are generally affixed to the tympan of a clamshell platen letterpress press.

gauze — (1) a loosely woven cloth material that is sometimes embedded in the perfect binding adhesive to increase strength. (2) an alternate name for crash, the material used on the spine of casebound books.

GB — the abbreviation for gigabyte, meaning billion.

GBC binding — the General Binding Corporation trade name for its plastic comb binding. *Related Term:* comb binding.

GCR (gray component replacement) — a technique for reducing the amount of composite cyan, magenta and yellow in a given area and replacing them with

an appropriate level of black. *Related Term:* under color removal (UCR).

GDI (graphics device interface) — the part of computer operating systems that draws all the graphical objects, including common elements such as scroll bars and lines, to the screen. In Windows, GDI functions are handled by a program called gdi.exe that loads automatically when you start Windows.

gear streaks — parallel streaks of ink or dirt appearing across the printed sheet caused by the gear teeth on a printing cylinder.

gelatin — the vehicle into which the sensitive silver salts are mixed when making photographic emulsions. It hardens when cold and softens when immersed in developer with an elevated or inappropriate temperature.

GEM (Graphics Environment Manager) — an obsolete computer environment manager developed by Digital Research Corporation. Largely supplanted by Windows and the Mac OS.

general trade items — generally, a term which refers to items not specifically made for one customer, but that are generic in nature. For example, the printing of a general calender in bulk with the intent to imprint individual corporate names on them prior to distribution.

generation — the number of steps of distance between the original and the reproduction. A first-generation image is the original; second-generation is made from the original; third-generation is made from the second generation. Typically, using conventional reproduction methods, printing is fourth generation: copy (first), negative (second), plate (third), print (fourth).

GEnie — a minor online service run by General Electric.

genlock — process of synchronizing two video systems by using the sync signals from one to drive the other. Used in video editing.

genre — a category of a certain type of writing, such as horror, romance, mystery, science fiction, and so forth.

GEO — *Related Terms:* geosychronous; geostationary.

geostationary — describes a satellite in orbit 22,300 miles above the equator and which revolves around the earth with an angular velocity equal to that of the earth's rotation about its own axis. The satellite's position relative to the earth's surface is constant (geosynchronous), so little or no ground antenna tracking is needed, and it appears to hover over one spot on the earth's equator.

geosynchronous orbit — in broadcasting, a satellite about the earth at 23,300 miles. At this distance, the satellite hovers over a fixed geography on earth and can be used for transmission of signals at all times. A number of these satellites placed around the earth provide worldwide communication channels for public, private and military use. *Related Term:* geostationary.

ghost halftone — a halftone that has been screened to produce a very faint image.

ghosting — (1) in computer monitors, shadows and streaking due to drastic changes in on-screen intensity. It is common to see white or black shadows to the right of a solid bar drawn on the screen. (2) in printing, the phenomenon of a faint image in a solid image area on a printed sheet where it was not intended to appear. Usually caused by insufficiency in the inking system which cannot recover in a single press cycle.

GHz — gigahertz. 1 billion cycles per second. Giga is a prefix meaning one billion.

GIF (graphic Interchange Format) — an image compression algorithm developed in 1987 by CompuServe that facilitates the transfer of high quality images over a network. A GIF image can be of any resolution but only has 8-bit (256) color. While both GIF and JPEG have particular uses, GIF is generally losing ground to the JPEG format for photographs. JPEG files are not as compressed and take longer to transfer across a network. GIF is a licensed format which must be licensed. For this reason a new format (PNG) which has all of the GIF attributes plus more colors and compression is being used by increasing numbers of users. Many people believe that PNG will be the ultimate replacement to GIF. While GIF images have limited colors, JPEG files can contain up to 16 million colors, and look nearly as good as a photograph. *Related Terms:* interlaced GIF; transparent GIF; JPEG.

GIF Interlaced — interlaced GIFs appear first with poor resolution and then improve until the entire image has arrived. The technique gives a quick idea of what the entire image will look like while waiting for the remaining parts. If the Web browser doesn't support progressive display, it will still display interlaced GIFs once them have arrived in their entirety.

GIF, transparent (89a) — transparent GIFs are useful because they appear to blend in smoothly with the user's display, even if the user has set a background color that differs from what the developer expected. They do this by assigning one color to be transparent; if the Web browser supports transparency, that color will be replaced by the a browser's background color, whatever it may be. In software packages such as Photoshop, Fireworks, etc., more than one color can be made transparent.

GIF89a/GIF animation/multiblock GIF — one type of GIF format which allows a series of images to be displayed one after another or on top of each other to simulate change or motion.

gigabyte (GB) — a unit of measure equal to 1,073,741,824 bytes. Derived from the term meaning billion. Prepress digital processes often require several gigabytes of storage due to the memory requirement of their sophisticated application and high resolution color images. Workstations were built specifically to meet these types of intensive storage requirements. *Related Terms:* bit; byte; megabyte; terabyte.

GIGO — a computer related acronym; garbage in, garbage out. Slang for bad input that produces bad output.

gilding — the powdered metal which may be used on the edges of book pages. The process involves coating the edges with a thin varnish and dusting with metallic powder.

glass screen — the traditional halftone screen formed from two sheets of glass that are ruled with grooved parallel lines filled with an opaque material. The tow sheets are cemented together so that the lines are at right angles. This produces a regular, even pattern of very small transparent squares which serve as apertures. When exposed to film through this device, photographic originals are reduced to halftone dots for reproduction. Now used infrequently for production, but still a preferred matrix for creating the more common contact screens generally used. *Related Terms:* halftone screen; contact screen.

glass, ground — the frosted glass component at the rear or on the film plane of view and process cameras onto which the subject is projected whenever the shutter is opened. Used, with a magnifying glass, to check sharp focus and position on the film, prior to exposure.

glassine — the glossy, translucent paper used for windows in envelopes and as release paper for labels and decals. *Related Term:* pergamyn.

glaze — a buildup on rubber rollers or blankets that prevents proper adhesion and distribution of ink.

glitch — originally in computing, a minor problem with the operating system, hardware or software. The term has been adopted and is now in general use for describing all minor problems in any field.

gloss — (1) in papermaking, the characteristic of paper or varnish that reflects relatively large amounts of light. A high gloss tends to increase the apparent density range of the image. (2) in presswork, lithographic, and letterpress, an ink type that will dry without penetration and that is often used for printing on coated papers. *Related Term:* gloss ink.

gloss ink — *Related Term:* gloss.

gloss paper — *Related Terms:* gloss; art paper; enamel paper; slick paper.

glossary — a list of definitions.

glossy — (1) a photographic print made on a shiny surfaced paper. (2) a paper with high surface reflectivity.

glue lap — areas on a signature or box blank that are provided for gluing to hold the product together when folded.

gluing off — in bookbinding, the process of applying glue and affixing the contents into a hard binding.

glyph — the word "glyph" is used differently in different contexts. In modern computer operating systems, it is often defined as a shape in a font that is used to represent a character code on screen or paper. The most common example of a glyph is a letter, but the symbols and shapes in a font like ITC Zapf Dingbats are also glyphs. *Related Terms:* character; character encoding; keyboard layout.

GNU (Gnu's Not UNIX) — a software writing project of the Free Software Foundation to create a free version of the UNIX operating system.

golden ratio — the rule devised to give proportions of height to width when laying out text and illustrations to produce the most aesthetically pleasing result.

goldenrod — in stripping, a specially coated masking paper (usually yellow or gold colored); sheet of paper or plastic that serves as a base attachment and precise positioning of film negatives to make flats used for making printing plates and prepress proofs. When exposure openings are cut through it to reveal the negative images, the remainder serves as an exposure mask that does not transmit actinic light. *Related Terms:* imposition; masking sheet; flat.

Gopher — named after a college mascot, and for its ability to "go for" information, Gopher is a text-menu based distributed information service that makes available hierarchical collections of information across the Internet. Once one of the most popular search engines, the Web is superseding Gopher for document retrieval. *Related Terms:* Archie; FTP; wide area information servers.

Gothic — (1) improper name often given to a plain, sans serif typeface with lines of unvarying thickness. (2) a general type class whose characteristics are typified by German Black Letter, Old English and similar faces which mimic the 14th and 15th century Gothic period letter forms. *Related Terms:* Black Letter; text.

GOTHIC, TEXT OR BLACKLETTER
This classification of type typifies the early letter forms.

gouache — a water-based opaque ink used to color areas where exceptional opacity is required.

Gouraud shading — in 3D graphics, a method of shading the wire frame polygons that make up the images to add realism. Gouraud shading is a complex process using algorithms to create a color gradient. *Related Terms:* flat shading, Phong shading, shading

.gov — a domain name suffix used in Internet addresses that denotes a nonmilitary government institution.

gradated colors — a graded series of colors that changes progressively from one color to another or from light to dark or dark to light within the same color.

gradation — a smooth transition between black and white, one color and another or color and the lack of it. *Related Term:* gradient.

grade — (1) a general term used to distinguish among printing papers, but whose specific meaning depends on context. Grade can refer to the category, class, rating, price, use, finish, or brand of paper.

gradient — in graphics, having an area smoothly

blend from one color or tone to another, or from black to white, from color to no color or vice versa, etc.

graduation — *Related Term:* gradation.

grain — (1) in paper manufacture, the predominant direction of the fibers in a sheet. When folding, the direction of the grain is important. A sheet folded with the grain folds easily; a sheet folded across the grain does not. (2) in printing, paper is said to be "grain-long" if the grain direction parallels the long dimension of the sheet. The paper is referred to as "grain-short" if it parallels the short dimension of the sheet. (3) in book binding, grain direction of all papers used must run parallel to the book backbone. (4) in photography, crystals of metallic silver in a negative, transparency or print. Artifacts present in extreme enlargements from photographic negatives. In "grainy" images, crystals are apparent, reducing image quality. There is a direct relationship between film "speed" (ASA or ISO) and size of grain. Faster film has larger and more obvious grain. That is, an ISO 400 film is grainier than an ISO 100 film.

grain direction — (1) in papermaking, the general alignment of pulp fibers in the direction of the travel of the Fourdrinier wire. (2) in printing, the reference to paper as "grain long," meaning the direction parallels the long dimension of the sheet. "Grain short" means that the prevalent grain direction is parallel with the short dimension of the sheet. (3) in book binding, the grain direction of all papers used must run parallel to the book backbone. *Related Terms:* grain; grain long or grain short.

grain direction, across — a method of printing at right angles to or opposite the paper grain direction.

grain direction, against — folding or cutting paper at right angles to the paper grain in the direction of the sheet's fibers.

grain long or grain short — paper whose fibers parallel the long or short dimension of the sheet, respectively. *Related Term:* grain direction.

grained paper — a paper embossed or printed to resemble material such as wood, leather, or marble. *Related Term:* embossed finish.

graininess — the visual impression of the irregularly distributed silver grain clumps in a photographic image, or the ink film in a printed image. Particularly noticeable extreme scale enlargements from relatively small negatives.

graining — a mechanical or chemical process of roughing a lithographic plate's surface; ensures quality emulsion adhesion and aids in holding solution in the nonimage area during the press run.

gram — a weight unit in the metric system of measurement.

grammage — the European basis weight of paper expressed in grams per square meter. *Related Term:* substance weight.

grammalogue – any word represented by a symbol such as & (and) or @ (at).

grant — a sum of money paid in the form of a gift to finance a particular project, such as a book.

graphic — any illustrative element in a page layout, such as a photograph, illustration, icon, ruled line, or any other nontext element.

graphic arts — (1) the creation and reproduction of images by any of the several printing classes and processes. (2) the crafts, industries and professions related to designing and printing messages. *Related Term:* graphic communications.

graphic arts film — a high-contrast film whose emulsion responds to light on an all-or-nothing principle without the ability to satisfactorily reproduce gray tones.

graphic arts magnifier — lens built into a small stand and used to inspect copy, negatives, plates and printing. *Related Terms:* "loupe"; linen tester.

graphic communications — the reproduction industries, including printing, publishing, advertising, design. All disciplines that participate in the production and dissemination of text and images by printed or electronic means. *Related Term:* graphic arts.

graphic designer — a professional who conceives a design, plans how to produce it and may coordinate production of a printed piece.

graphic papers — *Related Term:* fine papers.

graphical user interface (GUI) — computer systems or operating environments which rely on the use of small graphics, called icons, to direct user input.

graphics — art and other visual elements such as rules, screens, charts, tables, photos, drawings and other visual elements used to make messages more clear.

graphics accelerator card – an additional circuit card fitted into a computer to reduce the time to produce an image on-screen. It has additional built-in memory and processor chip(s) which relieve the data load on the main CPU, producing screen redraws at a higher rate than the CPU can do alone.

graphics, design — noninventive elements used to enhance appearance such as gradient screen tints, reverses, bleeds, die cuts, etc.

graphics, info — manufactured graphics such as charts, graphs, maps, tables, exploded views, etc., used to clarify complex information such as statistics in financial and government reports.

graticule — an extremely fine incremented ruler which requires magnification to be properly read.

grave — an accent above a letter "e" **(é)**.

graver — the scribing tool used to incise lines and images into metal intaglio class printing plates. Usually associated with the engraving process. *Related Terms:* plate engraving; intaglio.

gravity delivery — the process in which sheets simply fall into place on the press delivery stack. *Related Terms:* chute delivery; tray delivery.

gravure — a (most often) rotary intaglio printing process where all image areas are etched as small cells or dots into (generally) a copper cylinder which is then chrome plated for durability. The cylinder is rotated through a trough of fluid printing ink which fills the etched cells. The surface is wiped clean by a doctor blade leaving ink only in the etched areas. The paper is then passed, under great pressure, between the etched cylinder and an impression cylinder, drawing the ink out by absorption and capillary action, sometimes with an electrostatic assist onto the substrate. Gravure printing produces consistent quality in long production runs. *Related Terms:* rotogravure; photogravure.

gravure cell — the correct name for the sunken portion of a gravure cylinder that holds ink during printing. *Related Term:* gravure well.

gravure well — *Related Term:* gravure cell.

gravure, conventional — a method of intaglio printing in which the ink-receptive cylinder cells are etched to vary in depth but not in area, thereby transferring different apparent ink densities to the substrate.

gravure, halftone — *Related Term:* halftone gravure.

gray balance — the values for the yellow, magenta, and cyan that are needed to produce a neutral gray when printed at a normal density. When gray balance is achieved, the separations are said to have correct color balance. Gray balance is determined through the use of a gray balance chart. *Related Term:* color balance; gray balance chart.

gray balance chart — a printed image consisting of near-neutral grid patterns of yellow, magenta, and cyan dot values. A halftone black gray scale is used as a reference to find the three-color neutral areas. The dot values making up these areas represent the gray balance requirements of the color separations. The gray balance chart should be produced locally under normal plant printing conditions. *Related Term:* gray balance.

gray component replacement (GCR) — an electronic color scanning capability in which the least dominant process color is replaced with an appropriate value of black in areas where yellow, magenta, and cyan overprint. Color variation on press is less serious when GCR is used. *Related Term:* under color removal (UCR).

gray scale — the range or shades of black. (1) a reflection or transmission film strip showing neutral tones in a range of graduated steps. It is exposed with originals during photography and used to time development, determine color balance, to measure density range, tone reproduction and print contrast. Gray scales can also be used to check focus and resolution. (2) in prepress scanning, the range of luminance values for evaluating shading from white to black. Frequently used as a measure of ability to capture halftone images. More levels produce better results, but with correspondingly larger memory requirements. *Related Terms:* gray wedge; step tablet; step wedge; cameraman's sensitivity guide; Stouffer Gray Scale.

gray wedge — *Related Term:* gray scale.

grayness — in the Preucil Ink Evaluation System, the lowest of the three (red, green, and blue) densities expressed as a percentage of the highest.

greeking — (1) body text that is made illegible when viewed at 12 points or below, for the purpose of speeding screen redraw or creating a rough layout. (2) nonsensical character combinations and spaces used to simulate text copy blocks on mock-ups. (3) a method of overcoming the slowness of high resolution displays by replacing image detail with a fixed pattern.

Green Book — the compact disc standard created to work on CD-I players. It provides that CD-I players can play back audio CDs, but no other CD-playing platform can make use of Green Book CDs. *Related Terms:* CD-I; Red Book; Yellow Book.

grep — a UNIX shorthand expression for "get regular expression." A tool to find and replace variable names or look for patterns such as a Social Security numbers. While originally a UNIX tool, there are now grep tools on virtually every platform.

grid — (1) in computing, a pattern of nonprinting guidelines on paste-up board or a computer screen used to help align copy. (2) in design, a systematic division of a page into areas to enable and ensure consistency. The grid acts as a measuring guide and shows text illustrations and trim sizes.

grin through — in screen printing, the apparent loss of color in images whenever the fabric is stretched or distorted sufficiently to allow it to show through the printed image.

gripper — metal fingers or bars that clamp onto the leading edge of a press sheet to accurately guide it through the printing press.

gripper edge — the lead edge of a sheet held by the grippers, thus going first through a sheet-fed press. Allowance must be made for grippers when calculating sheet size, as this area cannot receive ink. *Related Terms:* lead edge; feeding edge; gripper margin/edge.

gripper margin — the unprintable blank edge on which the paper is gripped as it passes through a printing press, usually measuring a half inch or less.

gross weight — the total package weight, including product and wrapping or packaging materials. The weight of the produce, alone, is known as net weight.

grotesk — the European name for true Gothic type.

groundwood paper — any newsprint and other inexpensive paper made from pulp created when wood chips are ground mechanically rather than refined chemically.

groundwood pulp — paper pulp made by grinding trees into fibers without removing lignin. *Related Term:* mechanical wood pulp.

gsm (grams per square meter) — (1) in paper manufacturing and sales, the unit of measurement for paper weight in metric countries. Comparable to our substance # or "#" designations. (2) in communications, the abbreviation for Groupe Speciale Mobile, the global standard for mobile communications and widely used in Europe for cellular communications. It's audio encoding subset is used by computer users because its data compression and decompression techniques are also being used for Web-phone communication and encoding WAV and AIFF files. *Related Terms:* AIFF, codec, WAV.

guard — (1) a metal or electrical detection device on equipment to prevent operation whenever safety covers are not engaged. (2) the safety devices designed to prevent injury and minimize the possibility. (3) a narrow strip of paper or linen pasted to a single leaf to allow sewing into a section for binding.

guard bars — the bars which are at both ends and center of a UPC and EAN symbol. They provide reference points for scanning. *Related Term:* quiet zone.

guardian — an authorization and authentication scheme developed by Network Solutions that helps protect domain name records, contact records and host records from unauthorized updates. Guardian is available free of charge and helps support secure registration transactions in an automated environment. It also provides flexible security mechanisms that can accommodate changes in organizations, personnel, and security needs. *Related Terms:* authorization; authentication.

GUI (graphical user interface) — a screen controller that lets users interact via icons and a pointer instead of by typing in text at a command line prompt such as in DOS and UNIX. Popular GUIs are OpenWindows, Microsoft Windows 95/98 and Apple's Mac OS. *Related Term:* command-line interface.

guide edge — the side of a sheet at a right angle to the gripper edge that is used to control the lateral (side-to-side) position of the sheet as it travels through the press or folder. *Related Term:* guide side.

guide side — the control area on the press that controls the position of the sheet during printing. It is usually the side closest to the press operator. *Related Terms:* guide edge; jogger; jogging edge.

guillotine cutter — (1) in production, a manual or electronic device with a heavy, sloping blade that descends to a table or bed and slices through a stack of paper. The blade drops from above the material, similar to a guillotine. (2) in photoengraving and metal platemaking, a heavy steel knife, operated by electricity or foot treadle, that trims sheets of copper, zinc, or magnesium and is used to cut the excess from electrotype and stereotype casts.

gum arabic — a liquid used on presensitized lithographic plates to prevent oxidation and minor scratches.

gummed holland — a tape material with embedded water-soluble glue used to cover the stitched spine and backbone of a soft cover book.

gummed paper — *Related Terms:* label paper; moisture-activated paper; lick and stick.

gumming — applying a thin coating of gum arabic solution to the printing plate after it is developed or an image is placed on it. This is applied to prevent deterioration of the clear area of the metal plate via oxidation.

gusset — the expandable side and end portions of envelopes, bags, folders and portfolios enabling the insertion of bulky materials.

GUSSETED ENVELOPE
Note the expandable sides to allow for bulky items. The side pleats expand as needed.

Gutenberg, Johann — German goldsmith who is given credit for inventing printing by his invention of reusable, movable, metal type and letterpress printing (c. 1455). He developed the system of punches and molds which made possible the creation of many identical letters. He is credited with producing the first printed Bible, which had 42 lines per page. It is commonly referred to as either the Gutenberg Bible or the 42-line bible which, even by modern standards, is considered a typographic work of art. He never realized profit from his invention, as creditors foreclosed on his loans, taking type, presses, paper, ink and all other relevant materials.

gutter — (1) in bookbinding, the inside margin between facing pages or the margin at the binding edge. (2) in typography, the central blank area between left and right pages or between columns of type. *Related Term:* gutter margin.

gutter margin — *Related Term:* gutter.

GZip — a free compression program similar to ZIP compression, commonly available as a UNIX command for file compression. It is also available for MS-DOS, compresses files and appends either ".z" or ".gz" to the file name. In VRML 2.0, the file can be "GZIP'd," and the file will automatically decompress for rendering on the client machine. Compressed files reduce download time. *Related Terms:* VRML; ZIP.

Hh

H&J (hyphenation and justification) — an algorithm that determines line endings and spacing, used by typesetting systems and page layout programs. Sophisticated programs follow a standard dictionary or special dictionary created by the user for unusual or field-specific terms, and allow adjustment of the H&J parameters.

hacker — a person who claims to be neither dangerous nor a security risk and delights in gaining access and having an unauthorized, intimate understanding of the internal workings of any computer system, networks in particular. The term is often misused in a perjorative context, where "cracker" would be the correct term. The practice is a felony under federal law.

hacker ethic — an informal set of moral principles common to the first generation hacker community which champions the concept that all technical information should, in principle, be freely available to all and destroying, altering, or moving data in a way that could cause injury or expense to others is always unethical.

hacker jargon — a lingo used for expressiveness by members of the hacker community.

HAGO (have a good one) — a shorthand appended to a comment written in an online forum or e-mail.

hair space — the narrowest space used for separating letters or numbers when setting type. *Related Term:* thin space.

hairline — (1) the thinnest rule that can be reproduced in printing. Generally approximately 1/2 to 1/4 point or about the width of a hair (1/100 inch). (2) a nonspecific term for the thinnest strokes in a typeface. *Related Term:* hairline rule.

hairline register — to register color separations within one-half (0.5) point.

hairline rule — *Related Term:* hairline.

halation — (1) a photographic term for light that spreads beyond the sharp definition of an image. This phenomenon and the resulting noticeable halo may be caused by light that is scattered by the emulsion or by internal reflections within the film base. Halation diffuses image detail so that it reproduces erratically in printing. It is particularly prevalent in the highlight areas of film negatives. (2) poor contact between the negative flat and the negative-working plate during platemaking that distorts the image by allowing light to expose nonimage areas, causing a blurred effect. Dirt, masking materials, or tape prevents proper contact between the film negative and the plate. In positive platemaking, dot loss occurs. *Related Term:* antihalation backing.

half bound — a casebound book where the spine, and possibly the corners, are covered with leather and the remainder is covered with alternative materials such as cloth, or leather of a different color. *Related Term:* full bound.

half duplex — a mode for transmitting data in either of two directions but only in one direction at a time, e.g., a two-wire circuit. *Related Terms:* duplex; full duplex; KBPS; bisynchronous.

half sheet work — a type of imposition layout that uses one plate to print both sides of a sheet in two impressions. After the face side of the sheet is printed, the sheets are turned over (left to right) for the second printing. *Related Term:* work and turn.

half title — the title of a book found on the page that precedes the title page. *Related Term:* bastard title.

half up — artwork one and a half times the size at which it will be reproduced.

half web — a web press whose width and cutoff allow printing eight 8½ x 11 pages on one press sheet. *Related Term:* dinkey.

half-fold — a one-page brochure that is folded once, forming two halves.

halftone — a reproduction of continuous tone artwork or a photograph with tone value and detail represented by a pattern of equally spaced dots of variable size. The smaller dots are in highlight areas and the larger dots in the shadow areas. When printed, the dots merge to give the illusion of continuous tone. *Related Terms:* stochastic screening; crossline screen; FM screen.

halftone copy — any type of camera copy that has variations in tone with some areas lighter or darker

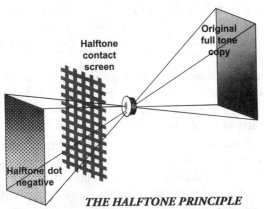

THE HALFTONE PRINCIPLE
A screen with tiny apertures is placed between the projected image from the camera lens and the film. It creates tonally reversed values in the form of small spots or dots on the film. Subsequent printing onto plates reverses the tones to their approximate original values when perceived from normal viewing distances.

than others and cannot be printed as simple line copy. Photographs, wash drawings, oil paintings, etc., are examples. *Related Terms:* continuous tone; halftone.

halftone dot conversion — breaking continuous tone images into high-contrast dots of various sizes, by use of a halftone screen, so they can be reproduced on a printing press. *Related Term:* halftone photo conversion.

halftone exposure computer — generally a hand-held, nonelectronic device used to assist in determining halftone exposures.

halftone gravure — an intaglio printing process in which the engraved cylinder cells vary both in area and in depth. *Related Term:* lateral hard dot process; standard gravure.

halftone grid pattern — a term sometimes used to refer to the shapes of halftone screen dots. Some common shapes are linear, elliptical and round. Different shapes cause different effects in the final output. Newer technologies have spawned the creation of different shapes; Adobe PhotoShop allows an operator a specified diamond-shaped dot which is supported by some of the newer imagesetters.

halftone photo conversion — the process of breaking continuous-tone images into high-contrast dots of varying sizes so they can be reproduced on a printing press. *Related Term:* halftone; halftone dot conversion.

halftone photography — the process used to reproduce a continuous tone illustration by use of conven-

tional film and a halftone screen or the use of autoscreen film. As the light rays pass through the screen apertures, they form a dot pattern on the negative film. This pattern creates the optical illusion of a continuous tone photograph when the plate made from the negative prints the image. *Related Terms:* halftone; halftone screen; halftone copy; vignette.

halftone resolution — the spacing of the dot matrix in a halftone image, usually measured in lines per inch (lpi). *Related Terms:* frequency; halftone frequency.

halftone screen — a fixed frequency vignetted glass plate or film matrix used to create a pattern of dots of

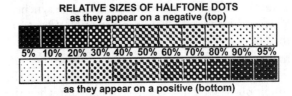

RELATIVE SIZES OF HALFTONE DOTS
as they appear on a negative (top)

5% 10% 20% 30% 40% 50% 60% 70% 80% 90% 95%

as they appear on a positive (bottom)

different sizes in negatives or prints to represent variable tonal densities in an image. Halftone screens with higher screen rulings (e.g., 200 lines/in., as opposed to 55 lines/in.) produce higher resolution images. Practical resolution is dependent upon the printing process and substrate to be employed. *Related Terms:* contact screen; crossline screen; stochastic screening; lpi; gradations.

halftone tint — an erroneous term applied to an area covered with a uniform dot pattern to produce an even tone or color. Halftone screens cannot produce the uniformity necessary for this effect. *Related Term:* mechanical screen tint.

hand composition — a method of setting type manually from a type case with a composing stick using precast letters and spaces.

hand cut film — screen printing stencil film that is hand cut with a knife. *Related Term:* hand cut stencil.

hand cut stencil — a screen printing stencil prepared by removing the printing image areas manually from base or support material. *Related Term:* hand cut film; direct photo stencil.

hand held scanner — an almost obsolete technology due to declining prices of conventional scanners. A small, usually limited resolution scanner which can be passed over stationary artwork while being held in the hand. Relatively inexpensive, but very limited in quality. Available in both color and black grayscale versions. *Related Terms:* flatbed scanner; sheetfed scanner; drum scanner; scanner; wand scanner.

hand laser gun — *Related Term:* laser scanner.

hand mechanical type — type composition prepared with template devices and by freehand techniques using a technical inking pen. *Related Term:* Leroy lettering.

hand work — any work which is more efficiently or economically accomplished by use of hand labor, such as some collating, pop-out assembly, book binding repairs, restorations, etc.

handle — in on-line chat the name you go by is called your handle. *Related Terms:* avatar; IRC.

handshake — an exchange of signals between communications devices to verify that they are synchronized and ready to exchange information. Occurs in such devices as modems and facsimile machines.

hang — (1) in typography, to place characters outside the left margin. (2) in telecommunications, the condition when a modem fails to hang up when the communications software is closed.

hanger sheets — the major part of the packing on a platen press.

hanging indent — the setting of type in paragraphs where the first line extends into the left margin and the remaining lines are indented.

hanging punctuation — text punctuation that extends beyond the normal type measure. An old technique for making the type characters appear to have better alignment. It was abandoned in favor of fast mechanical composition, but is reappearing in better typography with the availability of computing power.

hard bind — *Related Term:* case bind.

hard copy — the printed copy of electronic text transmissions that are to be typeset. Generally used as a backup if the electronic file is corrupted, or to doublecheck and spot errors.

hard copy printer — an impact, laser or thermal imaging machine that can be used to produce proofs of artwork and/or keyboarded copy for typesetting.

hard copy proof — a proof copy made on an impact, laser or other printer of the typeset file for the sole purpose of proofreading prior to film, plate or paper imaging.

hard disk — a rigid disk sealed inside an airtight transport mechanism. Information stored may be accessed more rapidly than on floppy disks and far greater amounts of data may be stored. Often referred to as Winchester disks. *Related Term:* disk.

hard dots — halftone dots that are fully formed and have no noticeable halation around the edges. They appear as opaque dark dots in halftone film negatives.

hard hyphen — a permanent hyphen that is manually inserted in a word, as opposed to a soft hyphen, which is inserted using a software command and which would go away if the text were to reflow.

hard proof — a proof on tangible output, such as laser paper, film, or photographic paper, as opposed to a soft proof, which is an image on a computer monitor. *Related Terms:* hard copy; hard copy proof.

hard returns — line end return codes in word processing programs that are usually mandatory, such as returns at the end of paragraphs.

hard sized — paper treated with a relatively large amount of sizing to make it especially water resistant, as compared to slack sized.

hardback — generally any book bound with sturdy cardboard. Specifically, a casebound book with a separate stiff board cover. *Related Term:* hardcover, hardbound, and case bound.

hard-bound — *Related Terms:* hardcover; case bound.

hardcover — a book bound with a case of binder's board.

hardcover — a book bound with boards. *Related Term:* case-bound, hard-bound, and hardback.

hardening — the stage in ink drying when the vehicle has solidified completely on paper surface and will not transfer.

hardware — the keyboard, monitor, chips and other equipment components of a computer system, as opposed to software (programs). *Related Terms:* software.

hardware abstraction layer (HAL) — a component of an operating system that allows programmers to write applications and game titles with device-independence. *Related Terms:* API; device independence; Direct3D; DirectX.

Hayes-compatible — an early term used for modems which used the AT command set. A more relevant and modern measure of compatibility is which international CCITT "V." standard your modem conforms to. *Related Terms:* AT commands; CCITT; V. standards.

haze — the milky appearance in improperly processed film caused by residual undeveloped emulsion. Most often caused in inadequate fixing.

HDTV — a digital standard for television which

will double the number of scanning lines to over 1000, and change the overall aspect ratio of the picture from the current 4:3 to 16:9. The system was mandated by the FCC and began broadcasting in larger cities in November 1998. The remaining cities will be phased into full operation by 2006. Conventional analog channels currently used will ultimately be surrendered to the federal government for reassignment. Older sets will be usable with the acquisition of a converter box.

head — (1) in book publishing, a line of display type signifying the title of a work or conveying crucial information. (2) generally, the top copy and/or margins of a page, book, or printing form. (3) in the digital world, a compact device that reads, scans, writes, or records data on a surface medium, particularly the fixed or moving electromagnetic elements used for reading and writing data on magnetic tapes, disks and drums. (4) in publishing, a shortened name for headline. The display-size text at the top of an article or story. *Related Terms:* foot; footer; head; head margin; headline.

head end — (1) in cablecasting, a head end is the originating point of a signal in a cable television system. Head-end equipment receives satellite and local broadcast TV signals and converts them to a form that can travel down coaxial cable to subscribers. (2) in papermaking, the starting end of the Fourdrinier machine where the stock is spread on the wire. *Related Terms:* head-end controller; router; modulation; demodulation.

head margin — the white space above the first line on a typed or printed page. *Related Terms:* foot; footer.

head stop — a mechanical gate or finger arrangement that stops the paper on the registration unit of a press just before the gripper fingers pull the sheet through the printing unit.

headband – the colorful beads of thread and reinforcing cloth used as decoration at the top and bottom of the backbone of a book.

header — (1) in data transmissions, the portion of a packet, preceding the actual data, containing source and destination addresses, error checking and other fields. (2) on the Internet, a header is the part of an electronic mail message that precedes the body of a message and contains information about the message originator and time stamp. (3) in publishing, a sometimes used slang term for a headline that appears at the top of a page.

heading — a hierarchical paragraph type that is displayed in a large, bold typeface. The size of a heading is related to its level: Heading 1 is the largest, heading 2, the next largest, and so on. Use headings to name pages and parts of pages.

headletter — any type designed especially for and used for headlines.

headline — in theory, the most important line of type in a piece of printing, designed to entice those who see it to read further or to provide a summary of the content of the copy which follows.

headline font — a font that has been designed to look good at large point sizes for use in headlines. Headline fonts generally do not contain a complete set of characters since they do not require a full set of special symbols and punctuation. *Related Term:* headline type.

headline schedule — a chart showing all the styles and sizes of headlines used by a publication.

headline type – *Related Term:* headlettert; headline font.

headliner — a manually operated photographic composition machine made exclusively to set display type, often on a transparent tape ready for pasting up.

head-to-head imposition — in prepress stripping, the assembly of negatives so that the top of each page butts up against another.

head-to-tail imposition — in prepress stripping, the assembly of negatives so that the tail of one page abutts the top of another.

heat sensitizing — one method of making a photo polymer plate very sensitive to ultraviolet light. Controlled heat is used to drive out the oxygen in the plate. *Related Terms:* carbon dioxide conditioning.

heat set ink — ink that dries by using radiant heat.

heat sublimation ink — a special ink used in heat transfer imaging systems. The ink turns directly to a vapor without a liquid state (sublimates) when heat and pressure are applied. It is most often used with textiles, particularly 50% cotton and synthetic blends.

heat transfer — *Related Terms:* heat sublimation ink; sublimation.

heat transfer paper — the substrate used in thermal transfer printing. The design is first printed on it with inks containing dispersed sublimate dyes. Under applied heat and pressure, the paper is placed in contact with another substrate to which the design adheres. *Related Term:* sublimation; thermal transfer printing; heat transfer ink.

heatset web — a web press equipped with ovens to make ink dry faster, thus able to print coated paper and at higher production speeds.

HelioKlischograph — a brand of cylinder engraving machine used to make halftone gravure cylinders, as opposed to chemical etching.

helium/neon laser — a red laser light source in which helium-neon gas that emits coherent red light at a wavelength of 633 nm to expose red sensitive film, paper or plates. Also used in some scanners for electronic dot generation. *Related Term:* He-Ne.

help — in computing, on-screen assistance screens which explain problems and propose solutions. In modern software, many such assistants are now interactive and by appropriate mouse clicks can be directed to more specific, rather than general, answers.

helper applications — special applications or applets that enable the Web browser to handle graphics, sounds, and multimedia. Browsers are basically HTML viewers, and anything else is a luxury that most browsers ignore. Helper applications are usually freeware or shareware but do not come with the browser software and users must hunt down the ones wanted, install and configure them so the browser will launch the particular helper application when it encounters a multimedia or other file. The helper applications' big advantage is that they can be added to the MIME standard to cover new file types, as necessary, without changing your browser. *Related Terms:* hypertext; MIME.

Helvetica — a popular sans serif typeface.

He-Ne — a scientific abbreviation for the gases in a for helium-neon laser.

Hering theory — a somewhat discredited theory of color vision proposed during the nineteenth century by the German physiologist and psychologist Ewald Hering. He regarded yellow and blue, red and green, and black and white as pairs of opponent colors, where one member of each pair is perceived at a time.

hertz (Hz) — the measure of frequency measurement. A hertz is equal to one cycle per second. Named in honor of Heinrich Hertz, first to detect such waves in 1883.

heuristics — a method of analyzing outcome through comparison to previously recognized patterns. For example, an antiviral program, familiar with behaviors typical of viruses (such as deleting files in sequence), could use heuristics to identify unknown virus strains by their behavior.

Hexachrome — the recently introduced six-color process printing system from Pantone, Inc.

hexadecimal — (1) a numerical notation system using a base of sixteen including numbers 0 through 9 and letters A through F. It works extremely well in computing environments. For instance, a single byte of information can be represented as eight bits (10011101), decimal numerals (913), or simplified to hex (9D). In hex, every byte can be shown as two hexadecimal characters. (2) as it relates to the Web, hexadecimal is the system used to specify colors in HTML: the hexadecimal equivalent of white is FFFFFF; black is 000000. *Related Terms:* bit; byte; RGB.

HFC network (hybrid fiber-coax network) —networks that combine both optical-fiber and coaxial cable lines. Optical fiber runs from the cable head end to neighborhoods of 500 to 2,000 subscribers. Coaxial cable runs from the optical-fiber feeders to each subscriber. Hybrid networks provide many of fiber's reliability and bandwidth benefits at a lower cost than a pure fiber network. *Related Terms:* optical fiber (fiber-optic cable), coaxial cable, head end, frequency spectrum

Hi-8 — an S-video tape format developed by Sony. Higher quality version of composite 8mm tape often used in consumer camcorders and superior to S-VHS.

HIBC (health industry bar code) — in optical reading technology, the symbology and label format for use by the health care industry.

hickey — a donut-shaped spot or imperfection in printing caused by small particles of ink or paper attached to the plate or blanket which appear on the which appear on the printed sheet as a dark spot surrounded by an paper-colored halo. It is most visible in areas of heavy ink coverage.

hidden directory — a folder in a FrontPage web with a name beginning with an underscore character, as in _hidden. By default, pages and files in hidden folders cannot be viewed from the FrontPage Explorer.

hidden field — in Microsoft FrontPage, a form field that is invisible to the user but that supplies data to the form handler. Each hidden field is implemented as a name-value pair. When the form is submitted by the user, its hidden fields are passed to the form-handler along with name-value pairs for each visible form field. You add hidden fields to FrontPage by clicking Add in the Form Properties dialog box.

hierarchy — a ranking system where each file is subordinate to the one above. In computing, directories and sub directories, etc. In USENET, the categorization of newsgroups and the way they are internally ranked.

hieroglyphics — a system of writing characters created by the Egyptian people using stylized pictures and symbols. Not an alphabet nor strictly phonetically based.

hi-fi color — process color reproduction using from six to twelve individual separations as opposed to the traditional four-color technique. It extends the tonal gamut of reproducibility for printing inks. Originally process specific, it now tends to be applied to any color specification and printing system that enhances the traditional four-color process system, such as Hexachrome.

high key photo — a photographic or printed image in which the main interest area lies in the highlight end of the scale.

high sierra format — a standard format for placing files and directories on CD-ROM, revised and adopted by the International Standards Organization as ISO 9660.

high-bulk paper — a paper made relatively thick in proportion to its basis weight, thus yielding fewer pages per inch, consequently higher volume in the finished product.

high-contrast — *Related Term:* high contrast image.

high-contrast emulsion — a photographic material that changes in developed density markedly with only slight changes in exposure. *Related Term:* graphic arts film.

high-contrast image — the relationship of highlights to shadows in continuous-tone or halftone photography, i.e., the darker portions of the image are uniformly very dark; the lighter portions are uniformly very light; and the midtone range is nonexistent or very small.

highlight — the lightest or whitest area of an original or reproduction, represented by the densest portion of a continuous-tone negative and by the smallest dot formation on a halftone negative and the printing plate.

highlight dots — the small dots on a halftone representing the lightest or whitest areas (highlight) of a photo or artwork; on a film negative, the darkest areas.

highlight mask — generally, a light negative image that is registered with a normal density continuous-tone negative for the purpose of enhancing highlight tone contrast.

high-resolution graphics — a system that provides a greater resolution than the standard 525-line video image; allows more detailed graphics to be seen clearly; often refers to a system with a resolution of 1,000 lines. *Related Term:* HDTV.

hinge bar — the part on a hand-operated screen printing unit that serves as the hinge support for the frame.

hinged cover — in perfect binding, a score placed on the cover, close to the spine, which forces the cover to fold away from the glue binding.

hints — the mathematical instructions added to digital fonts to make them sharp at all sizes and on display devices of different resolutions.

histogram — a visual frequency distribution bar graph that shows the history of what has happened at one point in a process. Useful in photo editing and adjustment programs.

history list — a drop-down menu in a Web browser that contains a log of the latest document titles and URLs you have visited during your Web session. It's a convenience feature that lets you jump back to where you've been without having to click repeatedly on the Back button.

hit — (1) a term erroneously used to indicate the number of visitors to a Web site. (2) properly used, a hit is a request made to a Web server. For example, if you look at a Web page that contains ten GIF files, one person visiting one page will make 11 hits on the server: one for the page, and ten for the graphics on the page.

HLS — a three-coordinate color model based on hue, lightness (or luminance) and saturation. *Related Term:* HLV.

hold — A status code in Web management for a domain name that is no longer included in the zone files for the top level domain. *Related Term:* deactivation.

holding fee — a supplemental charge made to clients who keep a photographer or photographs longer than agreed to.

holding lines — small red or black marks made on a paste-up to serve as guides for mounting halftone negatives in the stripping operation; are carried as an image on the film and are covered prior to platemaking.

holdout — *Related Term:* ink holdout.

Holland cloth — a material used as stamp.

Hollywood syndrome — the tendency to base one's video teleconferencing behavior on a model that includes a highly polished presentation rather than interaction, and the use of fast-paced visuals for effect rather than substance.

holography — (1) in printing, a technique using a laser or microcode to emboss images on a thin piece of film or foil. The resulting (usually foil) image has true three-dimensional properties. (2) in photography, a lensless process using a split beam laser. One beam directly exposes a photographic plate. The other reflects from the subject to expose the same plate. The two beams interact and cause interference patterns on the film, which when developed can be used to recreate a three-dimensional image of the subject. No discernable image is present on the developed film. It has the appearances of having swirls and blotches of exposure. In image re-creation, which (nearly) always involves a laser, the original scene appears as an image in space, behind the piece of holographic film.

holy war — cyclical arguments that involve certain basic tenets of faith, about which one cannot disagree without setting one off. For example: IBM PCs are inherently superior to Macintosh's (or vice versa).

home bases — the personalized start pages pioneered by the portals.

home page — the opening collection of text and graphics on a Web site and the site reference and navigation aid. Most sites have one or more home pages that you can use for orientation. A home page serves as the site's introduction, starting point, and guide.

homespun finish — *Related Term:* crash finish.

hood — in publication work, a 3-sided border (left, top, right) over a headline or column head. Often of a percentage screen tint.

horizontal bar code — a bar code or symbol presented in such a manner that its overall length dimension is parallel to the horizon. The bars are presented in an array which look like a picket fence. *Related Term:* vertical bar code.

horizontal blanking interval — the period of time when a scanning process is moving from the end of one horizontal line to the start of the next line. There are 525 lines in an NTSC picture and such an interval is included in the signals for painting each one. *Related Term:* vertical blanking interval.

horizontal camera — *Related Term:* camera, horizontal.

horizontal format — a page or image that is in landscape orientation, to be viewed horizontally.

horizontal line — not always, literally, a line but a horizontal graphic element on a World Wide Web page often used to separate sections of the page.

horizontal makeup — the arrangement of stories across multiple columns rather than vertically.

horizontal process camera — *Related Term:* camera, horizontal.

horizontal resolution — the specification of resolution in the horizontal direction, meaning the ability of the system to reproduce closely spaced vertical lines.

host — usually refers to a large computer which acts as the central connection for a network. May be either a server or a mainframe.

host (name server) — (1) on the Internet, a term which refers to name servers, the computers that have both the software and the data required to interpret domain names into Internet protocol (IP) numbers. (2) generally, computers (not terminals) that process data, act as data sources or destinations in a communications network. *Related Term:* name server.

host form (host template) — the electronic form used to register a new host (name server) or modify information for an existing host. Formerly referred to as the host template. *Related Terms:* host (name server).

hostmaster — a term commonly used by organizations for role accounts that handle e-mail related to network administration responsibilities. *Related Term:* role account.

hot list — a collection of site addresses or URLs; usually collected by Web enthusiasts and posted on their Web sites, often as hyperlinks, which lets the user simply "click on" the name of the site in order to be taken there. Most hot list sites share a common thread or they can be just a collection of Webmasters' favorite sites. *Related Terms:* URL; bookmark.

hot melt ink jet — a type of ink jet printing using plasticized hot-melt inks which are melted in minute quantities, ejected toward the substrate and solidify very quickly on contact. Because of their plastic nature and lack of absorption into the substrate, they produce excellent dot shape, contrast, edge definition, and holdout characteristics.

hot metal — type produced by casting molten metal into molds to form individual characters, slugs, borders or rules. *Related Terms:* cold type; Linotype; Monotype; Intertype; Ludlow; Elrod.

hot plugging — the ability to add and remove devices to a computer while the computer is running and have the operating system automatically recognize the change. It is supported in Windows 98 through two new external bus standards: Universal Serial Bus (USB) and FireWire. Hot plugging is also a feature of PCMCIA (PC cards).

hot type — the slang expression for type produced by casting hot metal types such as those from Linotype, Intertype, Monotype, and Ludlow and sometimes handset foundry type. *Related Term:* hot metal.

hot type composition — the preparation of any printing form used to transfer multiple images from a raised surface; examples include Linotype, Ludlow type, foundry type, and wooden type.

hotBot — a search engine which exploits parallel computing technology to achieve scalable, supercomputer-class performance from clusters of workstations and high-speed local-area networks (LANs). The technology enables a low-cost system, with unlimited scalability: it can increase the performance or database size simply by adding more machines, disks, or memory to the configuration to scale up with the rapid growth of the Internet.

hotspot — a graphically defined area in an image that contains a hyperlink. Images with defined hotspots are called image maps. In browsers, hotspots are invisible, but users can tell that a hotspot is present by the changing message of the pointer as it passes over the defined areas.

house list — a list of customers, subscriber or other names developed within an organization and not acquired from an outside list provider.

house organ — an in-house newsletter.

house sheet — any general-use paper kept in stock by a printer and suitable for a wide variety of printing jobs. *Related Term:* floor sheet.

house style — a publisher's adopted preferences for spelling, punctuation, hyphenation and indentation

used in all or specific works of that publishing house to ensure consistent typesetting.

HS (high speed) — a modem indicator light that tells you the modem is prepared to transfer at its maximum capacity.

HSB (hue, saturation, brightness) — a method of describing color by expressing relative levels of hue (the pigment), saturation (the amount of pigment), and brightness (the amount of white included) as percentages. *Related Terms:* HSV; HSL.

HSL — hue, saturation, and lightness.

HSV — a three-coordinate color model based on hue, saturation and value. *Related Terms:* HSL; HSV; HSB.

HTML (hypertext markup language) — the coding, or tagging, method used to format documents for the World Wide Web. The Web browser interprets the HTML commands embedded in a page to format the page's text and graphic components. HTML commands can interpret text formatting for bold and italic, bulleted and numbered lists, type color, and headlines in various sizes, etc. It also has the ability to include graphics and other nontext elements. *Related Terms:* browser, HTML 3.0, hypertext; URL; VRML.

HTML 4 — an extension of the original HTML specification, which was severely limited. HTML allows documents to contain only one column of text, very little formatting, and few graphics. HTML 3.0 enhances the graphics capabilities and permits type flow around figures, the inclusion of tables and frames, etc. HTML 3.0 must have a browser that specifically supports its enhancements. Both Netscape and Microsoft support it, although differently and in varying degrees. *Related Terms:* HTML; VRML; browser.

HTML attribute — a name-value pair used within an HTML tag to assign additional properties to the object being defined, for example <p> </p>. Most Web page authoring programs assign some attributes automatically when you create an object such as a paragraph or image map. You can assign other attributes by editing the file they create using a normal text-based word processor.

http (hypertext transfer protocol) — the letters which usually precede ".www" at the beginning of every URL. These tell the Web server what method the browser will use to communicate with it. On the World Wide Web, both sender and receiver use this protocol to transfer the document from the server to your system. *Related Terms:* hypertext transfer protocol; IP; TCP; FTP; Telnet; SMTP; CGI; server; URL.

HTTPS (hypertext transfer protocol secure) — a type of server software which provides the ability for "secure" transactions to take place on the World Wide Web. If a Web site is running off an HTTPS server you can type in HTTPS instead of HTTP in the URL section of your browser to enter into the "secured mode." *Related Terms:* proxy server; secure server; commerce server.

hub — (1) in networking, a type of hardware used to network computers together, generally using an Ethernet connection. It is a central, common wiring point from which information can flow to any other computer on the network. (2) in the case of an Internet environment, it is a focus of activity, not just a pass-through. Successful hubs are more narrowly organized and must focus on content and commerce appropriate to one particular audience's interests, i.e., a profession, recreational pursuits, arts, common interest issues, etc.

hue — (1) the attribute of color perception by means of which an object is judged to be red, blue, green, purple, etc. (2) the degree of black in a color; grayness.

hue error — in the Preucil Ink Evaluation System, the amount of the largest unwanted absorption of a process ink, expressed as a percentage of the wanted absorption content.

hue, primary — (1) in the additive transmissive system, any three hues, normally a red, a green, and a blue, so selected from the spectral scale as to enable a person with normal color vision to match any other hue by the additive mixture in varying proportions. (2) in the reflective subtractive color process, the primary hues are yellow, magenta and cyan—those transparent inks/colors that absorb only (or mostly) the additive primaries of blue, green, and red, respectively. *Related Terms:* physical primary hues/colors; physical color primaries.

human readable — the interpretation of bar code data, often printed immediately below the bar code, which is readable to humans.

humidity — moisture in the air. It affects printing by causing inconsistencies in platemaking, ink drying, the amount of static electricity generated, the size of the printing sheets, etc.

hundredweight (CWT) — 100 pounds in North

America, 112 pounds in the United Kingdom. From the Roman "C" for hundred + "WT," the abbreviation for weight.

hung initial — a large initial letter that is set in the left hand margin of the first line of a paragraph. *Related Term:* Initial letter.

hung punctuation — (hanging punctuation) A typographic style which allows certain punctuation characters to "hang" or extend beyond the left and/or right margins, giving a sharper line to the margins. *Related Term:* hanging punctuation.

Hurvich-Jameson Theory – *Related Term:* opponent process model.

hydrophilic — water-receptive; water-loving. The characteristic used to describe the nonimage areas of a lithographic image carrier. *Related Term:* oleophilic.

hydrophobic — oleophilic; water-repellent. The characteristic used to describe the image areas of a lithographic plate.

hydroquinone — a slow working developing agent in graphic arts photography, one main developing agent; a chemical that turns exposed silver halides to black metallic silver.

hygroscopic — the tendency of a material to absorb moisture.

hyperlink — in World Wide Web pages, hyperlinks are the primary way to navigate between pages and among Web sites. They are the underlined words or phrases you click to jump to another screen or page. Hyperlinks contain HTML-coded references that point to other Web pages, to which your browser then advances. *Related Term:* anchor; hot spot.

hyperlink view — a graphical view in the FrontPage Explorer that displays graphically the hyperlinks among pages and files in your FrontPage web along with the hyperlinks from that FrontPage web to other World Wide Web sites.

hypermedia — the integration of hypertext, hypermedia integrates text, images, video, and sound into documents with interactive links. The most common examples of hypermedia are the sound-laden pages of the World Wide Web. *Related Terms:* hyperlink; hypertext.

hypertext — originally, any textual information on a computer containing jumps to other information, now (1) a direct searching location method based on occurrences of specified terms in other text units containing the same term(s). (2) the second form is based on hypertext links that have been inserted in the text. These links can draw the user's attention to related sections even if the sections do not share the same terms and would not be found on the basis of term occurrences. *Related Terms:* meta tags; anchor; hot links; hypermedia; hyperlink.

hyphenation — *Related Term:* hyphenation and justification.

hyphenation and justification (H&J) — the grammatical and typographic end of line decisions that must be made by either the keyboard operator or the computer during the typesetting process. The practice of adjusting blocks of type so that they are both left and right aligned, with hyphenation occurring as appropriate word spaces, adjusted for good fit and overall appearance. Although hyphenation and justification are separate processes, they are often used together so that hyphenation may enhance the results of justification. Typesetting computers can be programmed to automatically hyphenate and justify lines. *Related Term:* H&J.

hypo — an abbreviation for sodium hyposulfite (sodium thiosulfate), the third chemical in standard process development; a chemical used to remove undeveloped silver from photographic emulsions; also a "fixer" component. *Related Terms:* fix; fixer.

hypo eliminator — a chemical used to neutralize the hypo from photographic films and papers and reduce washing time.

hysteresis — the ability of a wetted substrate to dry back into its original shape and dimensions.

Hz — in electrical work and electronics, the abbreviation for hertz; cycles per second.

I/O (input/output) — any socket in the back of a computer that you use to connect to another piece of hardware is called an I/O port. It is a generic term that refers to the devices and processes involved in reading and writing computer data. Input devices include any device capable of creating data (usually a keyboard or scanner). Output devices include printers, disk drives, the CRT and the typesetting photo unit.

i750 video processor — the name of the programmable video processor family from Intel, which brings DVI video to the PC.

IANS (international article numbering system) — a voluntary, nonprofit standards development association active in numbering and bar coding for products, services, utilities, transport units, and locations. Their basis symbology is a superset of UPC, and contains the same number of bars as UPC Version A, but it also encodes an additional digit which, along with the 12th and 13th digit, identifies a country by code.

IAS (image assembly system) — systems where multiple digitized images can be manipulated and combined prior to output on film or plates. *Related Term:* electronic makeup.

IBC (inside back cover) — the opposite side of the back cover in a magazine, booklet, or brochure.

IC — *Related Term:* integrated circuit.

icon — (1) in computing, a symbol shown on the screen to represent an object, concept or message, for example, a disk, folder, document, trash, etc. Icons are generally activated by an on-screen pointer controlled by a mouse or track ball. (2) in illustration, a general term for any form of drawing, diagram, halftone or color image that serves to enhance a printed piece.

ICR (intelligent character recognition) — mostly a marketing term, it has come to mean character recognition of hand printed data. It is a type of optical character recognition where letters and numbers written by a human hand can be read by a machine. *Related Term:* OCR.

IDE (integrated device [or drive] electronics) — a hard-drive interface that has all of its controller elec-

tronics integrated into the drive itself. The IDE specification handles hard disks up to 504MB in size. Because of its simple instruction set and the short route between controller and drive, it's a quick and easy type of drive to use. Because it's a limited specification, IDE is gradually being superseded by an enhanced version, EIDE. *Related Terms:* EIDE; SCSI.

ideogram — a picture or symbol used in a system of writing to represent an idea. Not an alphabet. *Related Term:* hieroglyph.

IEEE (Institute of Electrical and Electronics Engineers) — a nonprofit U.S. engineering organization that develops, defines, and reviews standards within the electronics and computer science industries.

IEEE 1394 — also known as FireWire. An IEEE standard for high-speed serial connections. It is designed for the exchange of information between PCs and consumer electronics devices that transfer large amounts of data, such as digital camcorders or VCRs. Currently, there is an extremely limited number of FireWire devices, but proponents of PC and TV convergence think the standard will be more widely adopted as Windows 98 supports FireWire-compliant devices.

IETF (Internet Engineering Task Force) — an open community of networkers who manage and shape the Internet.

IFC (inside front cover) — the opposite side of the front cover of a magazine, booklet, or brochure.

IIS (Internet Information Server) — a high-performance, secure, and extensible Internet server from Microsoft, based on Windows NT Server. IIS supports the World Wide Web, FTP, and gopher.

ILD (infrared laser diode) — a laser-producing light in the infrared region, used in optical bar code scanners as the reading light source.

illuminant — incident luminous energy specified by its spectral distribution.

illumination — (1) in photography, lighting a subject properly. (2) in book manufacturing, gilding book pages and manuscripts and decorating them with several colors. A popular practice in medieval times when books were rare and finished by hand.

illustration — a general term for any form of photograph, drawing, diagram, halftone, or color image that serves to enhance a printed piece. The term is commonly used to indicate drawings or diagrams to differentiate them from photographs in purpose and emotional impact. *Related Terms:* copy; image.

illustration board — a type of smooth, thick paper material used in paste-ups and mechanical art onto which all job elements are placed. Used for finished illustrations.

IMA (Interactive Media Association) — the industry association chartered with creating and maintaining standard specifications for multimedia systems.

image — (1) generally, any type, illustration or other original scene as it has been reproduced on computer screen, film, printing plate, substrate or by other means and that portrays the original in the proper form, color, and perspective. (2) in photography, a picture formed by light. The optical counterpart of an original focused or projected in a photographic device. (3) in data processing, an exact duplicate of an array of information of data stored in, or in transit to, another medium.

image acquisition — the process of scanning and digitizing images.

image alignment — in page layout or graphic manipulation software, the specification of how images and text are aligned with each other on the page.

image area — that portion of a mechanical, negative or plate corresponding to the ink on the substrate; the portion of paper on which ink appears.

image assembly — the second step in the printing process; involves bringing all pieces of a job into final form as they will appear on the final product and as they will appear on the film or plate. *Related Terms:* prepress; stripping.

image breakdown — a loss of image in a lithographic plate during the press run.

image carrier — the device on a printing press that carries an inked image either to an intermediate rubber blanket or directly to the paper or other printing substrate. A direct printing letterpress form, a lithographic plate, a gravure cylinder, and the screen used in screen printing are examples of image carriers.

image carrier preparation — the fourth step in the printing process; involves photographically recording the image to be reproduced on an image carrier (or plate).

image conversion — the third step in the printing process; involves creating a film image of a job from the image assembly step.

image depth — a reference to computer-generated image detail. In computer graphics, each pixel has three channels of color information–red, green, and blue–in various bit depths. In 24-bit display graphics adapters, there are 8 bits per color per pixel, but when the adapter has a 32-bit bus, the additional 8 bits are used as an alpha channel to control the color information of the other 24 bits of color. *Related Terms:* alpha blending; alpha channel.

image design — the first step in the printing process; involves conceptual creation of a job and approval by the customer.

image editing — the process of changing and manipulating photographs and other graphics, usually performed electronically using software applications such as Adobe Photoshop.

image form field — one type of form field that displays an image. By selecting the image, the user either submits or clears the form.

image guidelines — lines used in paste-up to position artwork and composition on illustration board; are drawn in light-blue pencil and are not reproduced as a film image.

image map — an image containing one or more defined regions, called hot spots or hot links, which are assigned hyperlinks. As the pointing device passes into these defined areas an associated hyperlink is available and can be accessed by selection. Usually, an image map gives users visual cues about the information to be made available by clicking on each part of the image and by having short identifying captions appear at the pointer location. For example, a building floor plan could be made into an image map by assigning hot spots to each room of interest to a visitor.

image plane — in digital video, display hardware that has more than one video memory array contributing to the displayed image in real time. Each memory array is called an image plane. *Related Term:* bit plane.

image recording — the output of digitized data onto film, plates, or RC paper.

image research — the use of an imaging system to pull up a payment check or other imaged account statement by customer, name, check number, or other item in the image index. Image research takes considerably less time than researching via microfilm or retrieving original documents from file cabinets.

image resolution — the fineness or coarseness of an image as it was digitized, measured in dots per inch (DPI). Typically 300-600 dpi for laser printers and in excess of 4,000 dpi for photographic imagesetters or film recorders.

image scanning — the process of converting art or photographs into digital files by use of a light-sensitive array. *Related Terms:* array; CCD; CMOS; flatbed scanner; drum scanner.

image statements — a statement system originated by the banking industry but which is gaining in general popularity because it includes electronic representations of checks or other scanned documents. Electronic image statements make it easier to reconcile accounts since the entire check image is visible, including name of the payee, etc., instead of just the date and check amount as on a traditional statement. *Related Term:* image research.

image station — a computerized workstation which allows interactive display and direct manipulation of digitized photographs or line art. *Related Term:* monitor.

image tool bar — a graphic display of icons in the form of a tool bar, that when selected issues execution commands to the computer program. Tool bar formats are somewhat uniform within operating system programs, although the precise nature of a specific program will dictate their functions. For example, most Windows programs provide for cutting and pasting, but only image editing packages will have the specific tools for image manipulation.

image transfer — the fifth step in the printing process; involves transfer of the image onto the substrate (often paper). Typically, the production phase in which images are printed on substrates and other products.

imagesetter — any device used to output fully paginated text and graphic images at a high resolution onto photographic film, paper, or plates. Usually exposures are made by use of a laser and resolution is 2,450 dpi or higher. *Related Terms:* PostScript; raster image processor; typesetting, digital; vectors; phototypesetter, fourth generation; area composition.

imaging system — a collection of components to capture, process, store, and retrieve images.

IMAP (Internet message access protocol) —a means of managing e-mail messages on a remote server, similar to the POP protocol. It offers more options than POP, including the ability to download only message

headers, create multi-user mailboxes, and build server-based storage folders, but is not supported by all ISPs. *Related Term:* POP; ISP.

IMHO (in my humble opinion or, in my honest opinion) — an acronym often used in e-mail, posting, and chat directly before an opinion which indicates that the writer is aware that they are expressing a debatable view, probably on a subject already under discussion. They may be honest but are rarely humble. *Related Terms:* IMO; YMMV.

IMO (in my opinion) — an acronym often used in e-mail, posting, and chat, this is usually more accurate than the alternative IMHO. *Related Terms:* IMHO; YMMV.

impact composition — composition by use of a device with a relief image set where individual letters sequentially strike output paper through a single-use carbon ribbon. Generally activated by a keyboard, or automated by tape control. A typewriter. *Related Terms:* strike-on composition; daisy wheel printer.

impact paper — *Related Term:* carbonless paper.

impact printer — any machine that produces characters by striking individually inked letters or pins through a carbon coated or cloth ribbon to strike a direct impression on a substrate. *Related Terms:* daisy wheel printer; strike- on composition; dot matrix printer; typewriter.

impact printing — any printing system where brute force is used to impress an image through ink on a plate surface, or even a microprocessor controlled hammer impacting against a ribbon and a substrate.

impose — *Related Term:* imposition.

imposing stone — a smooth metal or marble surface on which letterpress forms are arranged and locked into a chase or metal frame.

imposition — (1) in printing, assembling the various units of a page before printing and placing them on a form so that after the sheets are printed on both sides they will fold and trim correctly with proper page sequencing. (2) the plan for such an arrangement. *Related Term:* stripping.

imPRESS — a page description language (PDL) developed by Imagen Corp. One of the first commercially available PDLs.

impression — (1) the pressing of a substrate against type, plate, blanket, or die to transfer an image. (2) a single sheet of substrate passed through a printing press, measured by one press cycle.

impression cylinder — the cylinder on a printing press that provides a smooth, firm surface against which the paper (substrate) is held during impression. *Related Term:* back cylinder.

impression lever — the control on a printing press used to start and stop the printing (impression) action; independent of the actual press cycling, which may be continuous.

imprimatur — a published symbol indicating authorization or endorsement., i.e., from a church, charitable organization, government, etc.

imprint — (1) in presswork, to print new copy on a previously printed product. (2) in periodical publishing, the name and place of the publisher and printer required by law if a publication is to be published. Sometimes accompanied by data indicating the quantity printed, month/year of printing and an internal control number. *Related Terms:* colophon; half title; title page; overprint; surprint.

impulse jet — one type of ink jet printer where the ink supply is pressurized only when ink is needed for image formation. The pressure is not continuous.

in print — books that are currently available for sale from publishers.

in-band signaling — in ISDN, a separate wire called the D channel used to set up calls and handle their signals.

incandescent light — the original electric light bulb. Generally, all lights that use a hot wire in a vacuum to produce illumination; often used in photography because they are inexpensive but have the disadvantage of dimming with age.

incident light — light that illuminates, strikes, or falls on a surface.

InContext WebAnalyzer — a Windows software program for managing an Internet or intranet site for business or personal reasons. It is designed to provide diagnostic and statistical tools to manage an error-free Web site, detecting broken links and displaying what pages to fix.

incunabula — early printed books, specifically those done before 1500 AD, although some historians expand this to about 1540.

indelible ink — ink which tends to remain constant after exposure to cleaning and or laundering.

indent — in composition, setting type so that a portion of it aligns at a predetermined distance from the left or right margin of the column. A hanging indent is the opposite of a paragraph indent, with the first line of text set to the full measure and the following lines indented.

Indeo video technology — Intel's video technology which allows the creation of video messages and playback on any i486 (or higher) processor based PC through Microsoft's Video for Windows, etc. *Related Term:* i750 processor.

index — (1) one classification of heavier weight paper. (2) in microfilm/fiche and computing, a descriptive set of data associated with a document for locating that document's storage location and retrieval. In a more complex and demanding role, indices can be used to consolidate documents that may not be, at first glance, related or that may be stored in different locations or on different media. Index stored documents are the great intellectual challenge in document retrieval. Anyone can scan a piece of paper to microfilm. The hard part is devising an index scheme that describes every possible parameter of each document for later searches, comparisons and processing.

index Bristol — Bristol class paper made for products such as index cards and file folders.

index paper — a lightweight board paper for writing and easy erasure.

indexed color — an image which has been reduced to 256 or less colors. The index is a lookup table (LUT) containing up to 256 colors.

India ink — a special type of very black ink used for high-quality layout and illustration.

India paper — *Related Term:* bible paper.

indicia — postal permit information printed on objects to be mailed and accepted by the post office in lieu of stamps. Requires application and prequalification for use. Several types exist, the most common being bulk mail.

indirect color — *Related Term:* indirect screen color separation.

indirect contact print — a contact print in which copy is tonally and/or laterally reversed from the original. *Related Term:* contact print.

indirect photo screen — a method of preparing a screen printing stencil by using a special light-sensitive film, which is attached to the screen fabric after exposure. When developed, the image areas wash clear to allow screen inks to pass through to the substrate. *Related Term:* photographic stencils, sometimes referred to simply as the indirect process; direct process.

indirect relief — a printing process that involves transferring ink immediately from a relief form to a rubber-covered cylinder and then onto the paper; often called dry offset printing.

indirect screen color separation — a color separation method that first produces a continuous-tone separation negative from a positive print or transparency; requires additional steps to produce color-separated halftone negatives.

industrial papers — *Related Term:* coarse papers.

inferior character — A letter and/or number positioned below the baseline. The "2" in H_2O is a subscript or inferior character. *Related Terms:* subscript; superior character.

infinity — a limitless distance. The farthest distance at which a lens can be focused.

informal balance — a design characteristic in which images are placed on a page so that their visual weights balance on each side of an invisible center line, but their mechanical placement is unstructured.

information superhighway — an overworked term for the widening of computer-based information access. The overall plan to deregulate communication services and thus widen the scope of the Internet by opening carriers, such as television cable, to data communication. The term is often shortened to I-way, infobahn, etc.

information time value — the period of information validity; how long before it is obsolete, inaccurate or unuseful.

InfoSeek — an Internet search service that finds Web pages that mention a word or phrase you specify.

InfoSpace — a Web service that enable users to locate listings of people, businesses, government offices, toll-free numbers, fax numbers, e-mail addresses, maps and URLs.

infrared — that region of the electromagnetic spectrum that includes wavelengths from 780 nanometers to about 3,000 nanometers. Infrared radiation also serves as a source of heat.

infringement — when another party besides the copyright owner reproduces a copyrighted work, in whole or in part, without the copyright owner's permission. *Related Term:* copyright infringement; infringer.

infringer — any person who knowingly or unknowingly makes unauthorized copies of a work protected by a registered copyright.

in-house — functions performed within the company rather than by outside contractors.

initial — the first letter in a word set in a larger or more decorative face, usually at the beginning of an article, section, or paragraph. *Related Terms:* initial letter; illumination.

initial cap — a capital letter at the beginning of a paragraph that rests on the first baseline and rises above the x-height of the other letters. Sometimes incorrectly called a raised cap.

initialize — *Related Term:* formatting.

ink — the liquid or paste dispersion of a colored solid (pigment) in a liquid (vehicle) used to produce an image on substrates. Ink comes in different colors and forms designed for use on a wide variety of materials and by different printing methods.

ink brayer — equivalent of a press form roller, it is a hand tool containing an ink roller with a handle which is used for spreading ink on the face of relief type forms prior to pulling a proof. *Related Term:* ink form roller.

ink cell — minute dot-shaped areas engraved into a gravure cylinder to hold ink for impression. *Related Term:* ink well; cylinder well.

ink disk — a rotating round plate-like device at the top of a platen letterpress; used to distribute ink to the ink form rollers.

ink film thickness — the depth of a wet ink film in the ink train, on the ink form rollers or on the substrate.

ink form roller — *Related Terms:* form roller; brayer.

ink fountain — a trough-like device on a printing press, designed to hold a supply of paste or liquid ink and to transfer the ink to the ink train and subsequently to the form rollers of a printing press. The operator controls the ink volume with adjustment screws or "keys" on the fountain or from a remote console. *Related Term:* fountain; keys.

ink holdout — the characteristic of a paper that prevents ink from being absorbed into it, but rather to dry on its surface.

ink jet — a nonimpact method of printing by spraying

droplets of ink through computer-controlled nozzles to form characters. Early ink jets were prone to clogging, but were capable of delivering good print speeds and relatively high-quality text and graphics. Today's best ink jet printers have eliminated most of the clogging problems, rival laser printers for text quality, and now can produce color images of photographic quality and for less money than a low-end laser printer. The droplets' delivery can be either intermittent or continuous. *Related Terms:* continuous ink jet; asynchronous ink jet; daisy wheel printer; dot matrix printer; laser printer.

ink mist — threads of ink thrown into the air and caused by the ink splitting between rollers of a production press.

ink proofs — press sheets printed on special proof presses using the ink and paper of the final job; are extremely expensive and usually reserved only for high quality or long-run jobs.

ink train – all rollers of any type which transfer ink from ink fountain to ink form rollers.

ink trap — *Related Term:* trapping.

ink viscosity — a measure of ink's resistance to flow. A very important characteristic in runability for production. Viscosity varies by printing process from stiff to nearly liquid.

ink, sublimation — special inks which turn from a solid ink particle directly into a vapor, without a liquid state when heat is applied. The ink is most often used for heat transfer printing of textiles, particularly 50% cotton and synthetic blends. It is also popular with "iron on" elements which can be purchased for consumer use.

ink/water balance — in lithography, the critical adjustment which sets the appropriate amounts of ink and water required to ink the image areas of the plate and keep the nonimage areas clean.

inking system — all of the ink rollers, ink fountains and controls that apply and regulate ink to the plates on a printing press. *Related Terms:* ink train; ductor roller; form roller; inking mechanism.

inkjet printing — *Related Term:* ink jet.

inkometer — a device that measures the tack, or cohesion, of printing inks.

Inktomi — a research project in parallel computing technology at U.C. Berkeley which is a prototype for a commercial search engine.

in-line — the style of type in which a white line runs down the main stroke of the letter. A letter tracing which begins on the edge of the letter and goes in.

in-line finishing — the performance of finishing as part of the press operation. Can be accomplished by press attachments or auxiliary equipment at the delivery point. Can include numbering, slitting, punching, tipping, and similar operations.

in-line graphic — a method of keeping an illustration or photograph attached to specific test within a file if formatting is altered. A descriptor referring to a graphic image displayed within text, usually in a Web page or a document created in a desktop publishing program. The graphic will move with the text and will not be confined to a specific position on the page when the adjacent text copy is revised. *Related Term:* in-line image.

in-line image — *Related Term:* on-line graphic.

in-line press — a style of press that has two or more printing units positioned in a line so paper or another substrate can easily travel from one unit to another for subsequent image impressions. *Related Terms:* central cylinder press; McCall press; cylinder press; flatbed press; platen press; rotary press.

in-plant printer — *Related Term:* in-plant shop.

in-plant shop — a printing or duplicating facility within a manufacturing company or business. It does not usually perform work outside of the parent company or business. *Related Terms:* in-plant shop; in-plant printer.

input — (1) generally, a computer term used to describe entering information (data) into computer memory. (2) in computer composition, the data to be processed. *Related Term:* I/O

input device — mechanical or electronic devices for entering data. In electronic or desktop publishing, the keyboard, mouse, graphics tablet and scanner are the primary devices used. Entry and retrieval of data, as opposed to its processing.

input/output (I/O) — a generic term that refers to the devices and processes involved in reading and writing computer data. Input devices include any device capable of creating data (usually a keyboard or scanner). Output devices include printers, disk drives, the CRT and the typesetting photo unit.

insert — (1) an instruction to the printer for the inclusion of additional copy. (2) a separately prepared and specially printed piece which is inserted into another printed piece or a publication, e.g., newspaper circulars, advertising supplements, etc.

inserted signatures — unlike gathered signatures which are laid one upon another, inserted signatures are col-

INSERTED SIGNATURES
Note how the middle signatures appear to be sticking out further than the outside ones.

lected from the inside out. This is particularly popular when wire stitching is the binding method of choice. It seriously limits the practical number of total pages and con-tributes significantly to image shift and binder's gain.

inserting — (1) in bindery and finishing, an operation that involves placing one signature within another. (2) in newspapers, the placement of previously printed (preprints) into the current paper prior to delivery.

insertion order — an order form used by advertising agencies and ad sales reps to fulfill an advertiser's request to place an ad in a specific issue or series of issues of a publication.

inside margin — the space between the binding edge of the page and the text.

insoluble — those parts of the photo polymer plate which contain the image, have been hardened by the ultraviolet light, and will not wash away.

inspection — (1) applications that use bar codes or other forms of AIDC to identify and validate items. It is primarily used for the receiving or shipping functions of inventory control. (2) the final step in production prior to delivery to assure quality control.

inspection light — darkroom lights equipped with both safe and white lights; used to inspect film during and after processing.

instant image proof paper — proofing materials that create a visible image when exposed to light; does not require special equipment or chemicals. Often cannot be fixed and additional actinic exposure will cause the image to fade. Photographers use similar materials for studio proofs called "POP," or printing out paper.

intaglio — (1) any form of printing in which the image areas are engraved or etched below the nonimage areas on the image carrier to provide ink-retaining reservoirs or wells. Gravure is an intaglio printing process. (2) in papermaking, watermarking from countersunk depressions in the dandy roll to provide a whiter or denser design instead of increased transparency. *Related Terms:* gravure; halftone gravure; photogravure.

integral proof — a color proof of separation nega-

tives exposed in register on one piece of proofing paper. *Related Term:* laminate proof.

integral tri-pack — photographic film or paper with three main emulsion layers coated on the same base. Each layer is sensitive to one primary color of light. In processing, a subtractive primary color dye image is formed in each layer. Most color films are tri-packs.

Integrated Services Digital Network (ISDN) — a digital telecommunications channel that allows for the combined transmission of voice, video and data. *Related Term:* narrowcasting.

integrated technology — the general term which applies to combination appliances such as fax/copier/scanner units.

intelligent character recognition (ICR) — a sophisticated form of optical character recognition (OCR) in which the computer determines the probable meaning of a character not by looking for an exact match with a character pattern stored in memory but by analyzing the shape of the character, giving it the ability to learn new fonts and new symbols. ICR is, therefore, able to interpret a wide range of different typefaces and point sizes, thus differing from OCR, which is generally restricted to the specific face and point size combinations stored in memory. *Related Terms:* optical character recognition; IOCR.

intensification — the process which occurs to increase density and contrast through increased development of soft negatives which otherwise may have little useful value.

intensity — *Related Term:* color saturation.

interactive — the fusion of user programs and computer technology. An interactive program and a computer running in tandem under the active participation and control of the user. The user's actions, choices and input decisions affect the way in which the program unfolds to generate the appropriate response within context. *Related Terms:* linear; interactive video.

interactive color correction feedback (ICCF) — the reproduction of the original on a video display terminal to show the effects of color changes as they are being programmed.

interactive media — refers to telecommunications channels that allow the two-way exchange of information.

interactive video — the technique of involving the user in directing the flow of a computer or video program. A system that exchanges information with the viewer, processing input in order to generate the appropriate response within the context of the program.

interactive voice response system — an automated voice system that allows clients to make payments for services, make inquiries and place orders 24 hours a day, seven days a week, by calling a specified number and providing credit card information and/or account information to the system over the phone.

interbang — typographic symbol used to express a question and an exclamation at thesame time. A combination of an exclamation mark superimposed on a question mark.

intercap — the name given for the practice of having capital letters in the middle of proper names such as PostScript, QuarkXPress, etc.

intercharacter gap — in code and symbol reading, the space between two adjacent bar code characters in a discrete code The clear space between two characters.

interface — (1) the circuit, or physical connection, using hardware or software devices that allows information to be shared between two computerized machines, such as a modem interface for telecommunications. (2) the linking of two or more electronic devices so they can function as one unit; the place at which two systems or pieces of equipment meet and interact with each other.

interimage reflection — the passage of light between layers of ink and the substrate.

interlaced — a typical television display on some low-quality monitors. On TVs as on computer monitors, the whole screen is drawn line by line. Early display tubes couldn't draw the whole screen before the top began to fade, so engineers implemented a system called interlacing, which skips every other line on each pass. Odd numbered lines are scanned on the first pass and a second scanning sequence fills the even numbered lines. This system results in a little flicker, but it avoids having the bottom of the screen perpetually brighter than the top. Most normal-resolution monitors are noninterlaced (progressive), meaning that all lines are scanned with each scanning pass. Lower quality display adapters, however, when pushed into high resolutions and high color, sometimes are. *Related Terms:* noninterlaced; electron gun; refresh rate.

interlaced GIF — a feature of the GIF89a graphics standard; an interlaced GIF displays images in two passes of alternating lines, like normal television. Interlaced GIFs produce a "venetian blind" effect or blocky image that gradually sharpens with each scan. Interlaced GIFs let people see at least the outline of an image sooner and pages often appear to load faster than those with noninterlaced graphics. *Related Terms:* GIF; progressive JPEG.

interline spacing — a phototypesetting term for leading. *Related Terms:* line spacing; leading; set solid.

intermediate colors — the colors made by mixing a primary color with an equal amount of the secondary color next to it on the color wheel, e.g., yellow-green, red-orange, blue-violet.

intermittent inkjet — *Related Terms:* asynchronous inkjet; drop-on-demand ink jet; ink jet; solid ink jet.

internal hyperlink — a hyperlink to any file that is inside the current page of the web, often leading to specific portions of content as from a table of contents.

internal web — a World Wide Web site created within an organization and accessible only to members of that organization on an intranet. *Related Term:* intranet; extranet.

international article numbering — a voluntary, nonprofit standards development association and system active in numbering, bar coding, and EDI messages for products, services, utilities, transport units, and locations.

international paper sizes — the International Standards Organization (ISO) system of paper sizes is based on a series of three sizes: A, B and C. Series A is used for general printing and stationery, Series B for posters, and Series C for envelopes. *Related Term:* ISO paper sizes.

International Television Fixed Service (ITFS) — a nonbroadcast television service that is typically used for closed-circuit educational applications and requires special antennas and converters to translate the signals for viewing on ordinary television sets.

internegative — a negative made from a transparency or print for the purpose of making photographic prints. Without careful control, it often results in the new print having an undesirable increase in contrast.

internet — should not be confused with "The Internet" which is usually written with a capital I. This is a group of local area networks (LANs) that have been connected by means of a common communications protocol. Many internets exist besides the Internet, including many TCP/IP-based networks that are not linked to the Internet.

internet account — a service account with an ISP (Internet service provider) that allows a user to access the Internet.

internet address — an Internet protocol (IP) address that uniquely identifies a node on the Internet.

internet backbone — a superfast network spanning the world from one major metropolitan area to another is provided by a handful of national Internet service providers (ISPs). These organizations use connections running at approximately 45 MBPS (T3 lines) linked up at specified interconnection points called national access points, which are located in major metropolitan areas. Local ISPs connect to this backbone through routers so that data can be carried though the backbone to its destination. *Related Terms:* T3; internet account; internet service provider.

Internet Explorer — the Web browser from Microsoft. A rival of Netscape, it supports many of the Netscape extensions, as well as some innovative ones.

Internet or "the Net" — originated as ARPANet, a nuclear disaster proof communications network by the US Defense Department. The term now describes a system of linked computer networks, international in scope, that facilitates data communication services such as remote login, file transfer, electronic mail, and newsgroups. While on-line services offer Internet access, the internet does not offer on-line services. The Internet is a way of connecting existing computer networks that greatly extends the reach of each participating system. The World Wide Web runs on the Internet. While on-line services offer Internet access, the internet does not offer on-line services. *Related Terms:* internet; intranet; extranet; browser; ARPANet.

internet protocol (IP) — the network layer for the TCP/IP protocol suite. It is a connectionless, best-effort packet switching protocol.

internet protocol (IP) numbers (IP addresses) — a unique, numeric identifier used to specify hosts and networks. Internet protocol (IP) numbers are part of a global, standardized scheme for identifying machines that are connected to the Internet. Technically speaking, IP numbers are 32-bit addresses that consist of eight octets, and they are expressed as four numbers between 0 and 255, separated by periods, for example:

198.41.0.52. These are the actual addresses. The system will often translate them into word addresses.

internet relay chat (IRC) — a protocol that allows users to converse with others in real time. IRC is structured as a network of servers, each of which accepts connections from client programs. *Related Terms:* talk; chat.

Internet Scout — a project that provides current awareness tools and information about Internet resources and network tools. Operated by the University of Wisconsin at Madison, the project originated as an activity under the National Science Foundation cooperative agreement that created the InterNIC and was further developed as part of the InterNIC Information and Education Services cooperative agreement operated by Network Solutions. *Related Terms:* InterNIC; InterNIC Information and Education Services.

Internet security — all of the systems and techniques which help ensure that data passed over the Internet is not seen, copied or misused by any part of the system for which it was not intended. System firewalls, file encryption, call-back modems, passwords, PIN numbers, etc. are included as part of Internet security. All information on the Internet takes a circuitous route, not of the sender's choosing, through numerous intermediary computers to reach any destination. Any intermediary computer has the potential to eavesdrop and make copies or even deceive you by misrepresenting itself as your intended destination. These possibilities make the transfer of confidential information such as passwords or credit card numbers susceptible to abuse.

Internet service provider (ISP) — a person, organization, or company that provides access to the Internet. In addition to Internet access, many ISPs provide other services such as Web hosting, Web page design, domain name service, and other proprietary services. *Related Terms:* name service; ISP.

Internet Society — an organization dedicated to supporting the growth and evolution of the Internet comprised of companies, government agencies, and foundations that have created the Internet and its technologies as well as innovative new entrepreneurial organizations contributing to maintain that dynamic.

internetworking — communication between data processing devices on one network and other possibly dissimilar devices on another network.

InterNIC — the name given to a project that originated and operates under a cooperative agreement with the National Science Foundation (NSF), the InterNIC organization was formed in 1993 to handle domain name registrations. While Network Solutions manages the group, the National Science Foundation, AT&T, and General Atomics also play a part in how the organization is run. InterNIC maintains a database of domain names, to avoid duplication and provides domain name registration services in .com, .net, .org, and .edu; Internet protocol (IP) network number allocation; and information and education services. If you point your browser at InterNIC, click Registration Services, and then Whois, you'll find guidelines for researching the availability of domain names. *Related Term:* domain name.

InterNIC Directory and Database Services — a database and directory service developed by the National Science Foundation (NSF) with cooperative agreement awarded to and operated by AT&T. It provides: (1) a Directory of Directories, containing lists of FTP sites, servers, white and yellow page directories, library catalogs and data archives and (2) white and yellow pages-type Directory Services. *Related Term:* InterNIC.

InterNIC Information and Education Services — a provider of various tools and resources for the Internet community including: (1) InterNIC News, an on-line, monthly newsletter; (2) the 15 Minute Series, a collection of training materials for Internet trainers and others who support Internet users; and (3) Road map, an e-mail-based Internet tutorial. *Related Terms:* InterNIC; Internet Scout.

InterNIC Registration Services — a service that administers the registration of second-level domain names under the com, org, net, and edu top level domains. While Network Solutions was solely responsible for Internet protocol (IP) number allocation for the Americas, the Caribbean, and sub-Saharan Africa, it has transitioned this responsibility to the American Registry for Internet Numbers (ARIN). *Related Terms:* InterNIC; American Registry for Internet Numbers (ARIN).

interpolation — in digital imaging, a method by which actual optical data can be added to by either hardware or software approximation and manipulation of the pixels surrounding those actually recorded. The process increases the pixel information when scaling or performing any other transformation to an image. In reduction, pixels are averaged to create a single new pixel.

interPRESS — a page description language from the Xerox Corporation. It was the first such product of its type to be implemented, but has not been adopted commercially by any third party.

Intertype — a brand of hot-type casting machine with a keyboard that casts a line of type in one piece. *Related Terms:* Linotype; Monotype; Ludlow; Elrod.

intraframe coding — compression of video within each individual frame. The process results in less compression of a range of frames than interframe coding. *Related Terms:* interframe coding; compression.

intranet — a play on the word Internet, an intranet is a restricted-access network that works like the Web, but isn't on it. Usually owned and managed by a corporation, an intranet enables a company to share its resources with its employees without confidential information being made available to everyone with Internet access.

inventory control — in optical reading operations, the applications that use automatic identification to make sure the right material is in stock so it can be delivered for the right cost, to the right user at the right time. Generally, applications where bar coding and other forms of AIDC are used to add items or delete items from inventory with 100% item accuracy.

inverse address resolution (IN-ADDR) — the process used to resolve Internet protocol (IP) number(s) to their corresponding domain name(s), as opposed to the more familiar type of resolution that starts with a domain name and translates it into the corresponding IP numbers. Inverse address resolution can be helpful when using tools such as log files, because it enables the Internet protocol (IP) number(s) logged by the network and computers to be related to the corresponding domain names, which are more easily recognized by humans. *Related Terms:* domain name system (DNS).

inverted pyramid — (1) in makeup, a headline style in which each centered line is narrower than its predecessor; (2) editorially, a term that is also used for a news writing form where the facts are presented with the most important item first, and in order of importance, information of less relevance to the core of the story.

invoice — a bill for goods, services, etc. Invoices may be sent to the billing contact via both postal mail and e-mail.

ion deposition — *Related Term:* electrostatic.

IP (Internet protocol) — the required software protocol that divides data into packets for transmission between systems across the Internet. *Related Terms:* IP address; TCP.

IP (Internet protocol) address —the standard way of identifying a computer that is connected to the Internet, much the way a telephone number identifies a telephone on a telephone network. The IP address is groups of four numbers separated by periods, and each number is less than 256, for example, 197.241.72.55. For most users, their Internet service provider will assign your machine an IP address. *Related Terms:* IP; VAT.

IP packet — a chunk of data transferred over the Internet using standard Internet protocol (IP). Each one begins with a header containing addressing and system control information. IP packets vary in length depending on the data being transmitted. *Related Terms:* IP; IP address; ATM.

IPX (Internetwork packet exchange) — *Related Term:* IP packet.

IR coating — a liquid laminate coating bonded and cured with infrared light. *Related Term:* UV coating.

IRC (internet relay chat) — *Related Terms:* Internet relay chat; talk; client, server.

IrDA (infrared data association) — an organization whose goal is to establish standards for the exchange of data over infrared waves, a wireless technology that lets devices beam information from one to another. The devices, for example, beam a document to your printer instead of having to connect a cable. The standard developed by the group has been widely adopted by PC and consumer electronics manufacturers and Windows 98 supports it.

Iris — (1) in photography, the adjustable opening fitted into the barrel of photographic lenses which consists of a series of thin metal tongues overlapping each other and fastened to a ring on the lens barrel. The operator turns the ring backward or forward to make the aperture smaller or larger. *Related Terms:* diaphragm; aperture. (2) the name of one type of proofing system.

IRQ (interrupt request) — in the original PCs, a signal from a piece of hardware (such as a keyboard or sound card) indicating that it needs the CPU to do something. The interrupt request signals run along defined circuits to an interrupt controller that assigns it a priority and passes them to the CPU. The interrupt controller expects one signal from one device per line. The exact IRQ path was originally specified by the hardware installer by setting jumpers or switches. If more than one device sends signals along the same line, it will cause an IRQ conflict which

produces a protection fault and locks up the machine. Plug and play technology greatly improved the situation by having the machine recognize hardware and automatically assign nonconflicting IRQs. Even better is the newer USB system, which eliminates the need for them entirely. *Related Term:* plug and play; USB.

irregular sizes — paper which is cut to other than the basic sheet sizes customarily cut and stocked by paper suppliers or merchants.

ISA (industry standard architecture) — is the bus design that has been used in most PCs since IBM released the PC/AT. It's a limited 8-bit and 16-bit bus, but it's so widely compatible that it has outlasted technologically superior and much faster bus standards like PCI and USB. *Related Term:* PCI; USB.

ISAPI (Internet Server Application Program Interface) — a high-performance Web server application development interface for Intel servers. Developed by Process Software and Microsoft Corporation, it uses Windows' dynamic link libraries (DLLs) to make processes faster than under regular APIs. *Related Terms:* API; DLL.

ISBN (international standard book number) —a unique identification number that identifies the binding, edition, and publisher of a book. Assigned by the book's publisher using a system administered by the R. R. Bowker/Reed Reference Publishing Company in New York City. Usually found on the back of the title page or the back outside cover. The number is specific to the binding, edition, and other physical characteristics. When a publication is revised, a new ISBN number is assigned.

ISDN (Integrated Services Digital Network) — a planned hierarchy of digital switching and transmissions systems developed in 1984 to allow wide-bandwidth digital transmission on the public telephone network. The phone companies realized that its regular systems could not handle large quantities of data. ISDN specifications make it possible for a phone to transfer up to 64 kilobits per second of digital data. *Related Terms:* POTS; T1; Integrated Services Digital Network

ISO — (1) International Standards Organization. *Related Term:* ASA.

ISO 9000 — the International Organization for Standardization (ISO) registration standard to measure an organization's commitment to customer-defined quality. Establishes international standards to communicate clear process-based operations and facilitate the global exchange of products and services.

ISO envelope sizes — the matching sizes for envelopes to hold products printed on sheets using ISO (A, B, C, and D) paper sizes.

ISO number — (1) a number used to designate the light sensitivity or emulsion speed of photographic film. (2) the initials ISO stand for International Standards Organization. *Related Term:* ASA.

ISO paper sizes — printing paper sizes designated by the International Standards Organization and used throughout the world except in the United States and Canada. Expressed in millimeters. *Related Terms:* international paper sizes; ISO envelope sizes.

ISP (Internet service provider) — the front end to all that the Internet offers. Most ISPs have a network of servers (mail, news, Web, and the like), routers, and modems attached to a permanent, high-speed Internet "backbone" connection. Subscribers dial in to the local network, usually by modem, to gain Internet access, without having to maintain servers, file for domain names, or learn UNIX.

ISSN (international standard serial number) — a number assigned by the Library of Congress to newsletters, magazines and other serials requesting it. *Related Term:* ISBN.

issue — (1) in publishing, the end result of one production sequence such as the September issue or the Fall issue. (2) in management, a topic of concern which must be addressed.

italic — a forward-slanting version of a typeface with vertical slant, usually between 8 degrees and 20 degrees from the perpendicular of the character base line. It is a separate face from its roman counterpart, and is not merely the original face with a slanted angle and is more compact than the regular letter form of the face. In typeset copy, italic type is used to signify periodical titles and other special information. Originally commissioned and used by Aldus Minautius, italic figures are often modified individually by the designer to produce a better intraletter fit. *Related Term:* oblique type.

ITU-T (International Telecommunications Union-Telecommunication) — the new name for the international committee CCITT. *Related Term:* CCITT.

ivory board — a smooth, uncoated, high white board used for business cards etc. *Related Terms:* ivory Bristol; plate.

ivory Bristol — *Related Terms:* ivory board; plate.

Jj

jabber — transmission of meaningless data by either networks (along communication lines) or people (when communicating in chat).

jacket — (1) short for job jacket. A large plastic or paper envelope designed to hold mechanicals, art and other components of a production order. (2) in publishing, a short term for dust cover, or the paper outer wrap of a hardcover book. *Related Terms:* dust jacket; dust cover.

jaggie (jaggy) — the ragged edges or stair-stepped effect on a bitmapped image that are produced when a digital or circular line is scanned into a system. They are caused by the stacking of square pixels to form curved edges, and become more noticeable with enlargement. Jaggies cause an apparent lack of resolution; some say they resemble stair steps. They are less obvious in high-resolution work due to the increased number of pixels in a given image area. *Related Term:* antialiasing.

JAR (Java Archive) — a format to store compressed Java applets, developed by Sun Microsystems. Netscape Navigator can download JAR files and save them onto the local hard drive, so that the next time you access the applet it won't have to download it again.

Java — a general-purpose programming language created by Sun Microsystems. Java can be used to create small Java applications (applets) that can be safely downloaded to your computer through the Internet and immediately run without fear of viruses or other harm to your computer or files. The resulting Web pages can include functions such as animations, calculators, etc.

Java applet — a short program written in Java that is attached to a World Wide Web page and executed by the browser machine.

Java Virtual Machine —a program that interprets Java bytecodes into machine code. The VM is what makes Java portable—a vendor such as Microsoft or Sun writes a Java VM for their operating system, and any Java program can run on that VM.

JavaBeans — a component technology for Java that lets developers create reusable software objects. These objects can be shared—a database vendor can create a JavaBean to support its software, and other devel-opers can easily drop the Bean into their own projects.

JavaScript — *Related Terms:* Java; ActiveX; VBScript.

jaw folder — the final folder on a web press which folds the printed web into a final product.

j-card — the printed cover stock that fits snugly inside a clear plastic cassette case.

JDBC (Java database connectivity) —similar to ODBC, this set of application programming interfaces (APIs) provides a standard mechanism to allow Java applets access to a database.

jim dash — a small rule, usually used to separate decks in a headline or title.

JIT (just-in-time compiler) — when dealing with Java, it converts compiled Java source language statements into native code designed to run on a specific hardware and operating system platform. It compiles all the code at once before execution, so Java programs often run faster with it than with a Java Virtual Machine. *Related Terms:* bytecodes; Java Virtual Machine.

jitter — in video an artifact in the picture which is characterized by horizontal and/or vertical movement, making the image appear unsteady.

job estimate — a document submitted to printing customers that specifies a cost of producing a particular job under mutually agreed-upon conditions. An estimate can be changed to reflect a revision in specifications. It is different from a quotation, which is a formal offer of price based upon fixed specifications.

job lot merchant — a paper merchant that sells job lot paper. Also called clearance merchant and seconds merchant.

job printer — *Related Term:* commercial printer.

job schedule — the production schedule prepared for printing jobs as they pass through each production step; shows how long each step should take and the order of its movement through the shop. Job scheduling and production control departments of most printing companies direct the movement of every printing job through the plant.

job shop — a commercial printing plant, as opposed to a publication shop.

job stick — *Related Term:* composing stick.

job ticket — *Related Term:* work order.

jobber — an organization that buys books in large quantities for resale to retailers and libraries. *Related Term:* distributor.

job-lot papers — paper substrates that didn't meet specifications when produced, have been discontinued, or for other reasons are not considered first quality.

jog — to align flat, stacked sheets or signatures to one or more common edges, either manually or with a vibrating table or hopper. Some in-line finishing systems are equipped with a jogger/stacker that piles and aligns folded signatures as they are delivered.

jogger — a mechanical device on a printing press or other production machine that helps to position sheets of paper prior to or after the production step.

Jones diagram — a graph named after its inventor, Lloyd Jones, that presents steps in objective tone reproduction from the original to the separation negatives and the printed sheet. The relevant information at each stage is linked to the next by plotting the graphs on a quadrant in such a way that the influence of each successive step is displayed.

Jordan machine — a conical rotor and housing that refines fiber slurry in papermaking. It shortens the fibers and improves sheet formation.

journalism — the occupation that includes writing, editing and managing of publications that present information to a mass audience.

journalist — a writer or editor for a periodical.

journeyman — a person who has completed the apprenticeship requirements and who is a competent skilled worker in a technical trade.

joy stick — a hand-held lever-type device mounted permanently onto a base plate that when moved about its horizontal axis will alter the image qualities on a terminal in some predetermined manner, usually modifying the overall perspective being viewed. Often the joy stick will have other associated buttons, triggers, etc. to enhance the control capabilities. While most often associated with computer games, they are important control devices for many scientific computing applications. *Related Terms:* track ball; mouse.

JPEG (Joint Photographic Experts Group) — a file format for color-rich images developed by the Joint Photographic Experts Group. JPEG compresses graphics of photographic color depth better than competing file formats like GIF. It retains a high degree of color fidelity, and makes smaller files that are therefore quicker to download. Compression is scalable in JPEG, but it is a lossy format and the smaller you compress the file, the more color information will be lost. It is still image format with excellent compression for most kinds of images and is, along with the GIF format, the most commonly used on the World Wide Web for 24-bit color images. The benefit of using JPG images is the higher color and resolution you can have which is 16 million colors as opposed to the 256-color limitations of GIF files. It is also widely used in the banking arena to compress TIFF images. *Related Terms:* GIF; lossless; lossy.

jukebox — also called optical jukebox. A device designed to provide quick access to multiple CD-ROMs. Jukeboxes range from simple internal drives that hold a quartet of discs to massive external systems capable of holding hundreds of discs.

jump — the point at which a news story moves from one page to another, indicated with a "Continued on page _", or similar notation

jump head — a headline on the part of a story continued from another page.

jump line — a phrase indicating on what page a story is continued from or to. *Related Term:* jump.

jumper — an on/off switch used to alter hardware configurations. A jumper is made of wires and a small metal piece that can connect the wires to turn the jumper on. Jumpers are found on devices such as CD-ROM interface boards, bus expansion boards, controller boards, input/output cards, sound cards, graphics cards, modem cards, and mother boards. *Related Term:* controller.

jumpover — in publishing, used to describe text which begins above a photograph or illustration and continues below it.

junior carton — a case of five, eight, or ten reams of cut size paper. Junior cartons weigh approximately 50 pounds.

justification — the addition or subtraction of space from between words and/or letters when setting straight composition to cause the first and last letters of each line of type to fall in vertical alignment with the column edges. Hyphenation is sometimes employed but in other cases, only the spacing between words is adjusted. *Related Terms:* flush right; flush left;

ragged; ragged right; ragged left; quad left; quad right; quad center; word spacing; letter spacing; justified type.

justification range — the zone at the end of a line of type within which the computer or hot metal type casting machine will determine and make acceptable line breaks to cause justification.

justification, vertical — (1) the use of variable spacing between lines, or type elements, vertically in order to fill out a desired column or page depth. (2) modifying point size and intercharacter spacing instead of vertical space in order to force a given block of type to fill a desired depth.

justified — *Related Term:* justify.

justify — the alignment of text along a margin or both margins. This is achieved by adjusting the spacing between the words and characters as necessary so that each line of text finishes at the same point. *Related Term:* justification.

JVM (Java Virtual Machine) —an interpreter between the Java bytecodes and a computer's operating system. Using a JVM, you can run Java code on any number of different computer platforms, including Macintosh, Windows 95, and UNIX. JVMs read and execute Java statements one at a time, however, so they are often slower than a just-in-time compiler. *Related Terms:* microprocessor, bytecodes, JIT.

Kk

K — in printing, art preparation and color separation operations, the abbreviation for black.

KB — (1) the abbreviation for kilobyte. (2) letters representing the numeral 1024 (one kilobyte). (3) the number of bytes in a computer memory might be expressed as 64K, meaning 65,536 bytes.

Kb/s — kilobits (1000) per second. A standard for rating modem speed. *Related Terms:* KB; bit; BPS.

keep standing — a printing and publishing term indicating to hold type or plates ready for reprints.

kelvin (K) — a unit of temperature measurement starting from absolute zero, which is equivalent to minus 273.15° celsius. The measurement standard used to indicate the color balance of a light source.

Kenaf — a fast-growing tropical tree gaining in popularity for papermaking and now sometimes used to make pulp for newsprint. *Related Terms:* coniferous; deciduous.

Kermit — a popular terminal program and file transfer protocol developed at Columbia University and available for a variety of platforms. Because Kermit runs in most operating environments, it provides an easy method of file transfer and can be used to download files from a remote system to a personal computer. It is distinguished by its ability to transfer files over Telnet and other connections that would corrupt a binary transfer. Kermit is not the same as FTP. *Related Terms:* Xmodem; Ymodem; Zmodem.

kern — (1) generally, to selectively adjust the space between characters to improve readability or to achieve balanced, proportional type. (2) in hot metal, a kern was any part of a letter that extended into the space occupied by an adjacent letter. *Related Term:* kerning.

kerned pairs — *Related Term:* kerning pair.

kernel — the foundation on which the operating system rests. It provides low-level services, such as memory management, basic hardware interaction, and security. Without it, your system would not start, stop or perform any of its basic functions. *Related Term:* operating system.

kerning pair — a pair of characters for which tighter kerning is automatically applied. Kerning pairs are defined in kerning tables built into most fonts.

kerning table — a table built into most fonts containing kerning pairs.

kerning —to mortise. (1) the process of using negative letter space between specific character combinations so that they appear closer together. Typesetting characters are positioned within an imaginary rectangle of a specific width value. Typographic kerning defines characters by their shape as well as their width by optically reducing the space between certain letter combinations and programming them into the typesetting system. Kerned letters are common in italic, script and swash fonts. Commonly kerned letter combinations are AC AT AV AW FA LT LV LW OA OV OW PA TA TO VA VO WA WO YA YO Av Aw Ay Ta Te To Tr Tu Tw Ty Va Ve VO Vu Wa We Wo Wr Wo Wr Wu Wy Ya Ye Yo Yu T. F. Y. W. y. r. (2) that part of a letter which actually overhangs the body or body space of the letter itself. *Related Terms:* kern; kerning pairs; mortise; spacing; word spacing; tracking; kerned letters.

key — (1) in design, the relating of photographs or loose pieces of copy to their positions on a layout or mechanical using a system of numbers, letters or colors. (2) in composition, certain combinations of letters and/or numbers which cause a typesetting machine to perform specific functions, i.e., change font, size, etc. (3) in general computing, the buttons on the keyboard which are depressed to initiate an action. *Related Terms:* keyboard; direct input.

key lighting — in photography, the emphasis on lighter or darker tones in a print; high key indicates prevalence of light tones; low key, prevalence of dark tones.

key pal — a colloquial term for a person with whom you correspond that uses a keyboard to type e-mail messages rather than a telephone, pen, to write handwritten letters. Sort of an electronic pen pal. If correspondence occurs frequently or on a regular basis it could be said that the correspondents are "key pals."

key plate — the printing plate used as a guide for the registration of other colors.

keyboard — an input device which is used to cause typographic characters to be composed in the order in which the keys are struck, sometimes without an intermediate form such as magnetic tape storage or paper tape. The key layout most often resembles a standard typewriter.

keyboard layout, keyboard mapping — sometimes known as a character mapping, a keyboard layout or mapping. A table used by a computer operating system to govern which character code is generated when a key or key combination is pressed. *Related Terms:* character; character encoding; glyph.

keyboard template — a usually software-specific plastic or cardboard attachment to a computer keyboard which explains the function keys, and other key combinations, to cause its software to perform pre-determined tasks.

keying — in video systems, the process of inserting one picture into another picture under spatial control of another signal, called keying signal. Generally the type of electronic approach which places a person into an impossible scene, such as weather person in front of a satellite cloud photo, etc. *Related Term:* chroma key.

keyline — (1) guidelines inked on artwork to indicate the outline or limits of screens. (2) lines on a mechanical or negative showing the exact size, shape and location of photographs or other graphic elements. *Related Term:* mechanical.

keylining — the process of making a keyline.

key-ring — a pair of computer code keys that consists of both a public key and its corresponding private key. Key-rings are used in public key encryption systems such as Pretty Good Privacy (PGP). Data encrypted with someone's public key can only be decrypted with the corresponding private key, and vice versa. *Related Terms:* Encryption, Pretty Good Privacy (PGP).

keys — (1) the adjusting screws on an ink fountain used to control ink flow. (2) the decoding portion of encryption. The key(s) will provide the routine to translate the encrypted message.

keystroke — the striking of one character on a keyboard.

keyword — the subject word used by online databases and search engines. Usually entered manually to begin the search, and most engines provide methods of narrowing the results if they are too broad. *Related Term:* Boolean.

kick leg — the support bar that keeps the screen printing frame up while a printed sheet is removed and a new sheet inserted.

kicker — a small, secondary headline placed above an article's primary headline printed in smaller type, usually underlined and often italic if the headline is regular face, or vice versa. *Related Term:* teaser.

kid finish — a vellum finish on a soft bond paper made to feel like soft leather.

kill — (1) in editorial matters, the deletion of unwanted copy. (2) in printing production, to distribute or dump metal type from a form that has already been printed, or to destroy existing negatives or press plates.

kill fee — the money paid to writers and photographers for work done on assignment, then not used.

kill file — a file for USENET that lets you filter postings to some extent, by excluding messages on certain topics or from certain people.

kilobyte — *Related Term:* KB; K.

kilohertz — one thousand Hertz (cycles per second).

kiosk — a monitor screen mode that drops all the visual clutter of your browser: tool bars, menus, and borders. This leaves more room for the Web page content. Sometimes called presentation mode. *Related Term:* full screen view.

kiss — the minimum pressure which can be applied between two adjacent surfaces during production operations. *Related Term:* kiss impression; kiss cut; kiss die cut.

kiss cut — *Related Term:* kiss die cut.

kiss die cut — die cutting which cuts through only the upper layer of adhesive labels, leaving the carrier or backing undisturbed. *Related Term:* kiss cut.

kiss fit — a condition that exists when abutting colors in a knockout come together with no trapping, framing, or keylines.

kiss impression — a very light printing impression.

Kizan — in management, a Japanese term that describes the neverending effort for incremental improvement and customer satisfaction.

knife coating — *Related Term:* blade coating.

knife folder — a folding machine that uses a thin reciprocating knife blade to force a sheet or sheets of paper between two tight fitting, rotating rollers to produce a folded result.

knockout — a shape or object printed by eliminating (knocking out) its specific printing area, often revealing the blank sheet, in contrast to overprinting. *Related Terms:* surprint; dropout; reverse; mask out.

knockout film — *Related Terms:* masking material; Rubylith; Amberlith.

Kodalk — an alkali used in developing formulas as a replacement for carbonate. Carbonate developers are reactive to small changes and greatly affect the action of the developer. Kodalk is a more stable compound and produces minimal undesirable changes.

kraft paper — (1) strong unbleached ground wood paper used to make bags, large envelopes and for wrapping paper. (2) *Related Term:* sulfate paper.

Kromekote — the Champion Paper Company trade name for a high-gloss, cast coated paper.

KU band — frequencies in the 11 to 14 gigahertz band used to send and receive signals to and from satellites. The band nominally associated with small satellite dishes and direct satellite television systems.

LL

L1 cache (level one cache) — a term for the primary cache. *Related Terms:* cache; primary cache; secondary cache; L2 cache.

L2 cache (level two cache) — the term for secondary cache. *Related Terms:* cache; primary cache; primary cache; L1 cache.

L8R (later) — a departing shorthand appended to a comment written in an online forum or chat room.

label paper — a paper with an adhesive coating. Dry gum labels are activated by moisture, pressure-sensitive labels by peeling away from a release paper backing, and heat-sensitive labels by applying heat. *Related Terms:* gummed paper; lick and stick; crack and peel.

lacquer — (1) in platemaking, a solution placed on additive presensitized lithographic plates to make the image visible. (2) in production, a clear coating, usually glossy, applied to a printed piece for protection or special effect. More often called varnish.

lacquer inks — a type of fast-drying printing ink with a vehicle that is thinned by and can be dissolved in lacquer.

lacquer-soluble film — a screen printing stencil film that is hand cut and whose emulsion will dissolve in lacquer.

ladder code — *Related Term:* vertical bar code.

lagging — slowness in a time span. Movement which is retarded in obtaining maximum value.

laid antique — a paper finish that simulates the finish of original handmade paper which used screen segments which were sewn together. The seams between the screen segments created the watermarked characteristics. Laid lines are close together and run against the grain, as compared to chain lines. Today it is often simulated by use of a dandy roll. Used for higher quality stationary. *Related Terms:* laid paper; laid finish; laid lines.

laid finish — an attractive finish on bond or text paper on which grids of parallel lines run against the grain and simulate the surface of handmade paper.

laid lines — the closely spaced watermark lines running across the grain on laid paper.

laid paper — a substrate that shows the wire and chain marks when light passes through it. Usually used for high-quality stationery.

lamer — a less than flattering term used to describe any user who behaves in a stupid or uneducated manner; a description often applied to newbies.

laminate — a thin transparent plastic coating applied to paper or board to provide protection and give it a glossy finish. *Related Terms:* UV coating.

laminating — the process of placing a thin layer of transparent film on one or both sides of a printed sheet to provide protection, durability and give it a glossy finish.

LAN (local area network) — a computer network technology that is designed to connect computers and peripheral devices such as printers and mass storage units in a manner to enable shared access to data files within a single building or campus of buildings separated by a short distance. It can be connected to the Internet and can also be configured as an intranet. *Related Term:* WAN (wide area network).

lands — the reflective portions of all CD tracks; opposite of pits.

landscape mode — (1) horizontal format on a computer screen or printout that can resemble two newsletter pages side by side, or longest image direction is the longest paper dimension. (2) the orientation of tables or illustrations which are printed "sideways," i.e., top margin parallels the long dimension of the substrate. *Related Term:* portrait mode.

laptop — a portable computer, usually battery powered. *Related Terms:* hand-held computer; desktop computer; mainframe.

large format camera — generally, any camera which takes film and produces negatives 4 in. x 5 in. or larger. *Related Terms:* small format camera; medium format camera.

laser — (1) an acronym for Light Amplification by Stimulation of Emitted Radiation. A high-energy, coherent (single wavelength) light source. Unlike most light, laser light may be focused very precisely with little dispersion. The small spot of light produced makes it possible to expose light-sensitive and photoconductive materials at high speed and high resolution. (2) reference lines inserted in the body of an article; also called a refer or sandwich.

laser bond — a bond paper made especially to run well through laser printers. *Related Term:* laser paper.

laser communications — a system for transmitting coherent beams of light which are used as high-capacity communications links. *Related Term:* fiber optics.

laser cutting — (1) a process used in gravure printing to transfer an image to a gravure cylinder by means of a laser that cuts small wells into a plastic surface on the cylinder; the finished plastic surface is then chrome plated. (2) in finishing operations, a technique for cutting intricate designs and patterns into paper quantities without metal dies. A low-power laser cuts designs into the paper as it is directed by an optical scan of prepared line art.

laser engraving — the use of a laser to create an engraved printing plate or die.

laser facsimile platemaking — a system for making printing plates that consists of two laser stations, a paste-up reading station and a plate exposure station.

laser modulation — varying the light intensity from a laser to create an image.

laser paper — specially manufactured paper which is designed to work with the dry toners associated with laser printing.

laser platemaking — The use of lasers for exposing printing plates.

laser printer — a more complex and expensive common printer type. They are capable of producing extremely high-quality text and graphics (including color) at great speed. Current laser printers image at 600 dpi while some high-end newer models operate at up to 2000 dpi. At their most basic, laser printers apply an electrostatic charge to a drum inside the printer cartridge. A laser or a light-emitting diode then discharges portions of the drum to form the characters or graphics. Charged toner attaches itself to these discharged sections. A charged piece of paper is passed over the drum, transferring the toner. The toner is heated and fused to the sheet. *Related Terms:* daisy wheel printer; dot matrix printer; ink jet printer; page printer; nonimpact printing.

laser scanner — (1) in general digital graphics, a device that uses color filters, electronic circuitry and laser light to produce tone and color-corrected separations from color originals mounted on rotating drums. (2) in optical data reading, an optical bar code reading device using a low-energy laser light beam as its source of illumination. *Related Terms:* drum scanner; flat bed scanner.

laser typesetting — a technique whereby a laser light source directly imprints images onto paper, film or plates. *Related Term:* imagesetter.

laser writing — *Related Term:* laser printer.

LATA (Local Access Transport Areas) — local telephone service areas created by divestiture of the local Bell operating companies of AT&T. *Related Term:* baby Bells.

latency — (1) in networking, latency and bandwidth are the two factors that determine the speed of your connection. Latency is the time it takes for a data packet to move across a network connection. (Bandwidth is the capacity of data pipe that carries the data packet.) (2) in video, the length of time excited CRT phosphors require to return to a sleeping state; how long they glow after the electron beam is withdrawn. *Related Term:* bandwidth.

latent image — the invisible change made in film emulsion by exposure to light; chemical development makes a latent image visible to the human eye.

lateral — the end-to-end measurement of an object such as a press cylinder. The width of the press is measured laterally. The cylinder size or cutoff is measured circumferentially (around the outside diameter). *Related Term:* circumferential.

lateral hard dot process — a type of well design used in gravure printing in which a cylinder is exposed by using two separate film positives, a continuous tone and halftone.

lateral reversal — a positive or negative image transposed from left to right as in a mirror reflection of the original. Wrong reading. *Related Term:* tonal reverse.

lay — the position of print on a sheet of paper.

layered — an algorithm that is bandwidth scalable. Such a system will use more bandwidth if or when it becomes available. An example would be a modem transmission which, while transmitting or receiving, may fluctuate—become faster or slower dependent upon the varying traffic load on the system during the transmission.

lay-flat binding — a form of binding perfect-bound publications in which the cover spine is not actually glued to the edges of the bound pages so the book lays flatter when opened. *Related Term:* stay-flat binding.

layout — a guide prepared to show the arrangement and location of all the type, illustrations and line art that are combined together to compose the film flat. A "blueprint" of the job.

layout artist — a person who visualizes how the final printed product will look and makes the base layouts the production workers will use to execute the design.

layout table — a stripping table. Layout tables have translucent glass tops and are lighted from below. *Related Terms:* lineup table; stripping table; light table.

LBA (logical block addressing) — an improved addressing model for IDE hard drives which allow for drives up to 8.4 GB.

lc — a proofreading mark indicating the use of the lower case or small letters of a font of type. The margin notation looks like a script "lc." *Related Term:* miniscules.

LCD (liquid-crystal diode) — created by sandwiching an electrically reactive substance between two electrodes, LCDs can be darkened or lightened by applying and removing current. Large numbers of LCDs grouped closely together can act as pixels in a flat-panel display. *Related Terms:* active matrix; flat-panel display; passive matrix; TFT.

LCL — in the paper industry, an abbreviation for less than carload.

LDAP (Lightweight Directory Access Protocol) — it provides a simple protocol that allows access and searchability of data on the Internet that is often stored on a variety of incompatible systems. These disparate directory file types often contain information such as names, account numbers, birth dates, phone numbers, and addresses.

lead — (1) the thin line-spacing material used in hot type and hand composition; generally no more than 2 points thick. (2) a term sometimes used in place of leading to refer to the placement of spacing material between lines. (3) one of the basic metals utilized in foundry type and hot metal composition systems. The others are tin and antimony. *Related Terms:* leading; tin; antimony; hot metal composition; hand composition.

lead edge — that portion of a sheet that enters the printing press first; for sheet-fed automatic presses, the gripper margin is the lead edge. *Related Term:* gripper (edge).

lead in — the first words in a block of copy set in a contrasting typeface or all caps.

lead in/lead out — the insertion of white space between lines of type for the purposes of readability and fit. The amounts are measured in points or fractions thereof. The process is named after the strips of lead which used to be inserted between lines of metal type. *Related Terms:* leading; line spacing.

leader — a character string consisting of two or more dots, dashes, hyphens or other characters set in a row. Leaders are specified as 2, 3, or 4 to the em and are inserted between text on the left- and right-hand sides of a line. In lists, directories, tables of contents, etc., the leader guides the eye from the left-hand text to the right-hand text in the line. In fine typography they may be specified to align vertically. *Related Term:* dot leader.

leading — (line spacing) (1) in modern typography, the distance of the baseline of a line of type from the baseline of the line above it and less frequently measured from ascender to ascender, expressed in points or hundredths of an inch. Also called film advance and line spacing. (2) traditionally, the insertion of thin alloy metal strips to separate lines in hot metal composition. Pronounced "ledding."

leading edge — *Related Term:* gripper edge.

leads — (Pronounced ledds.) In metal type composition, the thin strips of metal (1 to 2 points) which are used to create space between the lines of type. Leads are less than type high and do not print.

leaf — one sheet of paper in a publication. Each side of a leaf is one page.

leaflet — a single printed sheet folded vertically in the center to produce four pages.

leased line — a dedicated private line that is rented or leased and used primarily to link two remote local area networks (LANs). Unlike frame relay, a leased line transmits data at only one speed, depending on the purchased bandwidth. Usually charged as a flat monthly rate. *Related Terms:* ISDN; frame relay; dedicated line.

LED (light-emitting diode) — in electronic terms, a diode is a semiconductor device through which current can pass in only one direction and that produces either visible or infrared light at a frequency determined by its chemical composition. LEDs require very little power and are often used as indicator lights, including (most likely) the drive access lights on your computer. LEDs are also found in some "laser-quality" printers and as the light source commonly used in wand-type bar code readers. *Related Terms:* diode; laser printer; semiconductor; LCD.

ledger paper — a category of printing paper generally used for printing bookkeeping record forms, statements, legal documents, etc.

left cylinder end guard — the covering over the left end of the cylinder of a vertical cylinder letterpress that serves as a safety device during operation.

left-justified — a paragraph of text in which the left edge is flush and the right edge is ragged. *Related Term:* ragged right.

legal paper — the North American term for bond paper trimmed to 8½ x 14-inch sheets.

legend — (1) in editorial matter, the descriptive matter printed below an illustration, usually referred to as a caption. (2) in design, an explanation of signs or symbols used in timetables or maps, etc.

legibility — one of the major factors in type selection, along with printability, readability and availability. The characteristic of copy having sufficient contrast with the paper on which it appears. It is determined by such features as typeface, size, leading, and quality of printing.

length — (1) a quality of ink that can be tested by tapping an ink with a corner of the ink knife and attempting to draw the ink out into a long string. A good lithographic ink has length. (2) in composition, the horizontal distance from the top margin to the bottom margin.

lens — an optical device usually consisting of precisely polished and shaped glass elements mounted in a barrel that collect and distribute light rays to form a focused image.

lens barrel — the cylinder tube or outside covering and support for the several glass lens elements of an optical device. *Related Term:* lens collar.

lens board — an accurate descriptor for the front case of a camera; the part of a process camera in which the lens is mounted and that holds the lens in alignment with the optical axis of the camera and allows it to move along that axis in connection with the reproduction percentage adjustment. *Related Term:* front case.

lens collar — the ring on the outside of the lens barrel that turns in order to focus the lens. *Related Term:* lens barrel.

lens scale — the aperture/reproduction ratio chart mounted above the lens barrel on a process camera to aid the operator in quickly and accurately setting apertures for any size enlargement or reduction.

lens, long — *Related Term:* telephoto lens.

lens, macro — a camera lens capable of focusing from approximately four inches to two feet. Often referred to as a "close-up" lens.

lens, short — *Related Term:* wide-angle lens.

lens, telephoto — a camera lens with focal length greater than 105mm that significantly magnifies objects.

lens, wide angle — camera lens with focal length less than 40mm whose field of view is wider than the eye normally sees.

lens, zoom — a camera lens that can be adjusted to various focal lengths along a continuum.

Lernout & Hauspie — a specialist in speech recognition, text-to-speech, and digital encoding of speech and music using subband coding, code book excited linear predictive coding and harmonic coding for a range of compression rates and sound qualities. *Related Term:* codec.

less carton — *Related Term:* broken carton.

less than carload — an amount of paper weighing less than 20,000 or 40,000 pounds, depending on the mill or merchant's definition of carload. Abbreviated LCL.

Letraset — a proprietary name for rub-down or dry transfer lettering used in preparing artwork.

letter — a graphic which, when used alone or combined with others represents in a written language one or more sound elements of the spoken language. Diacritical marks and punctuation marks used alone are not letters.

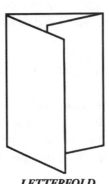

letter fold — two parallel folds which create a product which will fit into a #9, 10 or 11 business envelope.

letter paper — in North America, 8 1/2 x 11-inch sheets. In Europe, A4 sheets.

letter quality — type of the quality produced by typewriter keys, type element or daisy wheel.

LETTERFOLD
A sheet folded, usually into thirds, each fold being in the same direction relative to the face of the printed sheet.

letter spacing — (1) a hand or hot metal composition term meaning the adding of space between the individual letters in order to fill out a line of type to a given measure or to improve appearance. (2) in phototypesetting or imagesetting, either positive or negative (minusing) incremental adjustments to the space between letters to achieve the desired appearance of text and to aid justification. *Related Term:* kerning.

letter, primary — a lowercase letter such as "e," "m," "n," "o," or "c" that does not have ascenders or descenders.

letterform — the individual characters in a particular typeface.

letterhead — the imprecise terminology used to describe the stationery set system used by a business or professional organization. Although actually only the paper used for writing letters, it is often used to refer to the paper, envelopes, business cards, etc., which share common design features, colors, and paper and are used in a common environment.

lettering template — a device, usually made of plastic, containing letter and other image shapes that guide the technical inking pen when forming images.

letterpress — the method of printing in which the image, or ink-bearing areas, of the printing plate are in relief, i.e., raised above the nonimage areas. When ink is applied to the surface, only the high areas contact the inking mechanism. Paper applied to the raised, inked surfaces under pressure transfer the image. *Related Terms:* flexography; relief plate; relief printing.

letterset — an offset letterpress printing process that uses shallow flexible relief plates to print on an offsetting blanket as in a lithography press.

lettershop — (1) properly, an alternate term for a mailing service company. (2) generally, a term incorrectly applied to small print shops. *Related Term:* mailing service.

letter-size — an envelope that fits an 8 1/2 x 11-inch sheet of paper that has been folded twice. *Related Term:* #10 envelope.

letterspacing — the addition of space between the letters of words to increase the line length to a required width or to improve the appearance of a line.

lever cutter –hand-operated drop knife paper cutter. *Related Term:* guillotine cutter.

libel — written defamation that causes injury to another person.

library — (1) in general, a repository of knowledge; a place for books, periodicals and reference materials. (2) in computing, a collection of software, files and other electronically stored information including archives.

library corner — a special technique for folding book cloth around a corner of binder's board when handbinding a casebound book.

library edition — a book with reinforced binding to withstand extended use.

Library of Congress — the national library serving the United States Congress and the American public.

Library of Congress Catalog Card number — a unique number assigned by the Library of Congress to a given work for cataloging and identification purposes.

library picture — a picture taken from an existing library and not specially commissioned.

library rate — a discount postal rate for shipping books to or from libraries and educational institutions. *Related Term:* book rate.

lick and stick — a slang term for dry gum or moisture-activated paper.

LIFO — an inventory, **acc**ounting and accrual system. Last In, First Out.

lift — the number of sheets that a worker can conveniently pick up or handle at one time without the aid of a machine.

file flower office
file flower office

SOME COMMON LIGATURES
Ligatures reduce the number of individual characters in a line, and contribute to readability by producing more easily recognized word forms.

ligature — in metal or cast type, two or three characters joined on one body or matrix to enhance visual appeal and/or legibility, such as ff, ffl, m, Ta, Wa, Ya, etc. Not to be confused with characters used in logotypes cast on a single body, nor dipthongs which are combination characters which are phonetically based and will affect word pronunciation.

light — (1) a term used to describe body text, which is usually set in type that is less bold than the roman typeface in the same family and size. (2) electromagnetic energy with wavelengths (about 380 to 750 nm) that affect vision. (3) any radiation which produces change in the optical or physical properties of materials. *Related Terms:* light face.

light integrator — a device that measures the intensity rather than the duration of an illumination source. The system is computer driven and accumulates total units of light without regard to line voltage, surges or brownouts. Much more accurate exposures than with a regular timer.

light meter — more correctly it should be called a luminance meter, as it is an electronic device used to measure the amount of energy available for making an exposure on photographic film. It has three scales. One is set for the speed of the emulsion used and one is set to the meter reading of the illumination source. The correct exposure (in seconds or fractions of a second) is read from the third scale opposite the various apertures (lens openings) that can be used. *Related Terms:* exposure meter; luminance meter; photometer.

light pen — (1) in computing, an input device used in conjunction with a video display. The pen is touched to the display screen to identify the point to be processed. (2) in data acquisition and optical reading, a hand-held pen-like contact reader, which the user must sweep across the bar code symbol to read the code. *Related Terms:* mouse; track ball; joy stick; stylus.

light polarization — the reduction or elimination of reflections and glare by selectively eliminating the offending light waves travelling in specific directions and not affecting others.

light printer — one of the two halftones used to make a duotone; is usually printed with a light-colored ink and often contains the highlight to upper-middle tones of the continuous tone original.

light safe — a cabinet or room where light cannot enter and sensitive material such as film, plates and coated direct photoscreen printing frames can be stored without being subjected to light.

light source — the active illumination device in an observation instrument or in a visual observing situation.

light spectrum — the electromagnetic wavelengths (about 380 to 750 nanometers) that are visible.

light table — a glass-topped work area illuminated from underneath and used by production artists and for stripping negatives, opaquing and otherwise viewing images with transmitted light.

light trap — two or more connected passages leading into a darkroom through blackout curtains, revolving doors, etc., within the passages, prohibiting unwanted outside light from entering.

light valve imaging — in video and media, a powerful high-end projection process in which red, green, and blue halogen beams are used in place of the RGB CRTs and lens of conventional video projectors. The technique creates massive, bright displays of video signals for amphitheaters, convention halls, etc.

lightface — the part of a font family that has characters with strokes that have less weight than the normal or regular face.

lightfast — the desirable characteristic of some papers and other substrates to resist fading when exposed to bright light. Sometimes called lightfastness.

lightfastness — the ability of paper to resist fading or yellowing when exposed to light.

lightness — perception by which white objects are distinguished from gray objects and light from dark-colored objects.

lightness, dot — in production, the attempted balance between minimum and maximum dot gain.

light-sensitive emulsion (1) a liquid containing minute particles of light-sensitive silver halide material suspended in a gelatin solution. (2) a material that is chemically altered after it is exposed to light.

lightweight paper — any book-grade paper with a basis substance weight of 40 or less with high opacity for its weight. Used to reduce weight and bulk, thus keeping postage and space to a minimum.

lignin — the natural glue-like material that bonds cellulose wood fibers together in trees.

limestone — a porous rock, naturally formed from organic remains such as shells, used by Alois Senefelder, the inventor of lithographic printing. It is still used today by artists to make original lithographic prints.

limited edition — a special (usually) limited quantity printing of a book often signed by the author and sometimes in a special case or slip cover.

line advance — in phototypesetting, the space achieved by advancing the photographic material after composition of individual lines of type; commonly called "line" or "film" advance. Corresponds to the combined total of set size and leading in hand composition and/or hot metal composition systems. *Related Term:* line spacing.

line art — a piece of art or a plate exclusively in black and white, with no intermediate shades of gray. Examples are line work, type, rules, etc. *Related Term:* line copy.

line block — a letterpress printing plate made up of solid areas and lines and without tones.

line break — in word processing computer programs, a special character or command that forces a new line on the page without creating a new paragraph.

line conversion screen — a piece of film containing line patterns that break light into those patterns and through which light passes to render exposure.

line copy — any high-contrast copy that has only black and white image elements and no shades of gray. Type matter and drawings that can be reproduced without the use of a halftone screen. *Related Terms:* line art; line work; line drawing; line film.

line drawing — a black-and-white drawing with no shades of gray. *Related Terms:* line art; line copy.

line gauge — a ruler scaled in picas and points for a typographer's use. Line gauges come in different sizes, materials, and configurations and may include other scales such as inches, agate lines, etc. Used for copyfitting and layout measurements. Used for copyfitting and measuring typographic materials. *Related Terms:* pica rule; pica pole.

line length — the width to which a justified line of type is set. The length is usually expressed in points and picas in the U.S. and England or in Didots and ciceros in continental Europe. *Related Term:* line measure.

line measure — *Related Term:* line length.

line negative — a high-contrast photographic film made from line copy. Black areas correspond to white areas in the original copy and clear areas correspond to black or red areas.

line photography — the process of recording high-contrast images on pieces of film.

line provisioning — line provisioning is the way telephone companies set up ISDN to work with customer equipment. Since central office switching and hardware features are nonstandard, customers must specify how the ISDN line should be provisioned when they order it. Some hardware manufacturers have assigned codes to make ISDN configuration easier, but this procedure is necessary because of the multiple uses of ISDN and the fact that it is not yet fully standardized. *Related Term:* ISDN.

line screen — a reference to the number of dots in a halftone screen, per linear inch, horizontally and vertically. The screen ruling. Common screens are 65, 85, 100, 110, 120, 133, 150, 200 and 300 dots per inch. The higher the number, the more detail can be reproduced. Selection is based upon product, substrate, printing process, etc. Higher ruled screens require better paper surfaces to maintain their reproductive detail.

line spacing — a term used for leading in phototypesetting. Measured in points from one baseline to the next baseline. *Related Terms:* leading; lead; interline spacing; line advance.

linear — a motion sequence designed to be played from beginning to end without stops or branching, like film or audiotape. *Related Term:* interactive.

linear array — an arrangement of individual sensors arranged in a line which can be moved along a linear path and activated sequentially to capture data over an area. This contrasts with area array, where the en-

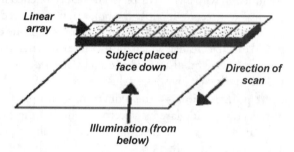

LINEAR ARRAY
Utilizes a small number of sensors in a row which are turned on and off as the array progresses down the length of the subject.

tire field of concern is covered and all are activated simultaneously.

linear editing — edits performed directly from videotape, requiring tape shuttling to access specific frames of video.

linear fill — a fill that is projected from one point to another in a straight line.

linearity — linearity refers to a monitor's ability to display shapes such as squares or circles in various places without any distortion. Poor linearity causes screen objects to be distorted. *Related Terms:* pincushioning; screen geometry; trapezoid error.

linecasting machine — a keyboard or tape-controlled hot metal device that sets complete lines of type using molten metal and intaglio matrices. *Related Terms:* Linotype; Intertype; Monotype.

linen finish — an embossed or impressed machine

paper finish on text paper that simulates the pattern of a linen cloth. Not to be confused with wove paper, which has a smoother appearance.

linen tester — a small magnifying glass mounted at a distance above its base equal to the focal length of the lens. Originally designed for counting threads in linen, modern achromatic linen testers are widely used to examine negatives, plates, and proofs. *Related Terms:* loupe; magnifier; graphic arts magnifier; printer's magnifier.

lines per inch (lpi) — the rating of mechanical tint and halftone screens indicating the number of rows of dots both vertically and horizontally per linear inch. Screens with a higher number, such as 120 or 133, have a higher resolution than screens with lower numbers, such as 65 lines per inch. *Related Terms:* screen ruling; halftone screen.

lines per minute (lpm) — a speed rating given to all typesetters. Expressed in terms of due number of lines of 8-point type, 11 picas wide, set in one minute. *Related Term:* pages per minute (ppm).

lineup table — a table with an illuminated top surface equipped with precision-geared ruling devices and used for preparing and checking alignment of page layouts and pasteups. *Related Terms:* light table; stripping table.

lining — a stiff piece of paper placed over the backbone of a book body before the case is attached.

lining figures — numerals that align on both the base line and top and are the same size as caps in a typeface: **1,2,3,4,5,6,7,8,9,10**. As opposed to oldstyle figures: 1,2,3,4,5,6,7,8,9,10. *Related Terms:* modern figures; oldstyle figures.

link — (1) in typography, the connecting line of a lowercase g between the loop and the upper bowl. (2) on the Internet, it is text you find on a Web site which can be selected with a mouse click which takes you to another area of that or another Web page. Links are also used to load multimedia video, movies and sound files. (3) a communication pathway between nodes. *Related Terms:* hyperlink; broken links.

Linotronic — a brand of PostScript imagesetter now manufactured by the Heidelberg Corporation.

Linotype — (1) originally, a Mergenthaler trade name for a brand of hot metal line casting machine that produced lines of type by casting all of the letters in a single piece of metal called a slug. Corrections to such composition are made by

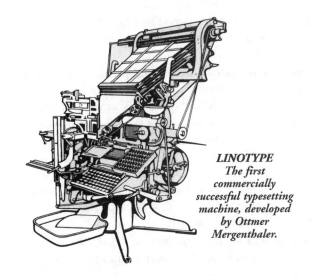

LINOTYPE
The first commercially successful typesetting machine, developed by Ottmer Mergenthaler.

merely replacing the slugs which contain errors as opposed to individual letters. (2) a manufacturer of a range of high-resolution phototypesetting machines capable of processing PostScript files through an external RIP and typesetting desktop publishing files direct from disk at 1,270 dpi and beyond. The company has been absorbed into the Heidelberg corporation. *Related Terms:* Intertype; Monotype; Ludlow.

Linotype slug — the solidified cast of molten metal delivered from a Linotype machine. Lines of relief type, ready for printing.

LINOTYPE SLUG

linseed oil varnish — a liquid material used as a vehicle or pigment carrier to make printing ink. *Related Term:* lith varnish.

lint — loose fibers not securely bonded to paper, thus causing shedding and picking during printing.

Linux — a freely distributed, Intel-processor-based alternative to UNIX, it is used by hundreds of thousands of people around the world. It has made some inroads into corporate life as an inexpensive substitute for high-priced UNIX Web servers. Linux is available from a number of vendors for several hardware platforms, including Intel x86, Compaq Alpha, Sun Sparc Stations and the PowerPC.

liquid crystal display (LCD) — a computer screen

in which images are formed by a liquid crystal material instead of the conventional cathode ray tube (CRT).

liquid laminate — any liquid plastic applied to a substrate then bonded and cured to a tough, smooth finish. Often cured with IR or UV energy sources.

list — (1) in word processing, a group of paragraphs formatted to indicate membership in a set or in a sequence. Most word processors and Web builder programs can create numbered lists or bulleted lists, menus, directories, or definitions. (2) in publication printing, a list of the titles a book publisher has in print and available for sale.

list broker — a person who sells and rents lists of names and addresses for direct mail campaigns.

list price — the suggested retail selling price of a product, as opposed to the net price or discount price, which is the price at which retailers or distributors purchase it from the manufacturer. *Related Term:* SRP (suggested retail price).

Listserv — software for setting up and maintaining discussion groups, automated mailing list distributions, etc. Many Listseerv discussion groups are a gateway to Usnet newsgroups.

liter — a unit of liquid measure in the metric system.

literary agent — *Related Term:* agent.

lithography — a printing technique in the planographic classification of printing processes. The largest segment of commercial printing, worldwide. Based on the principle that grease and water do not mix. Commercially, an image carrier chemically prepared so that the nonimage areas are hydrophilic (receptive to water), i.e., dampening or fountain solution. These areas repel ink while the image areas are made oleophilic (receptive to ink) and repel water. Ink is transferred directly to substrate or to a rubber-covered cylinder and then transferred (offset) onto the printing surface. The image carrier is said to be planographic, or flat and smooth.

lithol rubine — a reddish pigment used for making magenta inks. This pigment has relatively poor blue light reflection.

live — on the Web, designates an object file linked to another layer of information or describes when a particular Web site will be placed on the Internet, such as "it's going live next week."

live area — *Related Term:* image area.

live form — active work in the form of plates ready for printing. *Related Terms:* kill; dead matter.

Live3D — a method of extending browsers into the 3D realm. With this virtual reality modeling language (VRML) viewer, you can experience a rich new world of 3D spaces and interact with text, images, animation, sound, music, and even video.

LMP (Literary Marketplace) — a directory of people and firms in the book publishing industry. A listing of publishers and agents.

load — (1) on the Web, the process of transferring HTML documents and graphics from the web server into the local browser whenever a URL is accessed. (2) in computing, the process of sending data such as programs from the distribution format into the main memory or storage of a computer. You load programs.

loading — (1) in papermaking, the process of adding clay to pulps that will be made into uncoated papers. (2) in management, scheduling production hours by machine, i.e., press loading, bindery loading, etc.

local area network (LAN) — interconnected computers that can share programs and data files as well as the use of peripheral devices such as printers or CD-ROM drives. Each microcomputer connected to a LAN will typically require a network circuit board and software. A LAN allows many computers to access the same information files. *Related Terms:* WAN; internet; intranet; extranet.

lockup — the procedure used to hold a letterpress type form in a frame or chase. *Related Term:* chase.

log — *Related Term:* master roll.

log off — the opposite of log on, to disconnect from a network or remote system. *Related Term:* login or log in.

login or log in — (1) as a noun, the account name or username used to access a computer system. (2) as a verb "to log in," the term means the act of typing in your username and password.

login password — the required secret word used to log into a program, internet service provider, network or other controlled-access computing system.

LOGMARS (Logistics Applications of Automated Marking and Reading Symbols) — a U.S. Department of Defense program to place a Code 39 bar code symbol on all federal items.

logo — a corporate or organizational graphic identifier. *Related Term:* logotype.

logon — the act of connecting to a network or remote system. The opposite of log off. *Related Term:* login or log in.

logotype — a special ligature, symbol, trademark emblem, trade name or any other combination of art, characters, words, or phrases produced as a single graphic and used to represent the name of a publication, business, company, product or organization. *Related Term:* logo.

LOL (laughing out loud) — a shorthand term used in postings and on-line chat to show appreciation of a witticism in a previous posting. *Related Term:* ROFL; e-mail shorthand.

long grain paper — *Related Term:* grain long (paper).

long ink — thin ink with low viscosity that flows well on press. Typical of gravure and flexographic inks.

long lens — *Related Terms:* telephoto lens; lens, long.

Look@Me — software package that provides the ability to view another user's screen anywhere in the world. If they are running the companion software, you can watch their screen activity taking place, in real time, from within your Internet browser or as a stand-alone applet. It is a collaboration tool that allows the editing of documents, review of presentations and/or graphics, or providing training and support.

lookup table (LUT) (1) information stored in computer memory, that contains the dot sizes needed to reproduce given colors. The processed input signals of certain color scanners are used to search the table to find the values that will produce the color represented by the signals. (2) organized fixed reference data stored for recall in a computer. When accessed by an outside program, it will respond with appropriate input based upon the parameters established for the query.

loop – the lower part of the g, also called the tail.

loop knife — a cutting tool with a loop or circle-shaped blade attached to a pen-shaped handle; used to cut screen printing film and stripping film. *Related Term:* bean cutter.

loop stitch — a saddle stitch with staples that are also loops which slip over rings of binders.

loose proof — a proof of one color separation.

loose-leaf — *Related Term:* loose-leaf binding.

loose-leaf binding — (1) a nonrigid binding category that includes ring binding, post binding base, prong binding and action binding. (2) a method of binding which allows the insertion and removal of pages for continuous updating.

lossless — describes a compression system for digital photography and video files that, when decompressed, has no loss of information and no image degradation. *Related Term:* lossy.

lossy — a slightly degenerative compression method, and one of the two categories of compression techniques. Lossy techniques, such as JPEG, crunch files down smaller, but they throw out image quality in the process. Almost all graphic files are big, and most are larger than they need to be because their file formats are inefficiently coded. Lossless techniques throw away redundant bits of information without affecting the quality of the image and, generally, you can't see the difference unless you try to enlarge them significantly or output them to an imagesetter. *Related Term:* JPEG; LZW; lossless.

loupe — *Related Term:* graphic arts magnifier.

low finish – a paper finish that is relatively dull, as compared to high finish. *Related Term:* flat finish.

low key — a photographic or printed image in which the main interest area lies in the shadow end of the scale.

low noise amplifier — in satellite receivers, a special amplifier that boosts the satellite signal while contributing a negligible amount of noise.

low resolution — an image or screen in relatively coarse detail. In raster-oriented printing or displays, low resolution has to do with the number of pixels or dots used to reproduce the image. The fewer the pixels, the lower the resolution.

lower rail — the term originally used to designate the lower position on a matrix of a hot metal line casting machine which usually contained roman characters. *Related Term:* upper rail (for italics or bold face).

lowercase — (1) generally, the uncapitalized letters of the font alphabet. Originally called lowercase because the lead type versions were located in the lower portion of the printer's type case. (2) in proofreading, instructions to set lower case letters; the margin symbol looks like this: *Related Terms:* minuscules; uppercase; majuscules.

lpi (lines per inch) — the number of rows of halftone cells or dots per linear inch. The screen frequency. The larger the frequency number, the less noticeable the halftone dots. *Related Term:* lines per inch.

lpm (lines per minute) — an output rating for imagesetters. *Related Term:* pages per minute.

LS/2000 (local system) — a library automation system which provides catalog, circulation, serials and acquisition systems.

lucy — an overhead projector-type device to assist artists in drawing proportionally to scale. *Related Term:* camera lucida.

Ludlow — a hot type headline and display casting machine without a keyboard; specifically designed for larger type sizes, it casts a line of type from manually set brass matrices assembled in a special composing stick. The assembled matrices were injected with molten metal to produce the line of letters in one piece.

luminance — brightness. (1) generally, one of the three image characteristics coded in composite television, represented by the letter "Y." It may be measured in lux or foot candles. (2) in color management, the process of matching the amplitudes of red, green and blue signals so that the equal mixture of all three will produce an accurate white color or temperature.

luminosity — a value corresponding to the brightness of color.

lurker — the name for a person in a chat room or newsgroup who listens but does not participate. Lurking is encouraged for beginning users who wish to become acquainted with a particular discussion before joining in.

LUT — *Related Term:* lookup table.

lux — the metric unit of illumination. *Related Term:* footcandle.

luxometer — a photoelectric device used to control the operation of camera and other exposures according to actinicity and fluctuation of camera lamps, unbiased by time or duration. *Related Term:* light integrator.

LViewPro — a shareware graphics viewer/editor program for Windows that can be used as a helper application or program for Web browsers. It reads JPEG, TIFF, GIF, PCX, BMP, PBM, PGM, PPM, and Targa files and enjoys great popularity because of its ability to create transparent GIFs.

Lycos — a database of several million Web sites. Users can set custom search configurations to help find what they are looking for.

Lynx — a very fast, text-based World Wide Web browser. Unlike graphical browsers, it cannot display images or handle Java, but that is what gives it speed.

LZW compression —a lossless data compression technique that is an adaptation of two techniques by Abraham Lempel and Jacob Ziv. *Related Term:* STAC LZS compression.

Mm

M — (1) Roman numeral for 1,000. (2) In paper terminology, M refers to 1,000 sheets. It is prominent in referring to paper quantities, either when purchased or in product pricing, such as $25.00/M. (3) in computing, one megabyte or one million bytes. (4) in color reproduction, the abbreviation for the color magenta.

M weight — the weight of 1,000 sheets of paper in a specific basic size.

machine coated — any paper that is coated on one or both sides as part of the manufacturing operation as it comes off of the paper machine.

machine composition — a generic and very general term for the composition of metal type matter using mechanical means, as opposed to hand composition.

machine direction — the predominant direction in which the fibers in a sheet of paper lie, corresponding with the travel direction of the Fourdrinier screen on the papermaking machine. *Related Term:* grain direction.

machine finish — an off-machine method of smoothing a paper's surface by passing it through sets of calendering rollers on the papermaking machine. A grade of paper with a slightly less smooth finish than a grade of book paper.

machine glazed (MG) — paper with a high gloss finish on one side only.

machine language — the coded language understood and used directly by a computer. *Related Term:* machine readable.

machine readable — a general term for printed material which can be directly transferred to a data processing system. *Related Term:* human readable.

MacPaint — an obsolete format on the Macintosh computer for storing and transferring low-resolution, monochromatic bit-mapped images. It originated with the paint application of the same name. Now only supported by Photoshop and Streamline as a method of accessing old files.

macro — a series of instructions such as a typing a frequently used phrase, sentence, paragraph or other input which would normally be issued one at a time on the keyboard to control a program. A macro facility allows them to be stored and issued automatically by a single keystroke. *Related Term:* script.

macro lens — any camera lens capable of focusing from approximately four inches to two feet. Often referred to as a extreme close-up lens. *Related Terms:* telephoto lens; wide-angle lens; close-up lens.

magazine — (1) in hot metal composition, the storage case for matrixes (molds) in a hot metal line or slug typecasting machine. (2) in publication work, a printed booklet with a regular publication schedule, containing a variety of topics.

magenta — (1) a subtractive primary color created by a combination of the two additive primaries, blue and red. (2) one of the color printers used in four-color process printing. *Related Term:* process red.

magenta printer — (1) the plate that prints the magenta ink during four-color process printing. (2) the color separation film that will be used to produce the magenta printing plate.

magenta screen — a dyed contact screen used in making halftones, as compared to a gray contact screen. The magenta dye is designed to make contrast control easier for the cameraperson.

magnesium — an extremely strong, lightweight metal used in making photoengraving relief printing plates.

magnetic ink — used extensively by the banking and airline industries for checks and tickets, it is an ink with pigments that are treated with ferrous oxide and can be magnetized. Magnetically sensitive electronic equipment recognizes and reads specially shaped printed characters based upon footprint strength of the magnetic signal generated by each character of a special font of numbers and symbols. It is not an optical character recognition system. *Related Term:* MICR encoding.

magnetic media — any form of storage medium using a magnetic surface to capture and regain data. Floppy disks, cassettes, tapes and floptical disks are of this type.

magnetic printing — a printing method in which a magnetic print head transfers its image to a magnetized drum that picks up toner with the opposite magnetic polarity and transfers it to the substrate to form the printed image when the drum is demagnetized. *Related Term:* magnetographic printing.

magnetic tape — a Mylar tape or ribbon impregnated with magnetic material on which information may be placed in the form of magnetically polarized spots. Used to store data which can be recalled later.

magneto-optical — a type of storage disk that uses optical particles fused to magnetic particles. When heated by a laser and subjected to a magnetic field, the particle may be rotated on or off repeatedly. Permits an optical disk to be erased and used again.

magnification range — the amount of enlargement or reduction that a scanner or camera is capable of providing. A range of 20% to 3,000% are about the extremes of most commonly used devices.

magnifier — a convex lens system used to enlarge the appearance of objects and to examine photographic and printing detail. *Related Terms:* linen tester; loupe; graphic arts magnifier.

mail bomb — flooding of an e-mail address with angry or nasty messages.

mail filter — an e-mail sorting program that allows the user to classify e-mail messages according to header information.

mail fulfillment — to pick from inventory, pack, wrap and ship (fulfill) orders for merchandise via mail, UPS, FedEx, etc.

mail order — to place orders for merchandise via mail, usually from printed catalogs or a video presentation.

mail server — a software program that distributes files or information in response to requests sent via e-mail. Mail servers have also been used to provide FTP-like services.

mailbot — an e-mail server that automatically responds to requests for information. *Related Term:* bot.

mail-from — an authentication scheme where the e-mail address that is used to send a request is compared to the e-mail addresses that are on file. *Related Terms:* authentication; authorization.

mailing list — an e-mail address which expands to multiple e-mail addresses based upon participant subscriptions. Usually they are confined to specific topics of information.

mailing service — business specializing in addressing and mailing large quantities of printed pieces.

mail-to — an Internet protocol that is used to send electronic mail. You can create mail-to hyperlinks (mailto://) in the FrontPage Editor.

main exposure — the primary exposure made through the lens and a halftone screen; records detail from the highlights or white parts of a photograph to the upper-middle tones. Additional secondary (bump and flash) exposures may be made to improve detail. *Related Terms:* highlight exposure; detail exposure.

mainframe — a large computer originally manufactured in a modular fashion and centrally located, as in a data processing department. While still important, much of its significance has been eroded by the development of desktop computers.

maintenance — (1) in Internet computing, the processing, research, support, and administration involved in creating domain name records, contact records, and host records. (2) in production, the scheduled upkeep of machinery to ensure uninterrupted performance and equipment longevity.

majordomo — a mailing list processor which runs under UNIX. *Related Term:* Listserv.

majuscules — capital letters. *Related Term:* minuscules.

makegood — a free ad that is run in a publication to replace an ad that was previously run in an incorrect fashion (to make good), such as in the wrong publication page position or one with an error(s) caused by the publisher.

makeover — any printing plate that is remade.

makeready — in presswork, the physical prep-aration of a press for printing, putting on plates, adjusting impression, ink and image orientation on the sheet, i.e., all activities required to set up a printing press before production begins. *Related Term:* spoilage. Sometimes spelled as two words.

makeready sheet — a piece of paper containing affixed pieces of thin tissue paper in selected areas, which is placed under the form and used to equalize the impression of the type for letterpress printing.

makeup — assembling the typographic elements (type and engravings) and adding space to form a page or a group of pages of a newspaper, magazine, or book. *Related Term:* imposition.

making order — a custom order for paper that a mill makes to fill the customer's specifications.

making ready — (1) originally the specific time spent preparing make-ready sheets and associated packing to level a letterpress form on a press, prior to printing. (2) now, the time spent in performing all make-ready operations. Sometimes spelled as a single word.

malkinization — any form of chipping and/or scratching that occurs when two coats of solid ink have been applied to a cover stock but have not been protected by a UV or IR coating or lamination.

manifold bond — *Related Term:* onionskin.

manila paper — the strong, buff-colored paper used to make envelopes and file folders.

manual — (1) not automated; processed by hand. (2) a book of operating instructions.

manufacture execution system (MES) — the process of sharing shop floor data in real time across several departments using job site terminals and input devices.

manufacturer's ID — in UPC coding, the first 6 digits of the 12-digit UPC number applied by the UCC to uniquely identify a manufacturer or company selling products under its own name. *Related Term:* UCC.

manuscript (ms) — (1) the original from which type will be set and editors use to check. (2) traditionally, a reference to handwritten, as opposed to typewritten, material.

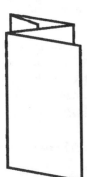

map fold — a type of fold which is used for map folding, although rarely in its basic form. It is often used as a combination or compound fold to accommodate larger sheets.

Mapedit — a graphical editor for Web-clickable image maps. With it and current release graphical Web browsers, you can use client-side image maps, which reside in your HTML page.

COMMON MAP FOLD

MAPI (Messaging Application Programming Interface) — an interface developed by Microsoft and other companies to enable Windows applications to access a variety of messaging systems, from Microsoft Mail to Novell's MHS. MAPI also works on a more everyday level: so-called mail-aware applications can exchange both mail and data with others on a network. *Related Term:* API.

margin — the white space on a page surrounding, beside, above and below text or illustrations.

Marionet — an Apple-based Internet scripting tool for the rapid development of customized interfaces that automate and simplify Internet tasks. Interfaces can be created in any authoring environment that supports both interapplication communication and provides a high-level interface to the standard protocols, such as those for e-mail (SMTP/POP3), file transfer (FTP), newsgroup (NNTP), HTTP/CGI, and Gopher.

markup — the process of indicating typographic specification such as measure, point size, face, etc., in preparation for text input. A set of production specifications.

masking equations — a set of linear formulas used to determine the mask strength required for color correction. They are based on the unwanted absorption of the colorants and assume perfect additivity and proportionality.

masking material — opaque material, often film, used in paste-up to outline photographs or in platemaking to withhold light from nonimage areas. *Related Term:* mask.

masking sheet — the opaque base material usually made of paper or plastic into which film negatives are affixed to make a flat for platemaking.

mass communication — communication to large numbers of people. Mass media include radio, television, magazines, daily and weekly newspapers, the Internet, etc.

mass tone — color of ink in mass as on an ink knife or in its container. Often differs from the printed color of the ink, which tends to be of lighter hue and saturation.

mass-market paperback — a smaller, lower priced paperbound edition of a previously published hardcover or trade paperback book. Usually printed on less expensive paper and sold in more mass-market outlets such as discount houses, etc.

master — (1) a common term for hard copy, digital files, paper or plastic offset printing plates. (2) paper carbon set used for spirit duplicating.

master flat — in multi-flat registration, the flat with the most detail; the one to which subsequent flats are positioned. The key flat.

master page(s) — a feature in many page layout programs that allows the user to specify repeating text and graphic elements that will appear on the pages of a publication.

master plate — in gravure printing, a frame with register pins that hold the film positives in correct printing position during exposure to the cylinder masking material.

master roll — the roll of paper as taken off the papermaking machine, before slitting into rolls for sheeting or winding for web presses. *Related Terms:* jumbo roll; log; reel.

mastering — etching the original CD-ROM disk using information from the premastering data.

masthead — (1) the block of information in a newsletter or newspaper that identifies publisher and editor, staff, owners, and subscription information for a periodical and tells how to contact them and may give subscription and business information. (2) sometimes erroneously, term applied to the banner headline or nameplate of a publication.

match color — in printing, the duplication of a specified color by using either multiple process colors or special flat colors. Match colors may be defined by supplied samples or by numbers from a color matching system.

Matchprint — originally a 3M (now Imation) trade name for their laminated color proofing material.

material — (1) any light-safe material used in photo processes to control light, such as goldenrod, goldenplast, Rubylith, studnite, etc. The opaque paper or plastic used to prevent light from reaching selected areas of film or printing plate. (2) the general term used to describe massive areas of leading used to support stereotypes and engravings in a letterpress form.

material safety data sheets (MSDS) — a product specification form used to identify chemical substances and their physical properties and access the potential hazards and medical treatments involved in their use.

mathematical signs — the typographic symbols used for equations and math composition such as = (equals), + (plus), - (minus), × (times), ÷ (divided by) and % (percent).

matrix — (1) in foundry type, an arrangement of typesetting characters in a hand composition case or hot metal font grid or magazine. (2) in hot metal typesetting, the individual molds from which relief metal type characters are formed by pouring or pressing hot metal. (3) in flexographic platemaking, the mold cast from a metal engraving into which rubber-like thermosetting materials are placed to form a flexographic printing plate. (4) in rubber stamping, the special fiber material used to create a sunken but right-reading mold for vulcanizing a rubber stamp.

matrix symbol scanners — special symbol scanners designed to evaluate data from a two-dimensional area, locating each printed element in both x and y coordi-

nates simultaneously. CCD camera technology is often employed to facilitate the process.

matrix symbols — in optical reading, a pattern that can appear as a checkerboard pattern, most likely square in shape, and which contains some form of unique "finder pattern" to distinguish it from others and provide a decoding reference for scanners.

matte — (1) in paper manufacturing, a dull surface that scatters the specula component of light, thus causing the underlying tone to appear lighter. Generally, a paper lacking gloss or luster. (2) the dull, clear coating applied to printed materials for protection or appearance. *Related Terms:* matte art; matte finish.

matte art — *Related Terms:* matte.

matte finish — (1) in photography, a textured, finely grained finish on a photo print as opposed to glossy. (2) a paper with an uncalendered, lightly finished or calendered surface. (3) A type of surface or coating that is dull, without gloss or luster.

matte ink — an ink that appears dull when dry.

matte print — a photograph that has a matte finish.

matter — a term applied to any composed text, especially body copy, i.e., type matter, set matter, etc.

mature — *Related Term:* condition.

Maxwell triangle — the equilateral color triangle devised by James Clerk Maxwell in 1851 to show the composition of the ranges of colors produced by additive mixtures of red, green, and blue light.

Mb — (megabyte) — one million bytes. *Related Term:* Meg(s).

MBone (The Internet Multicast Backbone) — a "virtual network" used for audio and video group conferencing, weather satellite images of the earth, U.S. Senate and House of Representative meetings and, "Radio Free Vat," and "Internet talk radio" to name a few. The MBone is a part of the internet. *Related Term:* MultiCast.

MBPS (megabits per second) — the abbreviation used to describe data transmission speeds of one millions bits per second, such as the rate at which information travels over the Internet. Factors that can influence transmission speed includes modem type and speed, bandwidth capacity, and network or Internet traffic levels. *Related Terms:* V.90; T1; T3; modem; bandwidth.

mean line — one of the typographic lines of reference. The line that marks the tops of lowercase letters without ascenders. *Related Terms:* base line; ascender line; descender line; x-height; waistline.

measure — denotes the line length of a type expressed in pica and pica fractions (points). *Related Term:* line length.

mechanical — the camera-ready assembly of all type and design elements pasted on art or illustration board in exact position and containing instructions, either in the margins or on an overlay, for the production operations. It can also be in the form of a digital print-out, or a digital file ready for high-resolution output. *Related Terms:* paste-up; final layout; camera-ready copy; keyline; artwork; mechanical layout.

mechanical artist — *Related Term:* production artist.

mechanical binding — a method of binding which secures pretrimmed leaves by the insertion of wire or plastic spirals through holes drilled in the binding edge. *Related Terms:* looseleaf binding; perfect binding; case binding.

mechanical color — color art copy separated for each color by hand rather than by electronic color separation methods. *Related Term:* mechanical separation.

mechanical composition — when material to be printed is drawn by the artist or pasted up from pre-printed sheets. *Related Term:* paste-up.

mechanical dot gain — *Related Term:* dot gain.

mechanical fabric treatment — the process of scrubbing screen printing fabric with a powdered cleanser and water to clean and roughen the fabric fibers.

mechanical layout — *Related Term:* mechanical.

mechanical lineup table — a special light table used in the stripping operation; comes equipped with roller carriages, micrometer adjustments, and attachments for ruling or scribing parallel or perpendicular lines. *Related Term:* lineup table.

mechanical pulp — *Related Term:* ground wood.

mechanical screen tints — (1) in stripping and plate-making prepress, a uniform size dot pattern on a clear base that is prepared mechanically and inserted by the stripper into areas of a flat to create a percentage dot value or other pattern. (2) in cold type com-position, an adhesive-backed preprinted sheet of dots, lines or patterns that can be laid down on artwork for reproduction. *Related Term:* mechanical tint.

mechanical separation — copy prepared by a designer where each individual color is on a separate

overlay. *Related Term:* mechanical color.

mechanical tint — *Related Term:* mechanical screen tints.

mechanical wood pulp — *Related Term:* mechanical pulp; groundwood pulp.

media — any means by which information is distributed such as print, broadcast, CD-ROM, World Wide Web, etc.

media conversion — (1) an alternate term for data conversion from one digital coding to another. (2) an indirect interface device which converts media from one unintelligible format into an intelligible format. *Related Term:* data conversion.

media kit — (1) in advertising, sheets of information about a publication's advertising rates enclosed in a presentation folder. (2) in promotion, a packet containing press releases, photographs, charts, corporate and company officer histories and backgrounds, etc., intended to entice publicity in editorial materials of a publication or other media outlet. If the intended media is audio or video broadcasters, video and audio tapes may be provided.

medium — a channel of communication such as radio, television, newspapers, and magazines. *Related Term:* mass media.

medium format camera — cameras that make 2 1/4 in. x 2 1/4 in. negatives.

medium screen — halftone screen rulings of 133 to 150 lines per inch.

meet-me bridge — a type of telephone teleconferencing bridge that can be accessed directly by calling a certain access number to provide dial-in teleconferencing.

meet-me teleconferencing — the result of the use of a meet-me bridge or using a dial-in conference number.

megabyte (Mb) — (1) 1,024 kilobytes. (2) mega is Greek for one million. In reality, however, a computer megabyte contains 1,048,576 bytes or slightly more than a "mega." *Related Terms:* Mb; bit; byte; kilobyte; gigabyte.

megahertz — one million Hertz (cycles per second).

memory — immediately accessible binary storage on a computer usually expressed in kilobytes or megabytes. Consisting mainly of RAM, which stores the applications software and data, and ROM, where permanent information such as the DOS bootstrap routines reside. *Related Terms:* RAM; ROM.

memory bandwidth — generally, bandwidth refers to data-carrying capacity and is expressed in cycles per second or Hertz (Hz). In the case of RAM, bandwidth is a function of its rated speed and the size of its data path. *Related Terms:* bandwidth; RAM; bus.

memory colors — common, unmistakeable colors around us, i.e., green grass, blue sky, white snow, etc.

menu — a selection list of programs, commands and/or functions displayed on a computer screen. The listing may be in alphanumeric format or in icon form.

menu bar — the area of most graphical user interfaces, such as Windows or the MacOS, at the top of a computer screen where the user can access the actual pull-down menus.

menu driven — programs which allow the user to request functions by choosing from a list of options. *Related Terms:* menu; menu bar.

menu list — a list of short paragraph entries formatted with little white space between them.

merchant brand — a brand or custom name of a paper assigned by a merchant, as compared to a mill brand. Also called private brand.

mercury vapor lamp — a once-popular light source for the graphic arts industry that operates by passing a current through a mercury gas in a quartz or glass envelope; this type of lamp emits massive peaks in the blue-violet and ultraviolet regions of the visible spectrum. At one time, mercury vapor lamps were the primary means of exposing sensitized printing plates. This is no longer true. *Related Terms:* metal halide lamp; pulsed xenon lamp; quartz iodine lamp.

merge copy — a command used within a typographic format to indicate that the next character to be processed is in the input stream and processing is to continue in the input stream until the next merge copy command is encountered (which transfers control back into the format).

merge/purge — the combining of two or more mailing lists (merge) then the elimination of duplicated addresses (purge), usually with the help of a computer.

mesh — the thread pattern of the woven fabric used in screen printing.

mesh count — a numerical system for identifying the fineness or coarseness of screen printing fabric.

meta tag — an HTML tag that must appear in the <head> portion of the page. Meta tags supply information about the page but do not effect its display. A standard meta tag "generator" is used to create meta tag code. Meta tags are a major way that search engines are able to categorize a Web page to offer it for display when queried.

metal halide lamp — a popular light source for graphic arts use, it is a mercury lamp with a metal halide additive and is rich in blue-violet emissions.

metallic ink — an ink with fine aluminum, bronze, or copper powders in it as pigment to simulate metallic colors.

metallic paper — a type of paper coated with a thin film of plastic whose color and gloss simulate metal. *Related Term:* foil paper.

metameric color — a color that changes hue under different illumination. If two colors match under one illuminant but differ under another, their spectral curves are different.

metamerism — the process where a change in illuminant will cause visual shift in a metameric color.

metamers — colors that are spectrally different but appear identical under a controlled viewing condition.

metaverse — like cyberspace, this term is from a book, in this case, *Snow Crash*, by Neal Stephenson. The term describes a virtual online representation of reality.

metric system — a decimal system of measures and weights with the meter and the gram as the bases.

metrics — all the information about how a font fits together, like kerning information and character widths. *Related Term:* font metrics.

mezzograph — (1) a mechanical screen with a grain formation instead of fixed frequency dots of a conventional halftone. It produces a grainy image such as would be achieved with a chalk or charcoal drawing on a pebbled surface. (2) a printing plate in which the details and tones have been transferred through a halftone screen with a grain formation. (3) an engraving produced on a roughened soft metal (usually copper) plate. After etching, highlight and midtone areas are hand-scraped to reduce ink retention, while shadow areas are burnished flatter to strengthen ink retention. *Related Terms:* mezzotint; mezzograph.

mezzotint — (1) an engraving produced on a roughened copper plate. The highlight and midtone areas are hand scraped to reduce ink retention, while shadow areas are burnished to strengthen ink retention. (2)

the effect of mezzotint engraving, produced by a mechanical screen tint.

mflops (mega-floating point operations per second) — a measure of computer power, it is a gauge of the capability of your system to deal with floating-point math instead of raw instructions. *Related Terms:* floating-point math; MIPS.

MG (machine glazed) — paper with a high gloss finish on one side only.

MHz (megahertz) — a megahertz is 1 million complete cycles per second. The unit of measurement used to measure the clock speed of a computer's microprocessor which handles all data-related tasks. *Related Term:* microprocessor.

MICR (magnetic ink character recognition) — the stylized printing on the lower left of personal and bank checks and uses a type of recognition that relies on detecting characters that have been machine-printed in magnetic ink to rigid specifications. The distinctive shape of each character and the amount of magnetic material in the ink are strictly defined by ANSI. The common classes of MICR font (E-13B and CMC7) are also readable using image-based recognition. MICR fonts can be read at high speed with an extremely high degree of accuracy by electronic data processing systems at very little cost per check. *Related Term:* magnetic ink.

MICR line — the row of characters printed in machine-readable magnetic ink at the bottom of each check. Part of this MICR line identifies the issuing bank and the check number and is usually preprinted on the check. The face value of the check is typed into the far right of the MICR line before that check is routed back to the issuing bank. This allows the check to be recognized cheaply and easily by the Federal Reserve in order to forward it back to the bank on which it was drawn. These are MICR numbers: 1 2 3 4 5 6 7 8 9 0

micro-channel architecture (MCA) — a computer bus architecture developed by IBM for its PS/2 computers. Designed for multiprocessing, it can function as a 16-bit or a 32-bit bus, but is incompatible with the original ISA bus. Although it was intended to replace ISA, the bus was never widely adopted and was largely overshadowed by the EISA bus, a 32-bit bus that worked with ISA cards. *Related Terms:* bus, EISA; ISA; PCI; USB.

microchips — integrated digital circuitry containing a very large number of computer components

on a single element. The highest densities are achieved in large-scale integration (LSI) or very large-scale integration (VLSI). *Related Term:* microprocessor.

microcomputer — a small, general-purpose computer with the ability to accept a wide variety of software and additional components to increase power; much smaller and less powerful than a mainframe and not dedicated like a minicomputer; also called a "personal computer" or a "PC."

microfiche — usually a 4 in. x 6 in. sheet of high-resolution film that contains reduced document pictures arranged as rows of micro-images. *Related Term:* microfilm.

microfilm — a roll film storage medium, for photographic storage of documents and information in greatly reduced form, similar to microfiche. Sometimes a term used generically for the several methods of storing images in extremely small spaces on light-sensitive film and/or by photographic means. *Related Term:* microfiche.

micrometer — an instrument for measuring the thickness of paper in microns. Pronounced "my-craw-met-or."

micron — one millionth of a meter (1/1,000,000). This unit is used to measure electroplating thickness, the height and width of gravure cells, and sometimes light wavelengths.

microprinting — an engraving technique for reproducing extremely small type which requires magnification to be perceived. Used extensively to confound counterfeiting of national currencies, stamps and other important financial or fiduciary instruments.

microprocessor — the microprocessor handles the logic operations in a computer, such as adding, subtracting, and copying. A set of instructions in the chip design tells the microprocessor what to do, but different applications can give instructions to the microprocessor as well. Chip speeds are measured in megahertz (MHz), so a 120-MHz chip is twice as fast as a 60-MHz chip. However, that doesn't mean your computer will run all tasks twice as fast, as speed is also influenced by other factors, such as the design of the software you're running, the operating system you're using, and so on. The first microprocessor, the 8080, was created by Intel. Other early microprocessors included Motorola's 6800 and Rockwell's 6502. The most popular microprocessors today are Motorola's PowerPC and Intel's Pentium. *Re-*

lated Terms: megahertz; operating system; microchip; kernel.

Microsoft Image Composer — the very competent image composing and editing application integrated into FrontPage.

microstorage — traditionally, the storage of data in extremely small form on photographic film. More recently, electronic storage in the form of CD's and other nonvolatile devices has assumed some of these storage duties. *Related Terms:* microfiche; microfilm.

microwave transmission — the transmission of information over distances using radio signals in the microwave spectrum (above 2 gigaHertz). These frequencies require line-of-sight transmission between sending and receiving antennas, but are impervious to normal weather disturbances.

middle tones — a tonal range between highlights and shadows in a photograph. Usually the tonal regions in a photograph or illustration about half as dark as its shadow areas and represented in halftones by dots between 30% and 70% full size.

middleware — software that manages communications between a client program and a database. Middleware allows the database to be changed without necessarily affecting the client, and vice versa.

MIDI (musical instrument digital interface) — a connectivity standard that allows users to hook together computers, musical instruments, synthesizers, etc., to create orchestrated digital sound. MIDI (pronounced "mid-ee") is used to describe the standard itself, the standard's hardware and the files that store information for the hardware. MIDI files contain descriptors, instructions for musical notes, tempo, and instrumentation and are widely used in game sound tracks and recording studios. *Related Terms:* sound; WAV.

midtone — *Related Term:* middle tones.

midtone dot — a point in a middle-gray area of a halftone. Its area equals or approaches the average of the nearby background areas. Together all exact midtone dots have a checkerboard appearance.

migration — a term describing how ink is absorbed and/or travels through the substrate, other inks and varnishes, etc.

mike — (1) in paper and printing, to measure the thickness of a sheet of paper using a micrometer. (2) an audio input device used in multimedia. (3) in audio and multimedia, the process of engineering an audio circuit to a microphone which is specifically set on or near a particular person. (4) a shortened name for a microphone.

mil — (1) 1/1000 in., the unit of measure often used to indicate thicknesses of paper. (2) the measure often used to quantify bar code printing and scanning dimensions.

mill brand — the name of paper assigned by a paper manufacturer, as compared to merchant brand.

mill order — supply orders filled from inventory at a mill, not from the inventory of a paper merchant.

mill swatch — a sample of paper provided by a manufacturer.

millisecond — a millisecond is one-thousandth of a second. The term is most commonly used to measure data access speeds, such as the amount of time it takes to retrieve data from a hard disk, a CD-ROM drive, or a floppy drive. Milliseconds are sometimes confused with MBPS (megabits per second), which describe data transmission speeds. *Related Term:* MBPS.

MIME (multipurpose Internet mail extensions) — a way to extend the power of Web browsers to handle graphics, sound, multimedia and anything but text which is handled by HTML. *Related Terms:* helper applications, HTML, viewer.

mimeograph — *Related Term:* mimeography.

mimeograph bond — a highly absorbent paper especially made for the mimeograph method of printing. The paper is characterized by high bulk, rough texture and great absorption properties, all of which contribute to the best reproduction possible with the mimeograph process.

mimeography — mimeograph duplication. An imaging process of the porous classification which makes printed copies from a stencil (image carrier). Ink passes through holes in the stencil. This method is sometimes called stencil duplicating.

mini skid — an informal term used in the U.S. and Canada to refer to a skid holding approximately 1,200 pounds of paper.

minicomputer — a small, dedicated computer usually used for only one task; smaller and less powerful than a mainframe computer. *Related Terms:* microcomputer; mainframe computer.

minimize — on a GUI computer system, the process of making one or more of the windows smaller or even to the point of being a recallable icon.

minimum line length — the shortest width of lines of type which maintain acceptable readability.

minus leading — the reduction or elimination of space between lines in excess of the point size of the type being set. As an example, 10 point type set on 9 points of leading. Not possible in hot metal and foundry type composition, the technique is possible only with photographic or electronic typesetting.

minus letterspacing — reducing the space between characters in excess of the typographic ideal; a technique that was possible, but infrequent, in hot metal and foundry type, but requires only keystrokes with photographic or electronic typesetting.

minuscules — lowercase characters; noncapital letters. *Related Terms:* majuscules; caps; lowercase; lc.

MIP mapping — a sophisticated texturing technique used for 3D game animation and architectural rendering walk-throughs. Through polygon manipulation, the system reduces the apparent aliasing on the screen or in the rendering. *Related Terms:* antialiasing; jaggies.

MIPS (million instructions per second) — a reference point in rating computing speed.

mirror — (1) writing the same data to two drives at once (RAID level 1); usually done when data integrity is paramount, though is not recommended for digital video. (2) in general, to maintain an exact copy of something. (3) Web sites or FTP sites that maintain exact copies of material originated elsewhere, usually to provide greater access from multiple locations.

MIS (management information systems) — usually, the department in charge of computing systems in a company; also called "data processing" or "information systems."

MISF (Microsoft Internet Security Framework) — a set of cross-platform, security technologies developed by Microsoft for Internet security.

misread — a mistake made by a recognition system in interpreting a character, field, barcode symbol or document. *Related Terms:* OCR; IOCR; bar code reader; scanner.

misregister — the inaccurate overlaying of ink from separate plates, causing a blurry or unusable reproduction. *Related Term:* mass media.

misting — fine airborne ink particles caused by the splitting action between rolling components of the inking system. *Related Term:* flying ink.

miter — the cutting of the ends of hot metal rule or borders at a 45-degree angle for making non-overlapping corners where horizontal and vertical sides meet.

mitography — the artistic application of printed images through a stencil pattern and screen fabric. *Related Term:* screen printing.

mixed fountain — a method of gaining a multicolor effect from one printing plate by slowing or limiting the oscillation of the distributing rollers and charging the fountain with multiple colors, each in a different section of the fountain. The colors will merge. After a limited number of impressions, 500 to 1,000 at most, the press must be completely washed up and recharged as before. *Related Term:* split fountain.

mixed mode CD-ROM — a compact disc which has both audio and data encoded according to the Red Book standards.

mixing — the combining of more than one style or size of type on the same line.

M-JPEG — motion JPEG. An M-JPEG file is the sequence of individual frames that have been JPEG-compressed. *Related Terms:* M-PEG-2; JPEG; JPG.

MM — Roman numerals for a thousand thousand, i.e., one million. 10MM means 10 million.

MMX technology — an Intel Corporation iteration of the Pentium Processor which utilizes a single instruction, multiple data (SIMD) technique. Current multimedia and communication applications often use repetitive loops that, while occupying 10 percent or less of the overall application code, can account for up to 90 percent of the execution time. The SIMD technique enables one instruction to perform the same function on multiple pieces of data, similar to a professor telling an entire class, "there will be an examination next week," rather than addressing each student one at a time. SIMD also allows the chip to reduce the compute-intensive looping common with video, audio, graphics and animation.

mnemonics — (1) in history, an ancient associative memory-aiding device. (2) in the modern era, a term used to refer to abbreviations of complex terms used in encoding computer instructions; symbolic names or letter combinations that permit one short set of codes (usually two letters) to represent a more complex, longer set or string of codes.

mockup — (1) in design and print production a

"dummy" or a full-size, experimental layout for study and evaluation; the rough visual of a publication or design. (2) in photography, product simulations used in some photo shoots. *Related Term:* dummy; comprehensive; comp.

model — an educational process whereby a computer-based learning system is used to represent another system or process. The person being instructed can change values and observe the effects of the change on the operation of the system. *Related Terms:* modeling; interactive.

model release — a photographer contract authorizing commercial use of subject matter that includes images of recognizable person(s) or, under some circumstance, of private property.

modeling — (1) the use of computer systems for simulation, testing or trial in advance of building or implementation. (2) the uneducated version of "mottle." This spelling frequently appears in poorly researched and edited material. *Related Term:* mottle.

modem (MOdulator/DEModulator) — an electronic device that varies the amplitude, frequency or phase of digital data from the computer terminal and converts it into signals that can be sent over regular telephone lines to another computer, network, electronic mail, etc. The power is measured in BPS (bits per second); the higher the BPS, the faster the information is transmitted and the shorter the connection time (and charges). *Related Term:* modulate.

moderated mailing list — a type of mailing list where messages are first sent to the list owner before they are distributed to all the subscribers.

moderator — the person, or small group of people, who manages moderated mailing lists and Usenet newsgroups. Moderators are responsible for determining which e-mail submissions are passed on to a list.

modern — a subclass of the roman serif type classification. The term generally refers to type styles introduced toward the end of the 19th century which tended to have generally straight serifs, thin and slightly rounded at the corners. *Related Terms:* old style; transitional.

modification — (1) in the computer arena, the process of updating an existing domain name record, contact record, or host (name server) record to reflect changes in existing information for an Internet site. (2) in production, any changes in job specifications which depart from the original specifications. Par-

ticularly significant if such changes are made after production has begun. *Related Terms:* domain name system (DNS); registrant; registration fee; AA; TA.

modular makeup — the arrangement of elements in rectangular units on a page. *Related Term:* Mondrian makeup.

modulate — a process imposing message information on a carrier by varying the amplitude, frequency or phase of a wave. *Related Term:* modem.

module — the narrowest nominal unit of measure in a bar code.

module check digit (or character) — a character within a barcode symbol data field calculated using modular arithmetic, which is used for error detection. The calculated character is determined by applying a code algorithm to the data field contents. *Related Term:* check character.

moiré — (pronounced "more ray") (1) in process printing, the objectionable result of superimposing improperly aligned fixed-frequency mechanical or halftone screens, thereby creating a pattern effect on the printed halftone or tint area. Normally detected in prepress during the proofing stage. (2) a similar pattern that results when a previously screened halftone is rescreened and printed. (3) the same undesirable pattern created whenever a previously halftoned original is scanned and bitmapped. *Related Term:* rosette.

moisture content — the relative amount of moisture in paper.

molleton — absorbent thick cotton flannel-like material used as a covering for lithographic press water form or dampening rollers.

monarch — a paper size (7 X 10 inches) and envelope shape often used for personal stationery.

Mondrian makeup — the arrangement of elements on a page into rectangles of various sizes and shapes. *Related Term:* modular makeup.

monitor — a high-resolution picture tube commonly used in video production and computing. It has no tuner and is unable to directly receive broadcast signals. It must rely on a separate input device.

monitor calibration — the process of correcting the color rendition settings of a CRT or monitor in an attempt to match selected colors of printed output.

monochromatic — (1) generally, refers to having one hue or color; monochrome. (2) in computers, a black and white display with no gray tones.

monofilament fabric — a type of synthetic material used for the fabric in screen printing. It has only a single filament or unit in each thread, such as nylon. *Related Term:* multifilament fabric.

monograph — a brief report on a particular subject.

monospace — a font in which all characters occupy the same amount of horizontal width regardless of the character's design width, i.e. "i" and "W" occupy the same escapement width. Typical of a standard typewriter.

monotonal — typefaces with strokes of equal thickness.

monotone — having uniform lines; a term used to describe square serif typefaces because their letter strokes are often the same width through the entire series of characters.

Monotype — a hot-metal typesetting machine which casts individual pieces of type in composed lines. The copy is first keyboarded to produce a perforated pa-

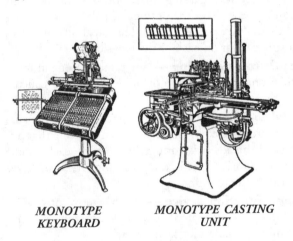

MONOTYPE KEYBOARD

MONOTYPE CASTING UNIT

per tape or ribbon. The tape is run through a casting unit, which produces the individual characters in justified form.

montage — (1) a group of related or similar subjects that have been combined to form a single, or composite, illustration. (2) mounting several color separation films that will print in the same color in register for subsequent transfer to the printing form. *Related Term:* gang.

moof – a term used in Net culture to explain the unknown. Especially used in IRC rooms to refer to unexplained disconnects.

MOPS (millions of operations per second) — a computer-based speed rating. In the case of interactive video technologies such as DVI, more MOPS

translate into better, smoother video quality. The i70 chip set can perform multiple video operations per instruction; thus the MOPS rating is generally higher than the MIPS (millions of instructions per second) rating.

morgue — *Related Term:* archive.

morphing — the transition of an elment into another through photographic or digital manipulation. A very common example is the transition of a normal character into an alien or monster in video and motion picture productions.

mortise — in hot metal makeup, an area cut out of a piece of art for the insertion of type or other art.

MOS (metal oxide semiconductor) — similar to a CCD but with very different methods of processing signals to create images.

motion video — linear imaging systems that display real motion by displaying a sequence of images or frames rapidly enough that the eyes see the image as a continuously moving picture. Because of vision response and a phenomenon known as persistence of vision, the eye is still perceiving a frame whenever the next one comes into view. The mind combines the two versions. When sequenced rapidly (above 15-20 frames per second), the illusion of motion is created. Motion pictures utilize 24 frames per second. Standard television uses 30 complete frames per second. Other standards vary. *Related Terms:* digital video; Indeo video; full motion video.

motivation — the driving force that makes a person want to accomplish one or more tasks.

mottle — pronounced "mott al." An undesirable, uneven color or tone; spotty, uneven ink coverage especially noticeable in large solids. Erroneously called "modeling" by the uninformed.

mounting block — (1) the specially shaped device, usually made of wood, to which a rubber stamp is fastened. (2) in relief printing the base upon which engravings and linoleum surfaces for carving are mounted to make their printing surface type high, .918 in.

mounting board — the thick, smooth piece of paper used to paste up copy or mount photographs. *Related Term:* layout board.

mouse — a hand-held pointing device using either mechanical motion or special optical techniques to convert the movement of the user's hand into movements of the cursor on the screen. Generally fitted with one, two or three buttons which can control specific software functions. May employ either optical or mechanical detection techniques to convert the movement into meaningful screen cursor instructions. Some models have wheels which can be spun with the finger which will cause the screen to scroll, etc.

movable type — the individual metal or wooden type characters that are taken from the type case, arranged to form words and sentences, and then returned to the case for reuse later.

moving beam bar code reader — a scanner where scanning motion is achieved by moving the optical geometry. A grocery store scanner is a good example, as it often uses a rotating polygon prism to make a stationary beam appear to be moving across a field. *Related Term:* fixed beam bar code reader.

MPC — *Related Term:* Multimedia PC.

MPEG (motion picture experts group) — a working committee under the auspices of the International Standards Organization (ISO) whose mission is to attempt definition of standards for digital signal compression and decompression for motion video/audio for use in computer systems.

MPEG2 — an evolving standard by MPEG to extend the standard MPEG compression and decompression capabilities for motion video. More comprehensive in nature it is a developing standard for unified, compatible, more robust video coding. *Related Term:* MPEG; JPEG.

MRD (minimum reflectance differential) — in optical reading, a measurement that is used to determine if there is an adequate difference between absorbed and reflected light.

MRP (material requirements planning) — in production, the actions required to anticipate need and place material supply orders based upon customer orders, production quotas, historical cycles, etc.

ms (manuscript) — (1) traditionally, the original written or typewritten work of an author submitted for publication. (2) the data file, diskette, tape, etc., with the manuscript codes.

MS-DOS — the Microsoft disk operating system. The operating system used in the original IBM personal computer as IBM-DOS and then marketed by Microsoft under the MS-DOS label to other competitive computer manufacturers.

MTBF (mean time between failure) — usually ex-

pressed in hours. The higher the number, the more reliable the device.

MTTR (mean time to repair) — usually expressed in hours. A lower number indicates the device can be repaired quickly.

MUD — *Related Term:* multi-user dungeon.

mug shot — a head and shoulders (only) photograph. Very popular with newspapers and magazines.

Mullen test — bursting strength test for paper. *Related Term:* pop test.

Mullen tester — a machine for testing the strength of paper.

multicolor — *Related Term:* multicolor printing.

multicolor press — two or more connected printing units (each with its own ink, image carrier, etc.), a feeder, a sheet transfer system, and a delivery. Normally two or more colors can be printed on one side of a sheet during a single pass through the press, but some presses have the optional capability of turning the sheet over between units. This capability produces a single color on each of the two sides.

multicolor printing — any printing done in more than one ink color; two or more colors used concurrently.

multifilament fabric — any type of synthetic or natural material used in fabric for screen printing that has several threads twisted together to form each strand in the fabric. Silk is one example. *Related Term:* monofilament fabric.

multifunction device — another term for devices which perform more than a single function, i.e., the combined fax/copier/scanner, etc.

multihosting — the ability of a Web server to support more than one Internet address and more than one home page on a single server. *Related Term:* multihoming.

multimedia computing — the delivery of computer information which combines different content formats such as motion video, audio, still images, graphics, animation, text, etc.

multimedia PC (MPC) — a specification for the minimum platform capable of running multimedia software. It is accompanied by a certification logo program. Computers with the logo are assured of the ability of running software which contains the MPC certification.

multiple master font — a single font capable of rendering a specific typeface in a number of weights, styles, and sizes, based upon hinting or processing instructions built into the specific font.

multiple submission — the submission of a manuscript for consideration to more than one publisher at the same time.

multiplex — transmission of two or more information streams over a single physical medium. The technique is used in both FM radio broadcasting and in television for implementing closed captioning and second languages.

multipurpose Internet mail extensions —*Related Term:* MIME.

multitasking — the ability of a computer to perform several operations concurrently, one in the foreground (active) and one or more in the background without conflict.

multi-user dungeon — simulations, adventures, and role-playing games played on the Internet. Players interact in real time and can modify the "world" in which the game is played.

Munsell Color System — a method of classifying surface color in a solid. The vertical dimension is called value, the circumferential dimension is called hue, and the radial dimension is called chrome. The colors in the collection are spaced at subjectively equal visual distances.

Murray-Davies equation — a formula for calculating printed dot area based on densitometer measurements. The resulting calculations are for total dot area, including the optical and physical aspects. *Related Term:* Yule-Nielsen equation.

mutt — a typesetting term for the em space.

Mylar™ — the DuPont trade name for its polyester film, but routinely applied to any clear or frosted stable base plastic material used for overlays in paste-up and for stripping together film positives.

Nn

NAK (negative acknowledgement) —a data packet negative reception response sent back to signal the sending modem. If everything is in order, the signal is an ACK, or acknowledgment. If some of the data is missing or corrupt, the modem sends back a negative acknowledgment, or NAK, which acts as a request to resend the data. *Related Terms:* ACK; CRC.

name server — a computer that has both the software and the data (zone files) needed to resolve domain names to Internet protocol (IP) numbers. *Related Terms:* host name server; zone file; resolve.

name service — a system for providing individuals or organizations with domain name-to-Internet protocol (IP) number resolution by maintaining and making available the hardware, software, and data needed to perform this function. Many Internet service providers (ISPs) operate name servers and provide their customers with name service when they register a domain name. Most individuals are not in a position to operate a name server on their own and will need to make arrangements for name service with an ISP or some other person or organization. *Related Terms:* resolve; Internet service provider (ISP); name server.

nameplate — that portion of the front page of a newsletter that graphically presents its name and subtitle and may include a logo, dateline and other information. *Related Terms:* banner; flag.

nanometer (nm) — measurement unit for wavelengths of electromagnetic radiation, each equivalent to 10-9 meters. Visible light wavelengths range from 400 to 700 nanometers.

nanosecond (ns) — one-millionth of a second.

narrowband — a low-frequency telecommunications system that includes telephone frequencies of about 3,000 Hertz and radio subcarrier signals of about 15,000 Hertz.

native file — an electronic file created in a specific native application file format such as Photoshop, Word and Excel, and which is not directly usable by another software configuration without translation or import filtering.

natural color — a pale off-white ivory, cream, or light brown shade of paper.

navigate — to move around on the Web by following hypertext paths from document to document in different domains.

NC (network computer) — a consumer product looking for a mass market. The under $500 Internet computer advocated by Larry Ellison, CEO of Oracle. The principle is based on the gigantic storage sites on the Internet which would, ultimately, contain data and applications. The original intent was to provide cheap computing and make the computer into another common appliance in homes. Several companies committed to the proposal, but weak public enthusiasm and understanding forced most to back away from the concept.

NCR paper — the accepted abbreviation for any no carbon required paper, a specific original brand name of carbonless paper of the National Cash Register Company. *Related Terms:* carbonless paper; no carbon required paper.

NCSA (National Center for Supercomputing Applications) — a scientific research center built around a national services facility, it is developing and implementing a national strategy to create, use, and transfer advanced computing and communication tools and information tech-nologies. These advances serve the center's diverse set of constituencies in the areas of science, engineering, education, and business. The NCSA was responsible for the development of Mosaic.

NCSA image map dispatcher — a program which handles server-side NCSA-style image maps when you are using the FrontPage Personal Web Server.

NDIS (Network Device Interface Specification) — a device driver for Windows that allows multiple network protocols to be used simultaneously with one adapter card. It was originally developed for use with Ethernet LAN cards, but is now used for many internal ISDN adapters. *Related Term:* ISDN.

near letter quality (NLQ) — an almost-obsolete term that was a descriptive characteristic given to dot matrix computer printers doing their best work (above 360 dots per inch).

negative — (1) a tonal reverse of colors or black and white. A film in which the dark areas of the images

**NEGATIVE/
POSITIVE IMAGE**
*A negative is tonally
the reverse of the
normal, i.e., the
background is black
and the image is
white or clear. It is the
first step in most
photomechanical
reproduction processes
and yields a positive
image from which to
print.*

or text appear light or clear and the light areas of the images or text appear dark.(2) any photographic film or plate that is exposed and processed to provide a tonal reverse image of that found in the original highlights and shadows or color values (i.e., white as black). *Related Term:* positive; negative film.

negative acting plate — a printing plate formulated to produce a positive image from a flat containing negatives.

negative film — film that converts black areas to white (clear) and white areas to black. When a contact is made of a negative onto negative film, a film positive of the original art is the result.

negative leading — a type specification in which there is less space from baseline to baseline than the size of the type itself (for example 40-point type with 38-point leading). *Related Term:* negative leading.

negative letterspacing — a type specification in which the space between characters is reduced beyond the default settings, either by kerning or tracking.

negative space — *Related Term:* white space.

neon-helium laser — a type of laser light used to expose red-sensitive photographic films, paper, plates or cylinders. It is also used in some scanners for electronic dot generation.

Nerd World — an Internet subject index with links to Web, USENET, and FTP resources and whose goal is to provide products and services to make Internet using easier.

nested — signatures assembled inside one another in proper sequence for binding.

nested indent — an indent whose indentation is measured from the margin of the last indent rather than the absolute margin.

nested list — a list that is contained within a member of another list. Nesting is indicated by indentation in most Web browsers.

Net abuse — abuse of network services, or violations

of netiquette. Types of Net abuse include (1) using too many of the system resources; (2) attempting to "hack," or break into other accounts; (3) using your account for any illegal activity; (4) sending unsolicited e-mail; (5) sending chain letters via e-mail; (6) advertising in inappropriate newsgroups; (7) off-topic posts to newsgroups; (8) "spamming" or inappropriate postings to many newsgroups; (9) disruption of newsgroups or IRC channels; (10) "flooding" someone with talk requests; (11) direct threats in newsgroup posts or e-mail; (12) sharing an account (in certain circumstances). *Related Terms:* Internet relay chat; Usenet.

net etch — a chemical technique to change the size of the dots over the entire halftone film image.

Net Nanny — a program intended for parents, guardians and teachers who wish to stop children from accessing pornographic and other undesirable material, while at the same time preventing the children's personal information such as names, addresses, telephone numbers, etc., from being circulated on the Internet.

Net Toob — a Windows multimedia player. It provides a single utility which plays all digital video standards, real-time audio and video via the Internet. It also enables playback of MPEG-1, Video for Windows (AVI), and QuickTime for Windows (MOV) via the Net. The utility has video screensaver capabilities, to enable users to create screen savers from their videos.

net.god — one who has been online since the beginning, someone who knows all and who has done all.

net.personality — someone sufficiently opinionated with plenty of time on his hands to regularly post in dozens of different Usenet newsgroups, and whose presence is known to thousands of people.

net.police — the derogatory term for those who would impose their standards on other users of the Net.

NetBuddy — Automatically checks a list of Internet Web locations which you want it to watch at a frequency you decide (once a minute, every two hours, etc.). If any of the sites have changed (have new information), NetBuddy lights up that site in its list to let you know something's different there.

netiquette — the internet pun on "etiquette," referring to proper behavior on a network.

Netnews — *Related Term:* Usenet.

Netscape Navigator — basically free Web browser. It was an early leader and is still a major browser, although market share has eroded in favor of Microsoft's

Internet Explorer. It allows for Gopher, FTP, and Telnet access as well as e-mail and newsgroup retrieval, management, and server software to create Web pages. The program is available for all platforms.

NetWare — software created by Novell to run on Intel-based computers, NetWare is the most widely used network operating system on that platform. *Related Terms:* IPX; NLM.

network — generally, a set of nodes connected via voice, data, or video communications to facilitate the exchange of information. Usually prefaced by an identifier such as television network, local area network, etc.

Network Information Center (NIC) — a provider of information, assistance and services to network users. The Internet Network Information Center (InterNIC) is a project administered by AT&T and Network Solutions, Inc. (NSI). AT&T provides directory and database services for registered Internet hosts, while NSI administers the registration process.

network location — in a URL, the unique name that identifies an Internet server. A network location has two or more parts, separated by periods, as in my.network.location. *Related Terms:* host name; Internet address.

Network News Transfer Protocol (NNTP) — a communications protocol for the distribution, retrieval and posting of Usenet articles through high-speed links available on the Internet. *Related Terms:* IP; SMTP; TCP/IP.

Network Solutions' Domain Name Dispute Policy — the policy established by Network Solutions to provide for actions Network Solutions may take when presented with evidence that the legal rights of a trademark owner are harmed as the result of a violation of the trademark owner's intellectual property rights. By submitting a domain name registration agreement, each domain name registrant agrees to be bound by this policy. *Related Terms:* domain name registration agreement; registrant; trademark.

networking — the ability to (1) exchange information between users or (2) share the resources involved in this process. A network consists of two or more information sources or destinations linked via communications media for the purposes of information exchange or resource sharing.

Neugebauer equations — a set of linear equations used for calculating tristimulus values. Equations of halftone color mixture combinations, when the dot areas of the contributing colors are known.

neutral — (1) generally, any color that has no hue, such as white, gray, or black. (2) in digital imaging, a pixel that has no particular color cast.

neutralized color — any hue dulled by the addition of white, gray, black, or some of the complementary color.

newbie — a slang term for a user who is new to the Internet or to computers in general.

news case — a container formerly used to store foundry type; upper case held capital (majuscule) letters and lower case held small (minuscule) characters.

news feed — the originating news source which supplies information, stories, data, etc., to local radio and television stations, newspapers, and internet ISPs. Each of them retransmits the information in their own format to their respective readers and subscribers.

newsgroup — on the Internet, a worldwide grouping of bulletin boards, organized more or less stringently around a specific topic. The topics are dictated by the subscription list. *Related Terms:* newsreader; Usenet.

newsletter — a short, usually informal publication presenting specialized information to limited audiences on a regular basis.

newsprint — an unsized low-quality grade of round wood or mechanical pulp paper containing about 85% ground wood and 15% unbleached sulfite. The weight is from 24 to 45 lbs. and the surface is coarse and absorbent. Used for printing newspapers and low-cost flyers or broadsides.

newsreader — a computer program that lets you selectively read, download, and reply to the newsgroup messages and encoding binary file attachments for you (the Internet was designed for text-only files. Binary or nontext files have to be encoded prior to transmission.) *Related Terms:* browser; newsgroup; Usenet.

Newton's rings — a pattern of concentric circles sometimes created when two glass surfaces come into contact, such as in the mounting frame of a process camera.

NFS (Network File System) — a protocol suite developed and licensed by Sun Microsystems that allows different makes of computers running different operating systems to share files and disk storage.

nib width — a corrupted version of nip width. *Related Term:* nip width.

nibble — four bits; one-half of a byte.

NIC — (1) network interface card. An adapter card that physically connects a computer to a network cable. (2) networked information center. Generally, any office that handles information for a network. The most famous of these on the Internet is the InterNIC, which is where new domain names are registered. (3) a handle. A unique identifier, which can be up to 10 alphanumeric characters, assigned to each domain name record, contact record, and network record in Network Solutions' domain name database. NIC handles should be used on registration forms whenever possible, as they save time and help to ensure accuracy in the records. *Related Term:* registration forms.

nicked corner — a special technique for folding book cloth around a corner of binder's board when hand binding a casebound book.

NID (network interface device) — a phone company term for a line junction box. Often only accessible by company employees with appropriate keys and combinations.

nip width — area of contact between two rollers, for instance the impression roller and plate (or image) cylinder on any rotary press, or the web entrance rollers on a publication press folder.

nipping — a part of the case binding method when the binding edges of the sewn signatures are pressed together to expel air under heat and pressure.

no carbon required paper (NCR) — *Related Terms:* carbonless paper; NCR.

no screen exposure — *Related Term:* bump exposure.

NOC (Network Operations Center) — the organization responsible for the day-to-day operations of the Internet's component networks.

node — any single computer connected to a network.

noise — any undesirable signal disturbance in an electronic or other medium which interferes with the effectiveness of the intended source material. Static in audio, voltage fluctuations in illumination, etc.

nominal — the exact (or ideal) intended value for a specified parameter. Tolerances are specified as positive and negative deviations from this value.

nominal basis weight — the basis weight of a specific paper as advertised or specified, which may differ from its actual basis weight.

nonbreaking space — in typesetting, a special space character placed between two words to keep the words from being separated by a line break.

noncontact reader/scanner — scanner or readers of any type which do not require physical contact with the original object or printed symbol. Most desktop and bar code scanners fit into this category.

noncounting keyboard — a keyboard that can capture operator keystrokes without character width knowledge or end-of-line decisions.

nonimage area — that portion of the mechanical, negative, or image carrier that will not print.

nonimpact composition — the use of a photographic negative and exposure or a laser in the imaging process to create type. Any system which does not physically impinge on the surface of the image carrier. *Related Term:* nonimpact printer.

nonimpact printing (NIP) — one of the five printing classifications. Any printing system which utilizes lasers, electrostatic charges, ions, ink jets or heat to transfer images to substrate. Usually a printing device that creates letters or images on a substrate without physically striking it. All nonimpact systems are electronically controlled, generally by computer. Laser printers, xerographic copy machines and dye sublimation printers are all examples of this classification.

noninterlaced — the design of monitors so that they paint images on a screen rapidly and repeatedly, sweeping their electron guns from top to bottom and left to right across the screen. A noninterlaced display paints every line on the screen each time it scans from top to bottom. Interlacing skips every second line on the first pass and fills them in on a second pass. That's acceptable on a TV, but it causes enough flicker to tire your eyes on a computer monitor. The modern computer preference is for noninterlaced monitors, and the new HDTV standards are also noninterlaced. *Related Terms:* electron gun; interlaced.

nonlinear editing — features immediate direct access to any frame of disk-resident video and audio source material. *Related Term:* random access.

nonpareil — a unit of type equivalent to one-half of a pica or approximately one-twelfth of an inch.

nonprocess printing — color printing in which the desired color is achieved by using an ink of the preferred color (such as a PMS color) instead of yellow, magenta, and cyan (process colors) in combination. *Related Term:* spot color.

nonread — the absence of data at the scanner output after an attempted scan due to no code, defective code, scanner failure or operator error.

nonread blue — a popular and unscannable color used as a writing guide for optical character recognition systems. *Related Term:* nonread color.

nonread color — (1) in printing, a light blue color that does not record on graphic arts film. Therefore it may be used to write instructions on mechanicals. (2) in optical scanning, a non-readable color which is often used as guidelines or boxes for handwriting. *Related Term:* read color; nonreproducible blue; nonrepro colors; nonscan ink.

nonreflective ink — generally a dense black ink used to print OCR characters which must be scanned. The dull surface makes good contrast with the background and improves the accuracy of the optical reading device.

nonrepro blue — a light blue ink that cannot be reproduced by orthochromatic films or platemaking cameras. It is used to mark up layouts and camera-ready artwork. *Related Terms:* nonrepro colors; nonrepro pencil.

nonrepro colors — *Related Term:* nonread blue.

nonrepro pencil — a light blue writing pencil that is used to mark up layouts and which cannot be reproduced by a platemaking camera. *Related Terms:* nonrepro colors; nonrepro blue.

nonscan ink — *Related Term:* nonread.

nonstandard cut — in paper cutting, the manipulation of positions and order of cuts to gain an additional sheet; always produces press sheets with different grain direction.

normal key — a photograph in which the main interest area is in the middle tones of the tone scale or distributed throughout the entire tone range.

North American paper sizes — standardized paper sizes used in the U.S. and Canada. Expressed in inches.

notch — a small serration, or indentation, along one edge of film or other photographically sensitive material that is used to identify the emulsion side of the materials and aid in positioning it in total darkness. Generally, when the notches are along the bottom right side (near the technician) of a substrate, and the long dimension is horizontal, it indicates that the emulsion is facing up.

notch binding — a part of perfect binding where small slits are cut across the spine of a book prior to gluing to hold more glue for adhesion and a produce a stronger overall bind.

notch mortise — a rectangle cut from a corner of a rectangular illustration.

notebook — a small extremely portable battery operated computer. *Related Terms:* laptop computer; desktop computer.

notification — the process of informing individuals and organizations who are listed as authorized points of contact on a domain name record that the registry has received a request to modify information in one of the records with which the authorized contacts are associated.

novel paper — *Related Term:* bulking book paper.

novelty — a classification of type that is really ornamental in design and does not display any strong characteristics. Intended to be used for single words or phrases, not text matter.

novelty printing — printing which is done on nonstandard formats such as pencils, balloons, key rings, and other advertising specialties.

nozzle — the orifice through which jets of ink are ejected to form an image in the inkjet printing process.

NREN (National Research and Education Network) — an effort to combine the networks operated by the U.S. government into a single high-speed network. *Related Term:* Internet 2.

NS 16550 (National Semiconductor 16550) —the modern UART (Universal Asynchronous Receiver Transmitter) chip found in standard PCs communications ports. *Related Term:* UART.

NSA line eater — paranoid chat phrase used to refer to the imagined supercomputer run by the National Security Agency that is assigned the task of reading everything posted on the Net.

NSAPI (Netscape Server Application Programming Interface) — a system designed to be a more robust and efficient replacement for CGI.

NSF (National Science Foundation) — an independent agency of the federal government, established in 1950, whose mission is to promote the progress of science and engineering.

NT-1 (network terminator-1) —an interface box that converts ISDN data into data a PC can understand

(and vice versa). It works similar to a cable TV descrambler, except it is exclusively for ISDN signals. It is often built into ISDN adapters.

NTFS (NT file system) — a unique 32-bit file system used by Microsoft's Windows NT. Only Windows NT can read and write NTFS-formatted drives.

nth name — the names selected from a mailing list in regular (nonrandom) increments, such as every 10th name, used to test the validity of mailing lists.

NTSC (National Television Standards Committee) — standard video encoding/decoding scheme used in Canada, Japan, South America, and the United States consisting of 525 lines/field and 30 frames/second. The NTSC standard combines blue, red, and green signals with an FM frequency for audio. NTSC is going away, with the arrival of HDTV, or high-definition TV. The FCC has ordered TV stations to start transmitting HDTV signals starting in 1998, and under certain conditions to abandon their analog wavelengths by 2006. *Related Terms:* PAL; SECAM.

NuBus interface — an obsolete Apple Computer interface designed as a one-piece socket with two rows of pins made for internal Macintosh cards. NuBus uses onboard ROM to configure itself. Power Macintosh computers use Intel's superior PCI standard.

null-modem cable — a special type of computer cable that lets you hook up two computers to communicate via their serial ports. It's called a "null-modem" cable because it eliminates using modems and phone lines for hooking together nearby computers.

number system — a coding system for identifying individual or groups of objects. One type is the significant digit system where each item is uniquely identified; another employs a nonsignificant digit where sequential numbers are assigned regardless of product or item description.

numbered list — the Web page paragraph style that presents an ordered list of items.

numbering – the hand or machine process for consecutively numbering elements of an order, such as invoices, tickets, etc. *Related Term:* imprinting; crash imprinting.

numeric — a character set that includes only numbers. *Related Terms:* alphabetic; alphanumeric.

numeric keypad — an auxiliary keypad adjacent to the normal alphabetic and function keys of a computer terminal. They are generally laid out similar to an adding machine comfiguration to facilitate speedy numerical input. In the Intel and Windows world, if used with the alt key they can produce an expanded character set including international symbols, etc.

nut — a nickname for an en quad.

Nvidia — an extremely sophisticated 3D application. A specific integrated circuit (ASIC) that adds a unique algorithm called quadratic texture mapping in addition to standard shading, texture mapping, Z-buffering, and acceleration. The process draws scenery using fewer polygons, making it easier and faster to render a scene. *Related Terms:* ASIC; quadratic texture mapping.

OO

object-oriented — (1) generally, a method of software development that groups related functions and data into reusable chunks. Properly handled, object-oriented programming can reduce development time on new projects. (2) in some circles, a term used to refer to a type of drawing that defines an image by mathematically describing lines, circles, text, and other objects (thus the "object-oriented" label), rather than as pixels in a bit map. *Related Terms:* vector drawing; vector graphics.

oblique — roman characters that slant to the right, as in pseudoitalics. *Related Term:* obliquing.

oblique stroke — (/)

obliquing — (1) electronically slanting characters by distorting an upright typeface so that each character is properly seated on the horizontal baseline while its upright axis deviates somewhat from the vertical in a forward or backward direction to create a letter reminiscent of true italic, but without the stroke weight adjustments, etc. (2) a top-forward-leaning stroke used as a separator in writing and computer keyboarding; sometimes indicates options such as "and/or." *Related Terms:* oblique stroke; pseudofont.

obscenity — material of low moral value, arousing disgust or erotic feelings. *Related Term:* pornography.

occasional — an infrequently used name for the "novelty," or "decorative," type classification which includes all typefaces that do not fit one of the other five classes.

Occupational Safety and Health Act — the federal law enacted in 1970 to protect workers from work-related dangers by requiring employers to provide places of employment free from recognized hazards.

Occupational Safety and Health Administration (OSHA) — the enforcer of the OSHA act; the inspectors that may appear unannounced or at the request of an employee to examine any plant for violations of the safety and health standards, and to require employers to provide places of employment free from recognized hazards as set forth by the act.

OCLC (Online Computer Library Center) — a special-purpose nonprofit international computer network and online library database for cataloging and interlibrary loan services.

OCR (optical character recognition) — a type of scanning that provides a means of reading printed characters in documents and converting them into digital signals that can be stored on tape or disk or read into a computer as an actual text file. Such a file can be edited, formatted, etc., in a normal word processor or page layout program, just like keyboard text. The process eliminates the need for rekeyboarding previously typed or printed documents. Most OCRs work by using either pattern matching or feature extraction. With pattern matching, the software is given a "template" of possible characters. When the scanner sees a letter, it compares it to its library of pattern templates. If there is enough of a match, it safely assumes it has "recognized" the letter and sends the ASCII equivalent of the letter to the output device. Feature extraction is more sophisticated. Its "library" consists of groups of information regarding a character's features; i. e., the letter W has two sets of diagonal lines; the lines intersect at the top; it has a horizontal line that crosses from one of the lines to the other, etc. As the OCR scans, it compares features of the character to its feature library. Feature extraction is used to recognize handwriting in certain cases. OCR software further supports its "guesses" by knowing a little something about the language. A digit "1" is not likely to fall in between a group of letters; the letter "h" frequently follows the letter "t," etc. *Related Term:* optical character recognition.

OCR-A — an abbreviation commonly applied to the character set contained in ANSI standard of 1974.

OCR-B — an abbreviation commonly applied to the character set contained in ANSI standard of 1975.

octal — a numbering system with a base of eight.

octavo — (1) a book which measures 6 inches x 9 inches. (2) a sheet with three right-angle folds; produces a 16-page signature.

ODBC (Open Database Connectivity) — a set of application programming interfaces, created by Microsoft, that defines how to move information in and out of any PC database that supports the standard. *Related Term:* API.

ODMA — Open Document Management Alliance.

OEM (original equipment manufacturer) — (1) originally an adjective used to describe a company that produced hardware to be marketed under another company's brand. TAL company, for example, produced CD-ROM drives that dozens of other companies would label as their own. (2) modern usage more often uses it as a verb, as in this sentence: "This CD-ROM drive is OEM'd by TAL."

off contact — *Related Term:* off-contact printing.

off-contact printing — the technique used in screen printing when the stencil and screen frame do not touch the substrate except at the squeegee contact point. *Related Term:* off contact.

official envelope — *Related Term:* business envelope.

off-line — (1) particularly in computing, a reference to equipment not directly controlled by a central processing unit or to operations conducted out of the normal process stream. (2) in production, operations conducted out of process, rather than inline (on-press), such as die cutting or the application of special coatings via dedicated equipment. *Related Term:* online.

offloading — relieving the intensive amount of data processing associated with a specific application (e.g., graphics) from the CPU, by performing those calculations in a dedicated or specialized processor, such as a video accelerator card.

off-press proof — approval proofs made by photomechanical means, such as a cromalin or matchprint, color key and sometimes by digital means, in less time and at a lower cost than it takes to create press proofs. *Related Term:* prepress proofs.

off-press proofing — *Related Term:* prepress proofing.

offprint — a rerun or reprint of an article first published in a magazine or journal.

offset — (1) in principle, any printing process in which the image is directly transferred from an image carrier to an intermediate cylinder which, in turn, contacts (and images) the substrate. (2) in the popular vernacular, short for photo offset-lithography, or just "lithography."

offset duplicator — a small, simplified offset press. A basic offset lithographic press configuration typified by lack of cylinder bearers, minimum sheet size, ink and water chain rollers, paper guide systems and delivery mechanisms. Generally considered suitable for simple single or multicolor work which does not require either high resolution or close registration. *Related Terms:* offset press; offset lithography.

offset gravure — a combination of gravure with an offset blanket cylinder, to form a special-purpose printing device in which the gravure image carrier (usually a cylinder) transfers ink to a rubber blanket that deposits the ink on the surface to be printed. Offset gravure is used to achieve special printed effects on metal surfaces and, in combination with flexography, to print on flexible packages. *Related Terms:* gravure; offset; intaglio.

offset letterpress — a printing method using flexible relief plates on an offset lithography press. *Related Terms:* dry offset; letterset.

offset lithography — a member of the planographic classification of printing, and the world's most popular printing method. It utilizes the simple principle that oil and water do not mix. Image areas on plate are treated to accept ink (rendered oleophilic), nonprinting areas of plate have been treated to accept water (rendered hydrophilic) and repel ink. The image on the image carrier is right reading and prints onto an intermediate rubber blanket cylinder which, in turn, transfers (offsets) the image to the substrate which is backed up by the impression cylinder, causing ink transfer. *Related Terms:* offset; lithography; direct lithography.

offset paper — an uncoated book category of printing paper whose surface is generally smooth and somewhat resistant to moisture. Designed to be used for printing manuals, form letters, advertising, etc.

offset powder — a fine powder sprayed on freshly printed sheets to prevent transfer of wet ink (setoff) to the back of subsequently printed sheets as they accumulate in the delivery stack.

offset press — a press design in which an image is printed from its image carrier on to an intermediate rubber blanket cylinder that transfers or "offsets" the image to the press sheet; this principle allows plates to be right reading and are generally of better quality than direct transfers.

offset printing — *Related Terms:* offset; offset lithography.

offset/gravure conversions — the use of lithographic positives, such as might be made for deep-etch plates, to produce gravure cylinders.

offshore sheet — a term used in the U.S. and Canada for paper made overseas, not in North America.

off-square sheet — any paper whose corners are not

90°. Production machinery is designed to handle square sheets, and when off-square occurs, it will not feed or register properly.

OH (off hook) — a modem indicator light that tells you the phone line is open and ready for communications. *Related Terms:* AA; RD; SD; TR.

ohm — impedance. In electronics, the basic unit of resistance. Used to measure resistance to flow of an electrical current.

oil mounting — the covering of a transparency with a thin coat of oil before scanning to prevent Newton rings and help eliminate fine scratches, fingerprints and dust.

oil-base inks — printing inks with a vehicle that can be thinned or dissolved with a petroleum solvent.

OK sheet — an approved press sheet that is intended for use as a quality guide for the rest of the production run.

Old English — (1) a member of the Gothic type classification. (2) an erroneous name for the Gothic text or black letter classification of type.

oldstyle — (1) a roman typeface subdivision whose designs are based on earlier hand-drawn characters and distinguished in design from modern typefaces by their clear, strong features; the comparative uniform thickness of all strokes; the absence of hairlines; the irregularities among individual letters; and the diagonal serifs, curves and cross strokes. (2) a style of type characterized by stressed strokes and triangular serifs. An example of an oldstyle face is Garamond.

OLE (Object Linking and Embedding) — a means of integrating applications introduced by Microsoft. It lets users copy objects between applications, with each object containing enough information about its format and its creation application to work in a variety of OLE-enabled applications. For instance, clicking an OLE image in a word processing document activates the application the image was created in and allows in-place editing. Instead of starting an entire new application when an OLE object is activated, the user simply sees a new set of tools or menu items. *Related Term:* DDE.

oleophilic — oil receptive, as in the image areas of a lithographic printing plate. *Related Term:* hydrophilic; oleophobic.

oleophobic — oil repellent, as in the dampened non-image areas of a lithographic printing plate. *Related Terms:* hydrophilic; oleophilic.

OMG (Object Management Group) — a consortium of software vendors, developers, and users that promotes the use of object-oriented technology in software applications. The group also maintains the CORBA software interoperability standard. *Related Term:* CORBA.

omnidirectional — bar codes which can be read in any orientation relative to the scanner.

omnifont — an operating technique employed by OCR scanners to enable them to read nearly all normal characters based upon their general parameters of construction as opposed to a memory map of a specific type face. The technique greatly increases the number and types of documents which can be electronically scanned to create digital files.

on spec — the designing or writing of copy without being paid but with the intent to collect a contract for a particular job.

on-contact printing — screen printing in which screen and stencil contact material throughout ink transfer. Related Term: off-contact printing.

on-demand printing — one method of producing only the required number of documents at a given time using electronically stored information. The process makes it easier to update and modify documents more frequently and to print additional copies at a later date without incurring new start-up costs. All excessive inventories are eliminated. *Related Term:* demand publishing; print on demand.

one-line text box — a labeled, single-line form field in which users can type text.

one-up — imposition for printing of only one item at a time on a sheet.

onionskin — a translucent lightweight paper traditionally used in air mail stationery and for carbon copies, but also as a fly sheet and/or overlay page in some quality printing and any application where product weight is a consideration. Used in animation, particularly the preliminary sketching phase, to product motion on separate layers. *Related Terms:* bank paper; manifold bond.

online — (1) in typography and rendering, a letter tracing which begins on the edge and continues on the edge, half in and half out. (2) in computing, being connected to the Internet via a live connection to an ISP. (3) an adjective used to describe a variety of activities that you can do on the Internet, for example, online chat, online shopping, and online games.

on-machine coating — the application of coating to

the paper in-line as it is made on the paper machine. *Related Term:* off-machine coating.

Onyx plates — a 3M Company polyester plate designed to be imaged in imagesetters for computer-to-plate (CTP) applications.

OOP (out of print) — the condition when a book is no longer available for sale from a publisher.

OOS (out of stock) — the term used when a book is still available for sale from a publisher, but temporarily unavailable from a particular retail outlet, the publisher, or the distributor.

opacity — a material's lack of transparency. (1) in photography, it is defined as the reciprocal of the fraction of light transmitted through, or reflected from, a given tone. (2) in printing ink, it is defined as the ink's ability to hide or cover up the image or tone over which it is applied. (3) in paper, a property of paper or other substrate that minimizes show-through, which is the amount of ink printed on one side of a sheet that shows through the other side. *Related Term:* show-through.

opaque — generally, the property of paper that makes it less transparent. A material that will not permit the passage of light; not transparent. (1) in photography, the nontransparent pigment (usually red or black) applied to pinholes or other areas of film negative to prevent light from passing through during platemaking. (2) a verb meaning to cover flaws in negatives with paint or tape. (3) total opacity.

opaque color proofs — proofs used to check multicolor jobs by adhering, exposing, and developing each successive color emulsion on a special solid base sheet. *Related Term:* laminated proof.

opaque ink — (1) in printing, any heavily pigmented ink that blocks out color of underlying ink or paper. (2) in conventional prepress, a misnomer for the dense opaque used to remove blemishes and defects in negatives and positives. *Related Term:* process color inks.

opaqueness — the property of a substrate material that minimizes show-through from the back side or the next sheet. The ratio of the reflectance with a black backing to the reflectance with a white backing. Ink opacity is the property of an ink that prevents the substrate from showing through. *Related Terms:* translucent; transparent.

open shop — a shop that does not require craftspeople to join a union to maintain employment.

open up — the addition of negative space to a layout to make it appear less crowded.

open web — a web press without a drying oven, thus unable to print on coated paper.

operating system — the fundamental operating instruction set for any computer. It provides the means to direct peripherals, perform calculations, etc., through operator or program software input.

operator side — the side of the press which houses most of the controls and where the pressman stands to monitor production. *Related Term:* drive side.

opponent process model — a theory of color vision that has been refined by Leo Hurvich and Dorothea Jameson.

optical center — a point about 5/8 of the vertical distance from the bottom of a sheet, above the mathematical or true center of a page or area.

optical character reader — electro-optical equipment that scans, interprets, and converts any copy or graphic elements to a machine-readable format. An optical character or intelligent reader. *Related Term:* optical mark reader.

optical character recognition (OCR) — a technique in which any printed, typed, or handwritten copy or graphic images can be scanned by an electronic reader and converted into information that can be read, interpreted, edited and displayed by computers. *Related Term:* intelligent character recognition.

optical density — the light-stopping ability of a photographic or printed image expressed as the logarithm of its opacity, or the reciprocal of the reflection or transmission.

optical disk — in certain formats, a plastic disk on which data may be written. In all formats a disk from which data may be read using a low-powered laser. When writing, a laser is used to burn a small pit or bubble onto the disk's reflective surface. In the reading mode, when the laser beam hits the pit or bubble, the light reflects differently than the shiny surface; this determines if the bit is a one or zero. Permits high-density, cost-effective storage of data. Used for software distribution, reference works such as dictionaries, encyclopedias, games, etc. *Related Terms:* CD-ROM; floptical; DVD; CD-ROM-RW; CD-ROM-W.

optical dot gain – *Related Term:* dot gain.

optical fiber — fiber cables consisting of thin filaments of glass (or other transparent materials), which can carry beams of light. A laser transmitter is often used to encode frequency signals into pulses of light and send them down the optical fiber to a receiver, which translates the light signals back into frequencies. Less susceptible to noise and interference than other kinds of cables, optical fibers can transmit data greater distances without amplification. But because the glass filaments are fragile, optical fiber usually must be run underground or in hardened protective cases rather than overhead on telephone poles. *Related Terms:* HFC network; fiber optics.

optical gain — an effect caused by light reflecting at various angles from a rough-surfaced paper. Hence halftone dots appear larger than actual size, resulting in image degradation.

optical mark readers (OMR) — electronic equipment that scans, interprets, and converts bar coded information into a machine-readable form. OMRs are used in inventory control systems, libraries, and grocery stores. *Related Terms:* bar code readers; optical bar readers; optical character readers.

optical resolution — a reference to the number of dots per inch an image scanner is able to gather with its CCD, PMT or CMOS sensors.

optical scanner resolution — the true hardware capabilities of a scanner based upon the number of reading elements and the sampling rate during a scan.

optical storage technology – any high-capacity storage device with removable media that employs laser technology to read and/or encode digital data by causing interference on the surface of a highly polished disk. Optical storage is estimated to have a 30+ year shelf life.

optical throw — (1) in data scanning, the distance from the scanner face to the closest point at which symbols can be read. (2) in photography, it is the difference between range and depth of field. (3) in media presentations, the distance from the projector to the reflective screen.

optics — a branch of physics that deals with the properties of light and its interaction with devices used to redirect it and alter its composition, i.e., lens, prisms, etc.

optimum color reproduction — a reproduction that represents the best compromise within the capabilities of a given printing system.

optimum format — a format in which the width of the type columns (line measure) is within the range of maximum readability, traditionally about 1 1/2 lowercase alphabet lengths of the type face and size being set.

optimum line length — making the width of the type columns within the range of maximum readability. Generally accepted as being approximately 1.5 lowercase alphabet lengths (a-z + a-l) of the font being used. *Related Terms:* alphabet length; measure.

option — (1) alternate methods. For example, in estimating, a comparison in pricing for different quantities, etc. (2) in publishing, the reserved right to purchase or sell an author's subsequent work(s), or the right to purchase or sell the subsidiary rights to an author's work during a specified period of time.

Orange Book — the format that enables CD-R drives to record discs that regular CD-ROM players can read. The main difference between it and other formats, such as Red Book and Yellow Book, is that Orange Book defines how CD-R devices can append index data to an existing disc's directory if you add more data to the disc in multiple sessions. *Related Terms:* CD-R; Yellow Book; Red Book; Green Book.

orange peel — (1) in image coatings, the effect of a defect in screen printing when IR or UV coating or thermographic resins crack or wrinkle when drying. The surface effect gives the appearance of an orange skin. (2) in small offset duplicator setup, the appearance of ink on the fountain roller which is sometimes sought by offset duplicator operators when initially setting up the press. It is achieved by manually rotating the fountain roller, with the ductor roller in contact, and opening fountain keys the correct amount to achieve an orange peel effect on the fountain roller. When this occurs it is believed to be a a good beginning point for setting the fountain for regular work.

.org — a domain name suffix used in Internet addresses that denotes an organization that doesn't fit into any of the other categories, such as a nonprofit, nongovernment organization.

organization chart — a diagram which outlines titles, responsibilities and pecking order within a group by compartmentalizing each position into a box and using a variety of solid or dotted lines to show the lines of authority in the hierarchy.

orientation — (1) in production, the predominant direction of flow of a design as in portrait (vertical) or landscape (horizontal) orientation. (2) in management, the process of familiarizing new employees or the newly promoted with the policies, procedures, responsibilities and management expectations of a position. (3) in optical scanning, the alignment of the symbol's scan path.

Two possible orientations are horizontal with vertical bars and spaces (picket fence) and vertical with horizontal bars and spaces (ladder).

original — any artwork mechanical, photograph, object or drawing that is submitted to be reproduced by a photomechanical or electronic process. Not pre- printed. *Related Term:* original copy.

original equipment manufacturer (OEM) — a company that develops, produces and sells equipment to others and applies the buyer's name brand, not their own.

ornament – typographic devices that are used to embellish text such as an initial capital letter (initial cap). Not to be confused with picture fonts or pi characters such as dingbats. *Related Term:* printer's ornaments; dingbat.

ornamented — a typeface that is embellished for decorative effect. *Related Terms:* decorative; novelty.

orphan — an element of type (such as a word or a line) which leads into a larger block of type but which has been left by itself at the end of a page or column. For instance, the first line of a paragraph or a section head or a single, short word or the end of a hyphenated word. Sometimes erroneously called a widow. *Related Term:* widow.

ortho safelight — a red light used during film processing that will not expose orthochromatic film. *Related Term:* safelight.

orthochromatic — photographic plates and films sensitive to all portions of the blue-green spectrum but not red light (electromagnetic radiation with wavelengths between 375 and 560 nanometers).

OS (operating system) — a software routine that knows how to talk to computer hardware and can manage a computer's functions, such as allocating memory, scheduling tasks, accessing disk drives, and supplying a user interface. An operating system eliminates the need for software developers to write programs that directly access hardware. With an operating system, developers can write to a common set of programming interfaces called APIs and let the operating system do the dirty work of talking to the hardware. *Related Terms:* API; kernel; DOS.

oscillator — gear- or friction-driven printing press rollers that not only rotate about their axis, but move from side to side, distributing and smoothing out the ink film and erasing image patterns from the form roller; designed to minimize or eliminate mechanical ghosting. *Related Terms:* oscillating drum; vibrator; roller.

OSP (online service provider) — any company that provides customer only content to subscribers of their service. Many OSPs offer Internet access, but maintain a private network that is only accessible to their customers. OSPs control the structure and content of their networks, and tend to be more logical and user-friendly, especially for beginners. Popular OSPs are: AOL (America Online), CompuServe and MSN.

Ostwald system — a system of arranging colors in a color solid. The colors are described in terms of color content, white content, and black content. The solid appears as two cones, base to base, with the hues around the base, and with white at one apex and black at the other.

out of print — any publication that is no longer available as a current release from a publisher and with no expectation that it will become available in the future.

out of register — the extremely undesirable occurrence when color separations for an image are not properly aligned, causing text and colors to appear fuzzy, blurry or with colored halos around image edges.

out point — the point during production where editing ends.

outdoor — the form of advertising used on buses, billboards, etc.

outline — (1) in typography, a typeface that is formed with only the outline defined rather than with a solid, filled face. (2) in creative writing, a manu-script or draft which contains only key points, without detail. *Related Term:* outline type.

outline halftone — a halftone in which background has been removed to isolate or silhouette an image.

OUTLINE TYPE
A typeface which comprises only the extreme edge areas of the design without fill.

outline type — *Related Term:* outline.

out-of-band signaling — a separate telephone network intended specifically for call signaling on a separate ISDN wire called the D channel. Its purpose is to do all the call setup and signaling. The existence of this separate telephone network intended specifically for call signaling is called "out-of-band." The actual call signaling is done independently of the data-carrying channels, thus allowing them to utilize their full bandwidth capability.

output — the results of the work produced by a computer or computer-controlled machine, e.g., the material produced by a typesetting machine. *Related Term:* I/O.

output device — any device that receives information from the microprocessor, most commonly the monitor or printer.

output resolution — the number of dots per inch (dpi) of the output device (high-end imagesetters can support various resolutions). The higher the screen frequency, the higher the output resolution.

output size — the size of film that is used for output with a particular scanner. Often times this will be referred to as 1-up, 2-up, 4-up, etc., with different sizes listed for each. It refers to the number of colors from the same transparency that can be separated, per film sheet. For example, a 1-up 16 in. x 20 in. on a 16 in. x 20 in. film necessitates rescanning for each additional color. Similarly 4-up 8 in. x 10" in. means that all four colors can be scanned at the same time on the same 16 in. x 20 in. film.

over — (1) the additional paper required to compensate for spoilage in printing. (2) in the plural, a term used to refer to a quantity produced above the number of copies ordered.

over the transom — a term used in publishing for unsolicited manuscripts or queries sent to a publisher by authors.

overcoating — the protective coating placed on most transparent-based films to protect film emulsion from grease and dirt.

overdevelop — the result of leaving sensitive materials in their reducing agent for too long a period of time. An example would be film which is left in developer beyond the prescribed time, temperature, agitation, or concentration recommended by the manufacturer.

overexposure — a photograph that appears washed out or too contrasty because too much light reached the film or print, resulting in an overly dense image.

overhang cover — a cover that is larger in size than the interior pages. *Related Term:* Yapp bind.

overhead — (1) in management, the cost of operating a business without considering cost of materials or labor; typically includes cost of maintaining work space and equipment depreciation. (2) in optical reading, the fixed number of characters required for start, stop and checking in a given symbol. For example, a symbol re-

quiring a start/stop and two check characters contains four characters of overhead. Thus, to encode three characters, seven characters are required.

overlap — the technique used by some electronic scanners to have consecutive scan lines created on the marking engine overlap by amounts as much as 40 percent.

overlay — generally, a sheet of tissue or acetate taped to a mechanical so that it covers the mounting board. (1) in layout, the overlay is used to protect copy. (2) in production, a sheet of translucent or transparent tissue, acetate, or Mylar attached over the face of the primary artwork (paste-up) or photos and used to indicate crops, surprints, knockouts, overlapping or buffing flat colors, placement of alternate or additional colors, or previously separated color art. (3) in letterpress printing, a sheet of packing added under the tympan sheet during make-ready; composed of built-up and cutout areas to increase or decrease pressure on the final press sheet.

overlay proof — thin, transparent pigmented or dyed sheets of plastic film that are registered to each other in a specific order and taped or pin-registered to a base sheet. Each film carries the printed image for a different process or spot color, which, when combined, creates a composite simulating the final printed piece. *Related Terms:* color keys; transfer keys.

overlay sheet — *Related Terms:* overlay; Rubylith.

overlay, acetate — a clear material base used to make color breaks. For instance, each color portion of an art object would be placed on a separate overlay which would be photographed independently and yield a negative with only that color's information. They are always registered with the base art with registration marks to facilitate stripping, platemaking and presswork.

overlay, tissue — a thin paper sheet on which the designer indicates production instructions about the base art and its acetate overlays. It also serves as a protective layer for the entire art piece. It is not photographed.

overline — a headline type heading placed above a picture.

overprint — (1) to print over an area that has already been printed using transparent inks, creating a third color, i.e., red overprinted on yellow would result in orange. (2) to print one image over a previously printed image, for the express purpose of obliterating the original image. (3) in lithographic plate-making, the exposure of a second negative onto an area of the plate previously exposed; a method of combining line and halftone images on the plate. (4) solid or tint quality control image elements that are printed over or on the top of previously printed colors. Overprint patches are used to measure trapping saturation, and overprint color density and may be measured from a color bar in the trim of a press sheet or from the printed image itself. *Related Terms:* knockout; surprint; reverse; double printing; multiple burns. *Related Term:* surprint.

overprint color — a color made by overprinting any two of the primary subtractive inks (yellow, magenta and cyan) to form secondary colors in the form of the additive primaries, red, green and blue.

overrun — (1) in production, the quantity of printed copies exceeding the number ordered to be printed. Trade custom allows a certain tolerance for overruns and underruns. (2) when an input device has speed which exceeds the output device's ability to perform. *Related Term:* over(s).

overs — *Related Term:* overrun.

overset — body type matter that has been set but exceeds the allotted space provided in the layout.

overspray — (1) in production, the excess misting from an anti-offset powder device on a press. (2) in NIP laser imaging, the appearance of flecks of toner in nonimage areas.

overstrike — a method used in word processing to produce a character not in the typeface by super-imposing two separate characters, e.g. $ using S and l of c and / to make a ¢ sign.

oxford rule — parallel heavy and light lines.

oxidation — (1) the process of ink drying by combining solvents with oxygen in the air. (2) the deterioration of unpreserved lithographic plates. The hydrophilic base metals tend to oxidize when left open to air. The oxidation which occurs causes nonimage areas to become ink receptive.

Ozalid — a trade name to describe a method of copying page proofs from paper or film.

Pp

package level indicators — individual items marked with UPC are frequently packaged in standard quantities of intermediate packs and shipping containers. The different package quantity for each product is assigned a "package level" and assigned a unique number.

packaging papers — (1) printing papers made for specific packaging applications, such as food, whose papers must meet government sanitary standards. (2) coarse and other inexpensive papers suitable for stuffing and wrapping packages.

packet — the unit of data sent across a network. It contains information about which computer sent the data and the data destination. If a packet runs into a problem during transmission, it can attempt to find another route. When all the packets are delivered, the recipient computer puts them together to recreate the original file.

packet switching — the process of breaking transmitted data into units, where each one has the address of where it came from and where it is going. The process enables data from many different sources to co-mingle on the same lines, and be sorted and directed to different destinations. By sharing data lines in this fashion, it enables greater numbers of users to access the network simultaneously.

packing — (1) on hand-fed platen presses, material used to control overall impression; top packing sheet holds gauge pins that receive press sheet during image transfer. (2) on cylinder and rotary letterpresses, the materials placed between the tympan paper and the impression cylinder to adjust impression. (3) in bindery and finishing, the placement of product in, and the materials used to protect a product as it is placed in, appropriate containers for delivery or storage.

pad — to bind by applying glue or other liquid adhesive along one edge of a stack of single sheets to produce a single unit such as a tablet or note pad. *Related Term:* padding.

PADDING SHEET COUNTER

pad counting tool — an adjustable device which is set to a caliper representing the number of sheets desired in each pad. Subsequent prodding into a pile of sheets of identical weight or bulk will yield amazingly accurate counts for subsequent pads.

pad printing — (1) the use of a soft rubber pad to transfer a relief image onto irregular substrate surfaces such as basketballs, masking tape, etc. (2) the production of notepads and similar.

page — (1) on the Internet, a single HTML document on a Web site. (2) in printing, one side of a sheet or leaf of paper. *Related Terms:* recto; verso; signature; imposition.

page buffering — the ability to spool an entire image to disk and print in a continuous motion.

page composition — the assembly of the elements on a page, including text and graphics. *Related Terms:* page makeup; page layout; area composition.

page count — the total number of pages in a publication. It includes blank pages and printed pages without numbers. *Related Term:* extent.

page description language (PDL) — in an electronic publishing system or phototypesetter, a standardized formatting routine in which all of the elements to be placed on the page, their *x-y* coordinates (respective position on the page), and the page's position within the larger document are identified. It is a special form of programming language which enables both text and graphics (object or bitmapped) to be described in a series of mathematical statements. The main benefit is that applications software is independent of the physical printing device. Without it, each output device would have to have its own unique translation software. Typical PDLs include interPRESS, imPRESS, PostScript and PCL. *Related Terms:* page descriptor; document descriptor; imagesetter; PostScript.

page layout —assembly of the elements on a page, including text and graphics. *Related Terms:* page composition; page makeup; paste-up.

page makeup — the assembly of the elements on a page, mechanically or electronically, including text and graphics. *Related Terms:* page composition; page layout.

page printer — the more general (and accurate) name

used to describe nonimpact printers which produce a complete page in one action. Examples include laser, LED and LCD shutter xerographic printers, ion deposition, electro-erosion and electrophotographic printers.

page proof — (1) in final prepress production, type output in page format, complete with headings, rules and numbers. (2) in makeup, the stage following galley proofs, in which pagination occurs.

page proportion — the aspect ratio or proportional relationship of a page width to its height.

page title — (1) in Web publishing, a text string identifying a page. (2) in conventional publishing, a page near the front which displays the title of the book, its author, and sometimes the name of the publisher or printer.

PageMaker — a software program for page layout from Aldus Corporation (now owned by Adobe) that everyone associates with desktop publishing due to its immense success on both the Apple Macintosh and PC platforms. A software program for the integration of type and graphics which metaphorically copies the steps in the conventional paste-up process. It is still used as a benchmark product although, for professional publishing, certain aspects of its design are coming under attack from other products, notably Quark Xpress.

pages per inch (ppi) — a measurement of the caliper of paper. Used to estimate the thickness of a bound publication.

pagination — (1) in conventional metal composition, the process of page makeup from galleys of raw type. (2) in cold composition, the making of mechanical art by makeup or paste-up. (3) in imposition, numbering pages with associated running heads and feet and sometimes including trim marks. (4) in electronic and digital prepress, the process of performing page makeup automatically through a computer program (computer-aided pagination) according to page parameters designated by the operator or by a database. *Related Term:* page makeup; CAP.

pagination dummy — a pattern used to indicate imposition that is prepared from a full-size blank press sheet according to the way the sheets will be folded during the finishing and bindery process.

paint program — a type of bit map graphics program that treats images as a collection of individual picture elements (pixels) rather than as a collection of shapes (or vector objects). *Related Term:* draw program.

Paint Shop Pro — one of the easiest, fastest and most powerful image viewing, editing and converting programs for the Windows platform. It supports over 30 image formats, contains several drawing and painting tools and effects and has the ability to use PhotoShop plug-ins. It is excellent for converting and preparing GIFs and JPEGs for use on the World Wide Web.

PAL (phase alternate line) format — the television broadcast standard throughout Europe (except in France, where SECAM is the standard). It broadcasts 625 lines of resolution, at 50 interlaced frames per second, nearly 20 percent more than the current U.S. standard, NTSC, of 525. *Related Terms:* NTSC; SECAM.

palette — a collection of colors or shades available to a graphics system or program. *Related Term:* gamut.

pallet — (1) a plastic or wooden platform onto which freight or cargo such as bulk paper or folded and gathered sheets are strapped for transport. Sometimes incorrectly referred to as a "skid." (2) A tool used for printing or gilding letters on book bindings.

pallet knife — a thin, flexible bladed knife used for mixing and spreading ink by hand. *Related Term:* ink knife.

pamphlet — an informational brochure that is usually only one or two pages.

pan — the process of changing a camera view by changing camera perspective through rotation about its own axis. For example, to begin with a view of a person on the left of a scene and, keeping the camera in its same relative position, turning through an arc to the right, capturing the entire scenic view.

panchromatic — any film emulsion which has sensitivity to the entire visible spectrum. *Related Terms:* orthochromatic; blue-sensitive films; panchromatic film; panchromatic material.

panchromatic film — a photographic film category that is sensitive to all three primary colors of light. *Related Terms:* orthochromatic; panchromatic material.

panel — one reveal of a brochure or other printed piece. It is only one side of the paper, the reverse being a second panel; thus a trifold brochure has six panels.

pantograph — (1) a tool or machine for quickly and accurately enlarging or reducing a design by tracing over an original. A series of interconnected arms holds a reproducing pen. The length of the arms and the points at which they are joined independently control the *x-y* axis of the reproduction. (2) in typography, it is used to produce a matrix for foundry type and duplicate engravings for high-quality printing such as currency and stamps.

Pantone Matching System (PMS) — the most commonly used ink mixing and color reference formula. It consists of a dozen or so carefully formulated colors which, when mixed in the proportions indicated on a reference chart, produce several thousand predetermined colors for spot reproduction.

PAP (password authentication protocol) — a moderately secure, commonly used, Internet access control; not as secure as CHAP, because it works only to establish the initial link. It is also more vulnerable to attack because it sends authentication packets throughout the network. *Related Terms:* ISDN, CHAP

paper — a substrate for printing and writing that is made of cellulose fibers from trees and/or other plants and which are sometimes combined with artificial fibers.

paper basis weight — in the United States, the weight in pounds per ream of paper cut to its basic size. The size varies with different grades and varieties of paper. *Related Terms:* ISO; gsm.

paper bind — *Related Term:* paperback.

paper distributor — a paper merchant selling wholesale to printers and other buyers of large quantities.

paper dummy — an unprinted sample of a proposed printed piece which has been trimmed, folded, and, if necessary, bound using paper specified for the printing job. *Related Term:* dummy.

paper grain direction — a characteristic mainly associated with presswork and/or bindery and is further identified as "cross" or "with." Refers to the general alignment of the majority of paper fibers as they were laid down from the headbox of the paper machine. "Cross" refers to the direction of fibers which run perpendicular to running direction of the Fourdrinier forming wire on the papermaking machine. "With" indicates that the predominant grain pattern is parallel to the direction of Fourdrinier. *Related Terms:* grain direction, cross; wire direction.

paper lines — lines drawn on paste-up board to show final size of printed piece after it is trimmed; they are usually measured from center lines and drawn in light blue pencil.

paper master — usually a term which refers to an offset lithography plate made of paper.

paper merchant — a wholesale merchant selling paper to printers and other buyers of large quantities.

paper mill — a manufacturing plant which produces paper substrates and other paper products.

paper plate — a short-run paper base offset printing plate on which matter can be typed, drawn or xerographically placed for reproduction.

paper ribbon — a narrow web of paper either one or two book pages in width that is created by slitting (cutting) the full width web of printed paper in the finishing section of a book or other web publication press.

paper stencil — a stencil formed by cutting the image in a sheet of paper and attaching it to the fabric of a screen printing frame with ink.

paper stock — the paper used for printing a particular job or order.

paper tape — a continuous band of paper (about 1 inch wide) that stores keystrokes by a series of codes, in the form of punched holes across its width; the holes can then be read and interpreted by typesetting or imaging machines to output characters.

paper, coated — a wood pulp or rag based printing paper with a layer of white clay or other pigments and a suitable binder applied to its surface. The coating improves surface uniformity, light reflectance, and ink holdout. Generally, coated paper is most often used in high-quality, four-color printing.

paper, NCR — *Related Terms:* NCR paper; carbonless paper.

paper, photographic — a paper substrate with a light-sensitive emulsion, used for making photographic prints.

paper, recycled — *Related Term:* recycled paper.

paper, uncoated — any paper that has not been treated or processed with a surface coating during manufacture. The rougher surface absorbs inks more readily and has less light reflectance than coated papers.

paperback — a perfect bound or burst bound book, as opposed to hard cover. *Related Terms:* paper bind; paperbound.

paperboard — a thick ground pulp paper product usually heavier than .010 inches in thickness. *Related Term:* board.

paperbound — A soft-cover book. *Related Term:* paperback.

papermaking — the name of the manufacturing process for creating paper from wood pulp, recycled paper or rags.

papeterie — (1) a soft, high-quality bond paper that accepts handwriting well, packaged in small quantities for sale as social stationery. (2) the stiff paper used for greeting cards. (3) a store selling high quality handwriting paper.

papyrus — a tall plant native to the Nile River region of Egypt, of which the leaves and pulp were sliced and pressed into matted sheets by the early Egyptians to produce the first writing material with many of the properties of paper. Provides the root for the word "paper".

paragraph indent — the negative or positive space created at the beginning of the first line of type in a paragraph when compared with succeeding lines.

paragraph mark (¶) — a type font symbol used to denote the start of a paragraph. Also used as a footnote sign.

paragraph openers — typographic elements used to direct the eye to the beginning of a paragraph. Often used when the paragraph is not indented.

paragraph rules — graphic lines associated with a paragraph that separate blocks of text. Rules are commonly used to separate columns and isolate graphics on a page. Some desktop publishing programs allow paragraph styles to be created that include paragraph rules above and/or below the paragraph.

paragraph style — a part of nearly every typographic manipulation software which stores a paragraph format that specifies the type of font to use in a paragraph, font size, etc. Paragraph style can also specify whether to use bullets and numbering, and controls indentation and line spacing.

parallax — the term used to identify the error in the image difference between that seen in the camera view finder and that covered by the lens.

parallel — (1) generally, side by side. (2) in the communications world, the sending of data over two separate wires in a single cable at the same time to speed transmission.

parallel fold — folds which occur on a sheet and are parallel to each other, as contrasted to a right angle fold where the folding directions are at 90° to one another; e.g., two parallel folds will produce a six-panel/page product.

parallel port — the 25-pin RS-232C connector, on the back of most PCs. Theoretically eight times faster than serial ports because it transmits all eight bits of a byte at once, parallel ports are also sometimes called the printer ports, but you can attach tape backup units, CD-ROM drives, scanners, and other devices as well. Parallel ports come in several configurations. *Related Terms:* EPP; ECP.

parameter — a variable that is given a constant value for a specific process used to refer to the limits of any given system.

parchment — originally, a fine, translucent paper made from the tanned hide of a sheep or goat. Today the artificial paper which simulates writing surfaces made from animal skins. *Related Term:* diploma paper; artificial parchment.

parchment, artificial — *Related Term:* parchment.

parchment, true — *Related Term:* parchment.

parent sheet — a paper mill term for sheets 17" x 22", A3 or larger. *Related Term:* folio size.

Pareto chart — a form of frequency distribution that displays categories from the most to least frequent.

parity bit — an obsolete error data checking process used in old modems because they transmitted data one character at a time. Each character had to have its own individual error check, and the usual rule was to add an extra bit (the parity bit) at the end of each character before the stop bit. This bit would be set to 0 or 1 based on the value of the previous data bits. Today, modems gather data into packets and send a larger 2- or 4-byte error-check value to validate data in the entire packet. The bit is no longer necessary. *Related Terms:* CRC; parity; start bit.

park — locking the heads of a hard drive into a non-contact space with the disk prior to moving the machine to prevent head crashes and disk or data damage or loss.

partition — a portion of a hard disk that functions as a separate disk drive. Any individual hard disk can be divided into several partitions, each of which functions as a separate drive and has its own volume name (such as D:, E:, F:, etc.). The purpose is to make the drive more efficient, as the computer can search smaller sections for a specific file rather than the entire drive. Partitioning refers to the process of dividing the hard drive into partitions. *Related Terms:* virtual drive; logical drive.

pass — the passage of a press sheet through all printing units.

passive matrix — an LCD (liquid crystal display, as opposed to an active matrix display). A flat-panel display created by laying a layer of liquid-crystal diode elements on top of a grid of wires. By applying current to the various intersections, the diodes can be excited; they light up and act as pixels. Passive displays simply apply current to the diodes at a specific refresh rate to maintain an image. Higher quality (and, therefore, more expensive) active-matrix or thin film transistor (TFT) displays control each diode with one or more transistors, making for sharper, brighter pictures. *Related Terms:* active matrix; flat-panel display; LCD; persistence; refresh rate; TFT.

password — a secret string of letters and/or numbers that allows a user access to a computer, Internet service account, etc., if the service requires it. The appropriate password may be decided by either the user or the provider, depending upon the nature of the system being accessed. Some systems assign passwords and periodically change them to prevent them from being compromised.

paste bind — to bind by adhering sheets with a thin bead of glue along the fold of the spine.

paste drier — a special combination of drying compounds used in ink-making to speed drying.

pasteboard — a chipboard base on to which another paper has been affixed to provide a better surface for printing, writing or drawing.

pastel colors — soft or light colors usually in the highlight to midtone range which lack saturation.

paster — (1) in printing, the device used to apply a fine bead of paste on either or both sides of the web to produce finished booklets directly from the folder without saddle stitching. The paste is applied from a stationary nozzle as the web passes. (2) an eight-, twelve-, or sixteen-page booklet that is pasted instead of saddle-stitched together. (3) automatic web splices on a press. (4) the rejected web with a splice in it. *Related Terms:* gluing; pasting; reel-tension paster; RTP.

paste-up — (1) in camera work and scanning, a noun which describes camera-ready mechanical art that includes all the necessary text and graphic elements. (2) a verb describing the process of preparing a camera-ready mechanical for final reproduction. *Related Terms:* mechanical; photomechanical.

paste-up artist — a person who arranges the type and illustrations to make camera-ready art according to a layout.

patch — the insertion of a few words or lines as corrections on original mechanicals, as opposed to outputting the entire page to adjust the error.

patching — a method of making corrections in repros or film in which the corrected "patch" is set separately and pasted into position on the repros or shot and stripped into film.

patent binding — *Related Term:* perfect binding.

path — that portion of a URL that identifies the folders containing a file. For example, in the URL http://that.place.site/howdy/earth/greetings.htm, the path is /howdy/earth/. Also used to define the location of a file by the computer operating system.

PBX (private branch exchange) — a computerized version of the telephone switchboard but with an expanded range of voice and data services; it serves a particular organization and has connections to the public telephone network; a multiline telephone exchange terminal with various features for voice and data communications.

PC — *Related Terms:* personal computer; microcomputer.

PC card — the credit-card-sized cards that plug into portable and desktop computers to add and remove RAM, modems, network adapters, hard disks, and other devices without requiring that you open the box. They conform to several standards set by the industry and not all slots will accommodate all cards because of the varying thicknesses permitted for specific purposes. Originally, the PC card was 3.3 mm thick, and used mainly to add RAM. A second type is 5.0 mm and used for RAM, modems and network adapters. Still a third type is 10.5 mm and is used for hard disks and radio devices. They all use a two-layer interface of instructions: card services and socket services. The first manages system resources the card requires, and determines IRQ and memory addresses. Socket services communicate with the card controller chips and act as go-betweens with card services. Socket services can be part of the BIOS or managed via software. *Related Term:* PCMCIA.

PCI (Peripheral Component Interconnect) — a self-configuring PC local bus. Designed by Intel, PCI has gained wide acceptance even by Apple, in its PowerPC series. It beats out the VESA Local Bus spec from a technical standpoint. If you have a Pentium, make sure any add-in board you buy is a PCI device. *Related Terms:* USB; ISA; EISA; microchannel; AGP.

PCM (pulse code modulation) — a system for converting analog sound signals to digital signals, so that a computer (and that includes CD and DAT players) can reproduce it. It is the most common technique for making this conversion. In its native format which is used by WAV and AIFF files, it is not well compressed. Sounds are often further encoded using a codec. *Related Terms:* ADPCM; AIFF; codec; WAV.

PCMCIA (Personal Computer Memory Card International Association) — formed by several modem card manufacturers in 1989 to define physical design, computer socket design, electrical interface, and associated software for PC cards. They used some of the Japanese Electronic Industry Development Association's (JEIDA) principles in developing their standard. Both organizations continue to support international standards for PC cards, as they are now called. The newest release of the standard incorporates both standards, a development which further enhances compatibility between products. The cards are now used in several types of RAM, preprogrammed ROM cards, modems, sound cards, floppy disk controllers, hard drives, CD-ROM and SCSI controllers, global positioning system (GPS) cards, data acquisition, network cards, pagers, etc. PC cards also provide the ability for hot plugging, meaning that the cards can be added or removed from a running computer without needing to shut down and reboot. Originally called PCMCIA cards, the association trademarked "PC card" and prefers its use. *Related Term:* PC card.

PCS — (1) in communications, personal communications services, a generic description of the cellular communication services that provide voice, data, and paging in a single unit. (2) in optical reading, print contrast signal, a measurement of the ratio of the reflectance between the bars and spaces of a symbol.

PCX — a graphics format, developed by Zsoft for its early DOS-based paint program PC Paintbrush. It was the de facto standard for bitmapped graphics before Windows. Even though the Windows BMP format is now more prevalent, PCX is more efficient in its use of disk space because it uses run-length encoding to compress graphics. *Related Term:* run-length encoding

PDF (Portable Document Format) — a file format created by the Adobe Acrobat Reader, Acrobat Capture, Adobe Distiller, Adobe Exchange, and the Adobe Acrobat Amber Plug-in for Netscape Navigator. It was developed in hopes of establishing a standardized format for documents that are used on the Internet. One of the benefits of using Acrobat and PDFs is that whether you're an executive using Microsoft® Office products, an engineer using a CAD program, or an art director using desktop publishing software, you can quickly deliver business documents to a colleague or to the entire company, without reauthoring or learning new applications. The other is that the integrity of the file is preserved between computing systems. PDF transfers exact fonts, color and position information. For this reason it is gaining acceptance in the digital publishing, advertising and service bureau industries, as the files are often smaller and easier to work with than traditional PostScript and, under the right circumstances, are editable.

PDF417 — a stacked or two-dimensional bar code with very high data density and a high degree of security.

PDL (page description language) —code generated by a typesetting or page-layout system that tells the output device, such as a laser printer or imagesetter, where to place elements on a page. Adobe System Inc.'s PostScript is an example of a PDL.

PDP (Programmed Data Processor) — a '60s and '70s computer series from Digital Equipment Corporation, before it developed the VAX minicomputer. The best-known families in the series were the PDP8, PDP10 and PDP11, which was the direct predecessor of the VAX .

PE (printer's error) — a mistake made by the typesetter or printer after the originals have been submitted by the client, as opposed to AA. These types of errors are not charged to the client.

peaking — an electronic edge enhancement effect that produces the appearance of increased image sharpness. *Related Term:* sharpening.

pebbling — an embossed texture placed into paper in its manufacture to create broken surface effect.

peer-to-peer network — any network where there is

no dedicated server. Every computer can share files and peripherals with all other computers on the network, when all are granted access privileges. Such a network is practical only for small workgroups of less than 10 or 12 computers.

PEL — a picture element. (1) in computer graphics, the smallest element of a display surface that can be independently assigned color and intensity. (2) elsewhere, the area of finest detail that can be reproduced effectively on a recording medium. *Related Term:* pixel.

pen name — non de plume. A pseudonym used by the author of a book or magazine article. *Related Terms:* wand scanner; fixed beam scanner; moving beam scanner.

pen scanner — a pen-like device, connected either by wire to a device or self-contained, used to read bar codes. Requires direct contact with the symbol. *Related Terms:* wand scanner; fixed beam scanner; moving beam scanner.

penalty copy – a very underused term. This is any original copy which is difficult to work with such as bad handwriting, badly typed or faxed copy or copy that is otherwise difficult to read. Should be charged out at a premium service rate.

penalty stock — a substrate that has undesirable or unusual characteristics that causes problems on press, making the printer charge extra to print it. Typically the problems are paper thickness or thinness, or difficult surface printability.

penetration — drying of ink by absorption into a substrate (usually paper).

Pentium — Intel brand name for the next generation microprocessor following the 80486.

pentop computer — a notepad-sized machine into which the user enters data with a stylus rather than a keyboard. When combined with a fax machine, users can send documents or diagrams back and forth with editing and notations.

penumbra — the soft edge of a shadow caused by the edge point where light passes over the sharp edge of a card while dodging or burning. As the card is moved closer to the enlarging lens the penumbra eliminated a sharp or harsh line.

percentage — the quantification measurement of the amount of enlargement or reduction of a piece of copy. Enlargements are shown as over 100%. Reductions are shown as less than 100%.

percentage scale — *Related Term:* proportion scale.

percentage wheel – *Related Term:* proportion scale.

perf — *Related Term:* perforation.

PERFECT BINDING

perfect binding — a relatively inexpensive method of binding in which the pages or signatures are folded, gathered, the spine edge roughened and sometimes sawed; a flexible adhesive is applied to the roughened area, a cover attached, the assembly is pressed together until the adhesive sets, and the product is trimmed. The process is less costly than sewing, but it is not quite as strong. Regardless of its shortcomings, the economies of its use cause it to be widely used for paperbacks, manuals, textbooks, and telephone books. *Related Term:* patent binding.

perfecting press — a printing press that prints both sides of a sheet in a single pass through the press. *Related Term:* perfector.

perfector — *Related Terms:* perfecting press; duplex.

perfector, convertible — a sheet-fed press with a special transfer cylinder in the gripper system that allows the sheet to tumble end for end between printing units so that the other side of the sheet is printed by the second unit. A two-unit convertible perfector can be configured to print two colors on one side of the sheet, or reconfigured to print one color on each side in a single pass through the press.

perforating — a finishing operation that cuts or punches a series of slits or holes in paper to facilitate tearing or ease folding by allowing air to escape from signatures or to prevent wrinkling when folding heavy papers.

pergamyn — *Related Term:* glassine.

periodical — A publication such as a magazine or newspaper that is published at regular intervals.

peripheral — (1) in computing, any input or output device used in conjunction, and usually connected by cable, with a computer; for example, a printer, modem, or disk drive. (2) generally, a piece of equipment that is attached to but not a direct part of a central unit.

PERL (practical extraction and report language) — for some, the programming language of choice for writing Web server applications. It is used for creating interactive forms and a slew of other CGI programs. It is a free-licensed language and comes in versions for Windows NT, Novell NetWare, and UNIX. PERL scripts are available free of charge all over the Internet. *Related Term:* CGI.

permanent paper — *Related Terms:* acid-free paper; archival quality paper.

permission — legal authorization from a copyright owner to quote or reproduce material from a copyrighted work. *Related Term:* infringement.

persistence — (1) in video display technology, the amount of time a phosphor or diode pixel stays lit after current has been applied to it. A pixel's persistence is what allows an image to remain on the screen between screen refreshes. Screens with short-persistence pixels may show flicker, while those with long-persistence pixels may show ghosting. (2) in vision, the length of time that the eye retains an image after it actually disappears. Motion pictures project a rapid series of still images. Persistence of vision causes the first image still to be in place when the second one is seen, etc. This overlap causes the appearance of motion, even though nothing is actually moving in the individual pictures.

perspective — the relative size and shape of objects as recorded on a plane surface, such as film. It provides the illusion of depth when a 3-dimensional image is imaged on a flat film surface.

perspective correct texture mapping — a realistic looking texture-mapping process, particularly in scenes rendered with large polygons. Without this correction, a hallway might appear to blend into the vanishing point. It looks good, but is not required for 3D gaming. *Related Terms:* texture mapping; perspective correction.

perspective correction — *Related Term:* perspective correct texture mapping.

petroglyph — an ancient or prehistoric drawing or painting done on a cave or rock wall. *Related Terms:* pictograph; ideogram; pictograph; hieroglyph.

PGP — *Related Term:* pretty good privacy.

PGP Public Key Server — a database of information that can be queried to find someone's public key. *Related Terms:* pretty good privacy (PGP).

Ph — a system that allows you to look up directory information, such as e-mail addresses at universities, research institutions, and governmental agencies throughout the world. To use it requires a Ph access program. When operating, all that is required is to identify which Ph server to use, and then enter a name you would like to search for.

pH scale — a numeric scale to measure of a liquid's acidity; it runs from 0 (very acid) to 14 (very alkaline, or a base); midpoint, 7, is considered neutral.

phase — the repeating peaks and valleys of a basic analog wave which starts at a zero point on a line, climbs to a peak a bit further down the line, goes past the zero point to a valley, and then finally returns to the zero point. *Related Terms:* amplitude; frequency.

Phong shading — a sophisticated way to shade polygons in 3D graphics. It works like Gouraud shading but requires more computer horsepower and yields better results. *Related Terms:* flat shading; Gouraud shading; shading.

phonogram — a symbol or character used to represent a word, sound or syllable in a system of writing. *Related Terms:* ideogram; hieroglyph.

phosphor — a material which converts radiated energy into visible light. The material on the inside face of TV and computer CRTs. When the electron gun fires and strikes the individual phosphor dots, they emit either red, green or blue. When they all are activated, the viewer perceives white.

photo CD — originally a proprietary system developed by Kodak for storing photographic images from film onto CD-ROMs, although certain parts of the specifications have been made public by the company. People use the term Photo CD to refer to two distinct entities: the Orange Book CD-ROMs that contain pictures, and the graphics file format with the file extension .pcd that stores the data, which can also be found on magnetic disks—floppy or hard. *Related Term:* Orange Book.

photo composing machine — not a photo-typesetting machine, but a device for repeatedly imaging a single negative over the surface of a printing image carrier. *Related Term:* step and repeat.

photo composition — *Related Term:* photocomposition.

photo conductive — any material that carries electricity in the light and serves as an electrical insulator in the dark.

photo conversion — the production phase in which

camera-ready copy is converted to photographic film or paper.

photo display typesetters — typesetting machines that set display sizes of type and operate either manually or automatically by keyboard, tape or online signal. *Related Terms:* PhotoTypositor; headliner; direct entry typesetters.

photo lettering — the manual setting of display-sized letters, usually from 48 points to several hundred points in size, for display headlines, etc.

photo masking — a technique used in color correction in which a special composite film mask is placed in contact with the unexposed film in order to selectively remove controlled amounts of color intensity. *Related Term:* masking.

photo processor — a machine for automatically processing photographic materials. *Related Term:* film processor.

photo release — a contractual document that grants permission and/or authorization for commercial use of a photograph or subject which is privately held. *Related Term:* property release; release.

photo resist — (1) in plate and cylinder engraving, a light-sensitive liquid material used in making relief and intaglio printing plates. When exposed to high-intensity light the material photochemically changes and becomes resistant to the action of acids and developers, while unexposed areas are etched or in some other manner affected. (2) generally, any material which hardens when exposed to light and changes its chemical properties. Often used by industry as a means of protecting specified parts of metals when they are processed with harsh solutions, such in the making of printed circuits.

Photo Typositor — a semiautomatic photo typesetting unit manufactured by the Visual Graphics Corporation for the creation of headline type on a 2 in. strip of photographic paper, that was ready for paste-up.

photocomposition — typesetting on photographic film or paper, usually in galley format. A term most generally associated with the first three generations of photo typesetting. *Related Terms:* phototypesetter, first generation; phototypesetter, second generation; phototypesetter, third generation; imagesetting; area composition.

photocopy — a duplicate image made from the original, using photography, xerography, electro-photography, etc.

photodiode device — a nonsemiconductor device; a solid-state material that responds to light and creates an electrical current. The current is translated by various means to activate an imaging or digital recording device such as a copier or a data file on a disk.

photoelectric densitometer — an instrument that produces density readings by means of a photo cell or vacuum tube whose electrical properties are modified by the action of light

photoengraving — a relief form made by photochemical process; after being exposed through a negative, the nonimage portion of plate is acid-etched away, allowing the image area to stand out in relief.

photogram — a technique for making silhouetted images by placing objects on photographic paper, exposing them to light and then developing the print paper. The result is a white image and a black background.

photograph — traditionally, a picture made using a camera, processing equipment, photo paper and chemicals. It is the most common kind of continuous-tone copy. Increasingly, it can also be applied to digital photography, which requires no processing equipment, chemicals or special paper, but merely a computer and printer. *Related Terms:* photographic print; digital camera; digital photography.

photographer's sensitivity guide — a strip of material with several shades of gray from white to black; often called a gray scale. *Related Terms:* gray scale; cameraman's sensitivity guide.

photographic — an image copy and transfer method in which an image is made by means of a light source, on a film negative and light-sensitive paper. The image on the light-sensitive paper is made visible with developing chemicals.

photographic emulsion – (1) the sensitive layer in photographic film often made of a combination of light-sensitive chemicals and/or metal salts held in suspension in gelatin. (2) also a liquid, light-sensitive material applied to screen printing fabric to make direct photo screens.

photographic print – a positive image made from a (usually) film negative or plate on photographic paper.

photographic processor — an automatic processing machine for photographic materials. *Related Term:* film processor.

photographic stencils — screen printing stencils produced using a thick, light-sensitive, gelatin-based emulsion that is exposed and developed either on a supporting film or directly on the screen itself; the three types of stencils include direct, indirect, and direct/indirect stencils.

photographic technician — a person who produces photographic film negatives and positives of image-set material from which printing plates are made.

photography — the use of actinic light to produce a latent or permanent image on sensitized materials.

photogravure — an intaglio printing process where the image is etched into or below the plate cylinder surface. The main advantage of this method of printing is the high-speed, long-run capability. Used mainly for mail order and magazine work. *Related Term:* gravure.

photolithography — the lithographic printing process that uses a plate made by a photographic process. *Related Terms:* lithography; photo offset lithography.

photomechanical — any process using photographic negatives or positives to expose plates or cylinders that have been covered with photosensitive coatings to prepare them for the platemaking process.

photomechanical proofs — a proof made by using light-sensitive emulsions to check image position and quality of stripped flats; are usually exposed through the flat on a standard platemaking device.

photomechanical transfer (**PMT**) — the Eastman Kodak trade name for their diffusion transfer process, most often used to make positive paper prints of line copy and halftones. *Related Term:* diffusion transfer.

photomultiplier tube (**PMT**) — a highly sensitive form of photocell for transforming variations in light into electric currents. Used in many color drum scanners for creating the input signals to the computing circuits. *Related Terms:* CCD; photomultiplier; CMOS.

photopolymer — a plastic material designed so that it hardens upon exposure to ultraviolet light. Used in some forms of dry offset and letterpress operations.

photopolymer coating — a protective coating applied to a printing plate that helps it resist abrasion; useful for long print runs.

photopolymer plates — plates formed by bonding a light-reactive polymer plastic to a film or metal base; the polymer emulsion hardens upon ultraviolet exposure and unexposed areas are washed away to leave the image area in relief.

photosensitive — the characteristic of paper, film, and printing plates coated with light-sensitive chemicals to react and change when exposed.

photosensitizing — a method of making a photo polymer plate more sensitive to ultraviolet light by first conditioning the plate in a high concentration of CO_2 before exposing the plate to ultraviolet light through a special filter. *Related Term:* carbon dioxide conditioning.

photostat — (1) originally a process used to make negative paper prints of line copy. (2) in modern world, it is often used in error as alternate term for PMT or diffusion transfer print. *Related Term:* stat.

Phototype — (1) properly, a brand-name, manually operated machine which sets type by a projecting imager, one character at a time. Usually utilized for headlines and larger sizes. (2) generally a name applied to any type of composition which is photographically based, regardless of equipment used to create it.

phototypesetter — (1) a machine that exposes and outputs imaged photosensitive paper or film according to the signals it receives from a computerized typesetting system. (2) a person who sets type using phototypesetting equipment. *Related Terms:* typesetter; photocomposition; composition; typesetting; compositor.

phototypesetter, first generation — photo composition machines that were patterned after hot-metal machines. Using conventional hot-metal recirculating matrix principles, a photo image was manufactured into each matrix, replacing the hot metal letter mold. The negative was used to expose a character on photographic material, by way of a camera, light and shutter system, which replaced the molten metal caster mechanism.

phototypesetter, fourth generation — *Related Terms:* imagesetter; laser imagesetter.

phototypesetter, second generation — the first machines actually designed with photography and photo output in mind; those phototypesetting machines with font masters comprised of photographic images on a film strip, a font disk, or a font grid matrix, which was mechanically positioned in the optical path, with a flash lamp behind the character master. Moving and fixed lenses and mirrors in front of the character master sized, focused, and positioned the characters on the image-receiving photographic emulsion.

phototypesetter, third generation — phototypesetting machines based on the CRT principle, where an imaging tube was used in place of the negative, light source and shutter. Images displayed on the high-resolution tube were then optically transferred onto photosensitive materials. Because it has almost no moving, revolving or spinning parts it was extremely fast. Upon high magnification, the images suffered from somewhat rough edges due to the phosphor glow at the end of the scanning strokes. *Related Terms:* typesetting, digital.

phototypesetting — a term erroneously given to cold type, it is actually any means used in the preparation of manuscript for printing by projection of images of type characters onto photosensitive film or paper as opposed to manually or digitally. *Related Terms:* photocomposition and (erroneously as) cold type.

phthalocyanine — an ink pigment available in a green shade or blue shade. A combination of the two is often used to create the cyan ink for four-color printing inks.

pi — (1) in hand composition, metal type that has become indiscriminately mixed, such as when a type form spills, so that it is unusable until it is put back in order. (2) the sixteen letter of the Greek alphabet. (3) in mathematics, p, representing the ratio of the circumference of a circle to its diameter.

PI (per inquiry) ads — advertising, in which the publication provides the advertising space for free but receives a percentage of each sale generated by the ad.

pi characters — type characters not among the alphanumeric set of a font, such as stars, ballot boxes, checkmarks, some fractions and any other characters not usually included such as accented letters, mathematical signs and reference marks, special ligatures, symbols and dingbats and extended diphthongs. *Related Terms:* sort; pi font.

pi font — *Related Term:* pi characters.

pibble — the word used to describe type when a word is set to look what it says.

pica — (1) a printing unit of linear measure, equal to approximately one-sixth of an inch. There are twelve points in a pica and approximately six picas in an inch. One pica is approximately 0.166 in. (2) a typewriter type with 10 characters per inch, (as opposed to the elite typewriter type, which has 12 characters per inch).

pick up art — to reuse art or text from previous work or editions, rather than outputting or regeneration.

picket fence code — *Related Term:* horizontal bar code.

picking — a press problem identified by small particles of paper torn from each press sheet and fed back into the inking system. Caused by ink being too tacky and lifting fibers out of the paper when the paper lifts toward the ink during printing. Shows up as small white dots on areas of solid color.

PICT — A standard file format that allows for the exchange of graphic images (usually bitmapped), often on the Macintosh.

PICT/PICT 2 — a standard data format in which data can be created, displayed on the screen, and printed by routines incorporated in the Macintosh operating system. A program needs no graphics-processing capabilities in order to incorporate PICT data generated by other software. The more recent PICT 2 format supports 24-bit color.

pictograph — (1) a chart which uses drawings or photos as replacements for common bars or other conventional indicators. (2) a symbol or character used to represent a word, sound or syllable in a system of writing; a letter or number.

picture font — spot illustrations that are installed in the computer and work like digital fonts or typefaces. *Related Terms:* dingbats; wingdings; webthings.

PIF (program information file) — a data file that contains settings such as memory allotments and program locations for DOS programs running under Windows.

pigment — the fine, solid particles derived from natural or synthetic sources used to impart color, body and opacity to printing inks. They have varying degrees of resistance to water, alcohol, and other chemicals and are generally insoluble in the ink vehicle.

pile board — a risible printing press platform on which the paper to be printed is loaded.

pile feeder — an automatic sheet-feeding mechanism on a printing press, folder or other machine that is designed so that individual sheets are moved from the top of the pile into the production process automatically in time with the production machine.

pile height — the maximum rise of the paper pile as

it is brought into the sheet-feeding mechanism of a sheet-fed press.

pile height regulator — the device that senses the height of a pile and controls its position.

piling — a printing problem that occurs when ink builds up (piles) on either the plate, blanket or both.

pillcrow – more commonly known as the paragraph symbol: (¶)

pin feed holes — the holes on the sides of continuous computer forms. Used to assure proper alignment and as a guide through the printing device. Generally the holes are on an outside perforated strip which is removed prior to the form being used.

pin register — the use of holes and pins through film and printing plates to ensure proper registration.

pincushioning — on computer screens and other video displays, when lines that should be straight look bowed or curved. Most common at the right and left edges of a displayed image, resulting in a screen that appears to be bowed inward at the centers. Better monitors include controls to help you compensate for this error. *Related Terms:* rotation error; trapezoid error.

PING (Packet INternet Groper) — an Internet program used to determine whether a specific IP address is accessible. It works by sending a packet to the specified address and waiting for a reply. PING is used primarily to troubleshoot Internet connections. It also reports how many hops are required to connect two Internet hosts. There are many freeware and shareware PING utilities available for personal computers. *Related Term:* TCP/IP.

pinholes — small clear openings in the developed film emulsion that pass light. Major causes are dust in the air during camera exposure, a dirty copyboard, or, sometimes, acidic action in the fixing bath; must be painted out with opaque in the stripping operation. *Related Terms:* opaque; opaquing spotting.

pipeline burst cache — a type of secondary cache installed on the mother board to boost computer performance. It replaced the older Sync Burst cache, which works less efficiently with fast processors. *Related Terms:* SRAM; cache; level 2 cache.

pipelining — the ability of a program to automatically flow text from the end of one column or page to the beginning of the next. An extra level of sophistication can be created by allowing the flow to be redirected to any page and not just the next available. This is ideal for U.S.-style magazines where everything is "Continued on page __." *Related Term:* autoflow.

pitch — (1) in composition, the number of equally spaced characters per inch that a typewriter or computer printer will produce. Standard pitches are 10, 12 and 15. (2) in bar code symbolism, the rotation of a bar code symbol about an axis parallel to the direction of the bars.

pits – the laser-etched holes in the CD tracks that do not reflect light. *Related Term:* lands (opposite).

pixel — short for picture element, which refers to a part of a dot made by a scanner or other digital device. It is also a location in video memory that corresponds to a point on the graphics screen when the viewing window includes that location. The term refers to the smallest sample of an image, or one Pel on a computer monitor. Digitally, each pixel has an associated bit depth that defines its color. A one-bit-deep image has each pixel set as a foreground and background color, usually black and white. 8 bits per pixel represent 1 of 256 unique levels, most often a grayscale, but also used to represent an indexed color image that has an included pallet. 24-bit deep represents 1 of 1.67 million colors in additive color theory or a 32-bit pixel, which represents the equivalent color in subtractive color theory. In monochrome displays, each pixel is either black or white, and represented by a single bit; thus, the display is said to be a bit map. In color or grayscale displays, several bits in RAM may represent the image. In a high-resolution display, each pixel is represented by either two or four bits. Thus, the display is a pixel map instead of a bit map. *Related Terms:* PEL; electron gun; resolution; texel; texture element.

pixel depth — the number of bits of data per pixel in a digital image. The greater the number of bits per pixel, the greater the number of distinct colors or levels of gray can be produced in an image.

PKZip - PKUNZip — a shareware utility for compressing and decompressing files developed by PKware in 1986. There are versions for Windows- and DOS-based systems. *Related Term:* WinZIP.

plagiarism — the copying of work of another with a claim as being your own.

plain paper — (1) generally, any paper lacking a thermally receptive, photoreceptive, or dielectric coating

(2) in paper manufacture, paper made from one grade of stock such as plain chipboard and plain straw board.

plain paper copier — an electrostatic or electrographic copier which makes copies on regular, untreated (plain) paper.

planer block — a block of wood used to press all relief type down against the imposing table surface.

planographic — (1) one of the five major classifications of printing. (2) a term used to describe a flat image carrier, such as a lithographic printing plate, which has no relief or intaglio images and has image and nonimage areas on the same level (or plane). *Related Terms:* lithography; chemical printing; spirit duplication; collotype.

plasma display — operating under the same basic principle as a fluorescent lamp or neon tube, a plasma display consists of approximately one million hair-thin fluorescent tubes set on a substrate; the light emitted by each tube is controlled to form a video display. As plasma displays can be produced in large size with a thin profile, they are expected to come into wide use as picture screens for the wall-mounted TVs of the future. Plasma displays benefit from all the advantages of flat panel displays over CRT displays. They are approximately one-tenth the thickness of a CRT and one-sixth the weight. This becomes a major consideration as the display gets larger. In addition to the advantages of bulk and weight, plasma displays have a field of view in excess of 160 degrees horizontal and vertical, are not affected by magnetic fields and have stable images, as opposed to the flicker which characterizes conventional CRTs.

plastic coating — a more expensive image coating process than press varnishing, but offers greater protection and higher gloss. The coating is most often applied by rollers. Each sheet is then passed through an oven on a conveyor belt so the coating will dry. As with varnishing, only high-gloss papers benefit from plastic coating. *Related Terms:* UV coating; varnish.

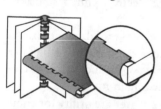

PLASTIC COMB BINDING
Premade plastic strips are inserted in specially punched slots in covers and internal pages. The "combs" are expanded to allow insertion of pages and when released they return to their closed state, holding the contents securely.

plastic combs — a round, fingered or comb-like device used to bind a group of single sheets together. *Related Terms:* mech-anical binding; coil binding.

plastic grip binding — binding by a slide-on friction strip. Suitable for very small page count work. Not a serious commercial binding process.

plastisol inks — the classification of screen printing inks containing resin and plasticizer that will not air-dry on a screen or printed product. To dry (cure), the ink is subjected to intense high heat such as that of infrared rays.

plastisol inks — the classification of screen printing inks containing resin and plasticizer that will not air-dry on a screen or printed product. To dry (cure), the ink is subjected to intense high heat such as that of infrared rays.

plate — a printing surface; an image carrier. The master device that bears the image to be printed. It may be made of thin metal, plastic or paper.

plate bend — that part of the lithographic plate that is bent to fit into the clamp that it on the plate cylinder of the press.

plate clamp — a device that grips the lead and tail edges of a printing plate and pulls it tight against the cylinder body. *Related Terms:* plate; image carrier; plate bend.

plate cylinder — the cylinder of a press on which the image carrier is mounted.

plate engraving — an intaglio-class printing process in which images to be reproduced on paper are cut into (below) the surface of a plate with a graver or etched with acid. *Related Term:* graver.

plate finish — in paper classifications, a very smooth finish on bond, cover, or Bristol paper similar to supercalendered finish on book paper.

plate ready film – *Related Term:* flat.

platemaker — (1) in quick printing, a camera device that makes plates automatically after photographing a mechanical. (2) in commercial printing, a machine used to make (usually) metal plates from negatives. (3) the person who prepares plates for printing.

platemaking — the preparation of a printing plate or other image carrier from a finished film or flat, including sensitizing the surface if the plate was not presensitized by the manufacturer, exposing it through the flat, and developing or processing and finishing so that it is ready for the press.

platemaking baths — solutions used for electroplating, etching photoengravings and developing flexographic rubber plates.

platemaking sink — a sink area used to hand process printing plates; usually has a hard, flat surface to hold the plate and a water source to rinse the plate.

platen — the flat area on a platen letterpress machine where the paper is positioned for printing. Commonly used in some types of letterpress, screen and intaglio printing.

platen dressing — the sheets of paper used to cover the platen of a letterpress platen press; often called packing. *Related Term:* make ready.

platen press — the traditional design used almost exclusively for relief-class operations including printing, die cutting, numbering and foil stamping; a flat

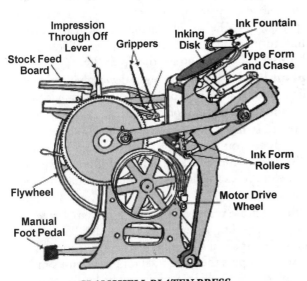

CLAMSHELL PLATEN PRESS
The platen and bed continuously open and close as the operator removes printed sheets and replaces them with new ones. Originally operated by a foot pedal and later revised to be driven by an electric motor.

type form is locked into place and moves into contact with a hard, flat surface (called a platen) that holds the paper or other substrate. Ink and pressure are applied to create the image.

platen safety guard — a metal bar and attached heavy canvas that raises each time the platen press closes. Designed to keep hands from being caught between the platen and type form.

plates — *Related Term:* image carriers.

platform — the type of computer or operating system on which a software application runs. Some common

platforms are PC, Macintosh, and UNIX. When someone knows more than one of these platforms or when a program or data file can be used on more than one of these platforms, it is known as cross-platform.

plating bath — the solution used for electroplating.

playback head — a bit of a misnomer, as it is the device in a magnetic diskette system which both reads and writes to the magnetic medium. It converts the weak magnetic pulses on the disk into electrical current and signals which direct the software to execute, display or otherwise process data.

pleasing color — color which satisfies the customer or client. Is accepted as satisfactory even though it doesn't match original samples, scenes, or objects.

Plessey code — a pulse-width modulated bar code commonly used for shelf marking in grocery stores.

plotter — an output device that translates computer data into a graphic form on paper by use of a pen, drawing device or knife which is moved laterally as the substrate moves to and fro beneath it. By moving the pen or knife along one axis as the media is moved along the perpendicular axis, the combined moves make the pen or knife draw or cut any shape desired. The main output device of draftsmen, architects, stencil cutters, etc. Capable only of line art; no tones.

PLU (price lookup) — the process by which a price is looked up in a database, based on information decoded from a bar code. Very commonly utilized in retail stores.

plug and play — peripheral standards developed by Microsoft, Intel, and other industry leaders to simplify the process of adding hardware to PCs. It is designed to conceal unpleasant details, such as IRQ and DMA channels, jumper settings, etc., from people who just want to add a new board or other device to their system. Even devices that don't seem that difficult to install also take advantage of PnP standards. A PnP monitor, for example, can communicate with both Windows 95/98 and the graphics adapter. It will automatically set itself at the maximum refresh rate supported for a chosen resolution and ensures that the monitor will not be driven beyond its capabilities. *Related Terms:* DMA; IRQ; refresh rate.

plugged up — the undesirable characteristic in printing when halftone dots on the image carrier have run into each other causing loss of shadow detail, and a sharp optical jump from some percentage of coverage to full coverage.

plug-in — a smaller "add-on" computer program which works in conjunction with a larger application. Usually a plug-in enhances the capabilities of the program it is "plugged" into. One of a set of software modules integrate into Web browsers and offer a range of interactive and multimedia capabilities. Larger applications must be designed to accept plug-ins, and the software's maker usually publishes a design specification that enables people to write plug-ins for it. Two notable applications designed around a plug-in architecture are Adobe PhotoShop and Netscape Navigator. Notable examples of plug-ins are Kai's Power Tools for PhotoShop and Shockwave for Netscape Navigator.

PLV (production level video) — the highest quality motion video compression algorithm. Compression is achieved offline in nonreal time, while playback or decompression is real time. The offline technique always produces better image quality than real time symmetrical compression because of the deliberateness of the process and the higher power processing employed.

ply —in paper specifications, each single layer of paper in multi-ply papers such as duplex cover stock.

PMK — a relatively new compound developer using Pyro, metol and Kodalk to produce fine grain, high sharpness and acutance. It adds print density through its natural staining effect and is popular because of its image-rendering capabilities and extreme shelf life.

PMS — (1) the abbreviation for Pantone Matching System, a standard trademark for color reproduction and color reproduction materials owned by Pantone, Inc. This abbreviated term is no longer used. (2) one of the most commonly used systems for identifying and specifying ink colors.

PMT (Photo Multiplier Tube) — (1) in drum scanners, an electronic sensor that takes in an analog pulse of light and turns it into a digital value for measuring intensity of the light. Original device for making scanning digitizers. (2) in camerawork, PMT, a Kodak trademark, is often misused to refer to any diffusion transfer processes for making positive or negative prints directly from original copy. *Related Terms:* diffusion transfer; stat.

PNC (personal network connection) — an obsolete SLIP/PPP connection type that used to offer accounts but stopped due to the large amount of support resources required to maintain them.

PNG (portable network graphics) — a lossless graphics file format that combines many of the benefits of GIF and JPEG. It has many features that GIF does not, including 254 levels of transparency vs. GIF's one, more control over image brightness, support for more than 256 colors, supports progressive rendering and tends to compress better than a GIF. *Related Terms:* GIF; interlaced GIFs; JPEG; lossless.

point — (1) in typesetting, composition and typography, the smallest English/American unit of typographic linear measurement, equal to 0.0138 in. (approx. 1/72 in.). One-sixth of a pica. Type height, size and line spacing are measured in points, height being .918 in. and the standard sizes being 6, 8,10,12,14,18, 24, 30, 36, 42 point. (2) in paper and papermaking, a unit of measure that indicates the caliper of paper in thousandths of an inch. (3) in some editorial and proofreading environments, an alternative term for the punctuation mark called a "period." (4) the act of placing your cursor onto a link to direct it to another Web page. "Point your browser to" means "go to" that Web site.

point light source — a source device that produces light from a single spot or point, such as a single bulb as opposed to a group of lights.

Point Listing — a database of popular Web sites that will direct you to areas of interest. Some Web sites display the "top 5%" seal indicating the site was selected for its excellence in content, presentation, and experience.

point sample texture filtering — the simplest technique for adding texture to a 3D model. Point sampling selects a single texel to place on an appropriate pixel, then puts the same texel on pixels with similar properties. Since this technique can make a scene look blocky, it's unacceptable for true 3D gaming. *Related Term:* texel.

point size — The measurement of type, generally from the top of the highest ascender in the typeface to the bottom of the lowest descender. However, because of variances in type design, the designated point size of a particular font might be somewhat different from the actual measurement. *Related Term:* set size.

point system — the English/American system of measuring by points and picas in typographic composition. It has been in use since 1878. Most of the rest of the world uses the European Didot system, with the cicero as the smallest unit. *Related Term:* point; pica; Didot system.

pointer — an arrow or other graphic on a computer

screen that can be moved by dragging the mouse and that gives commands by pointing at icons. *Related Term:* cursor; mouse.

point-of-purchase display — a box or rack, sometimes made of sturdy cardboard, that displays merchandise near the location of the sales transaction. *Related Terms:* POP; dump.

point-to-point protocol (PPP) — a method for transmitting packets over serial point-to-point links. *Related Terms:* serial line Internet protocol.

polarity — the condition of being electrically charged, negative or positive. Electrostatic image transfer uses negative and positive charges to make the image.

Polaroid — a transparent material containing embedded chemical elements to control, modulate or eliminate certain wavelengths of light, based on wave direction.

poly bag — the plastic bags used to enclose publications, etc., to protect the contents and, if additional items are included, obtain the preferred single piece rate for the entire contents.

polychrome printing — a term (sometimes) applied to multicolor printing.

polymerization — the chemical reaction for drying epoxy inks when two or more smaller molecules combine to form larger molecules that contain repeating structural units of the original molecules. *Related Term:* evaporation.

poor trapping — an undesirable printing condition in which different color inks do not properly register, causing thin white lines in between colors.

POP, PoP — (1) point of presence (PoP). In networking, the combination of modems, digital leased lines, and multiprotocol routers, etc., used by an Internet access provider to offer a local dial-up number for access to the Net. The provider usually either maintains or leases PoPs throughout the areas it serves. (2) point of purchase. An advertising description for product displays which are positioned close to the transaction site, i.e., cash register, automatic teller, etc. (3) Post Office protocol (POP). In electronic mail, a current popular Internet e-mail mailbox access standard, but it has limitations. Basically, you connect to a server and download all your messages, which are then deleted from the server. This technique limits flexibility of selecting what messages will be downloaded, and when. Some clients let you leave all messages on the server, and/or refuse to download messages above a certain size but, as messages become longer and larger and contain multimedia features such as sound or video, objects, etc., flexibility in what we retrieve and when we retrieve it becomes increasingly important. That's where the Internet message access protocol (IMAP) has significant advantages. The current version of POP is POP3, but later versions are incompatible with earlier ones. *Related Term:* IMAP.

pop-up — that part of a printed piece which folds toward the viewer when the pages are opened, creating a 3D effect.

pork chop — small head shot of a person, usually half a column wide.

pornography — printed material that is intended to cause sexual excitement. *Related Term:* obscenity.

porosity — a rating of how readily paper or other substrates allow liquids and gasses to pass through them. While not always significant for normal work, it is extremely important in papers used in the packaging industry. *Related Term:* porous.

porous — the ability of paper to allow air to permeate and ink to penetrate. The more porous the paper, the higher the capability for ink penetration. *Related Term:* porosity.

port — generally, a circuit in a electronic device for the input or output of signals. (1) in computing, a receptacle where information goes into or out of a computer, or both. E.g., the "serial port" on a personal computer is where a modem would be connected. (2) on the Internet, one of the network input/output channels of a computer running TCP/IP. It refers to a number that is part of a URL, appearing after a colon (:) right after the domain name. (3) the term also refers to translating a piece of software code to bring it from one type of computer system to another, e.g., to translate a Windows program so that is will run on a Macintosh.

portable — (1) generally something which can be moved and remain operational without special installation, tools, personnel, etc. (2) most generally in graphic comunications, a small laptop or notebook computer. (3) in software, the ability of a program to be used and/or write files which will function on several different platforms.

portals — an Internet gateway that passes you through to other destinations; an on-ramp where you would stop to look at a "map" to find a route to what you want. Portals will be very important for a long time but soon the emphasis will switch to the next phase, hubs. Portals are general interest because they help you find anything, anywhere without particular focus.

portfolio — (1) an expansion envelope or presentation folder. (2) the collected best works by an individual, used for show. Especially in advertising and art, as it gives prospective employers a synopsis of the applicant's quality levels, diversity, variety and types of work experience.

portrait — a mode for images which are upright on a page and where the height is greater than the width. *Related Term:* landscape.

portrait mode — the vertical format of a computer screen or printout that resembles one newsletter page. The vertical dimension is larger than the horizontal dimension. *Related Term:* landscape mode.

POS (point of sale) — a reference to the electronic management of sales in a store through a centralized and integrated system of networked printers, cash registers, card verification systems and bar code readers.

position stat — a photocopy or diffusion transfer print made to size and pasted to a mechanical showing how to crop, scale, and position loose art or photos, but not for use as reproduction art.

positioning — in advertising, the placement of a print, video or audio ad or the strategy devised for presenting a product to the market.

positive — a photographic reproduction with the same tonal values as those in the original scene. (1) the characteristic of an image on film or paper in which blacks in the original subject are black or opaque and whites in the original are white or clear. (2) the image itself. *Related Term:* negative.

positive transparency — a positive image on film, containing a tonally correct right-reading image. *Related Terms:* positive; film positive.

positive-acting plate — a printing plate formulated to produce a positive image from a flat containing positives.

POSIX (Portable Operating System Interface for Unix) — a U.S. government creation, POSIX-compliant programs are designed to be easily ported, and run on any POSIX-compliant operating system, including many Unix variations and Windows NT.

post — to put up a message. Subscribers to newsgroups and mailing lists often take part in discus-

sions by sending, or "posting," their articles or comments online. *Related Term:* posting.

POST BINDING
Cover and inside pages are drilled and then placed on an internally threaded collar and a screw is affixed from the cover side, holding the contents securely in place.

post binding — a method of binding in which two or more bolt-type posts are used to fasten sheet of paper together. *Related Terms:* loose leaf binding; comb binding; coil binding.

post mill waste — a paper manufacturing term identifying any waste generated after the paper has left the mill, including pulp substitutes and pre- and post-consumer waste.

Post Office Protocol — *Related Term:* POP; IMAP; postnet code.

post-consumer waste — finished material that is recycled or disposed of as solid waste after its product life span is completed.

poster board — a water-resistant paper. A heavy cardboard whose caliper is over .024 inches and which usually has a smooth laminated or coated surface, with the other being slightly rough.

poster makeup — an arrangement of a newspaper's front page that usually consists of large art and a few headlines to attract attention. Especially popular with tabloid format publications.

poster paper — *Related Term:* poster board.

posterization — the deliberate constraint of a gradation into visible steps to create a special effect. Generally the high-contrast reproduction of a continuous-tone image made without benefit of a halftone screen; usually consists of two, three, or four tones and is reproduced in one, two, or three colors. The variations in content of the colors is controlled by utilizing different filters and/or exposure times to maintain or drop out different areas of the original.

posting — (1) in computing, a single message entered into a network communications system. (2) in business, the recording of transactions into ledgers, inventory or other data bases, either electronically or manually. *Related Term:* post.

postmaster — the administrator responsible for resolving e-mail problems, complaints, answering queries about users and other related duties at a site.

postnet code — the code developed by the U.S. Postal Service to assist in automatic sorting of mail.

post-press — all operations after ink is placed on the substrate, i.e., folding, binding, packing, delivery, etc.

PostScript — a page description language (PDL) developed by Adobe Systems. It is a universal page description language which enables high-quality printing of text and graphics in digital format. Files created with PostScript can be used with any PostScript-compatible output device regardless of resolution; it is resolution independent. The PostScript file includes (encapsulates) a low resolution (generally 72 ppi) low color (usually 256) image for display and rendering when used with applications or devices which lack PostScript interpolation capabilities. In its Macintosh form, the encapsulated image is a PICT format which prevents cross-platform capability. In its PC form it includes a TIFF preview which supports cross-platform usage. Generally, what this means is that PostScript can support the highest resolution of any device, and that a user can reasonably preview PostScript on a low resolution screen. Widely supported by both hardware and software vendors, it represents the current 'standard' in the market, because it can interpret electronic files from any number of personal computers (front ends) and use off the shelf software programs. Text and graphics can be controlled with mathematical precision. *Related Terms:* imagesetter; page description language; raster image processor; RIP.

PostScript compatible — any software program that translates statements written in the PostScript page description language, but is not a licensed Adobe client. Sometimes called a "PostScript clone."

PostScript, encapsulated (EPS) — a file format used to transfer PostScript image information from one program to another.

potassium bromide — a chemical used as a restrainer in developers to prevent fogging.

potassium carbonate — an alternate to sodium carbonate, used in developers as an accelerator.

potassium ferricyanide — used in combination with hypo as an image reduction chemical and sometimes as a bleach in the sulfide toning process.

POTS (plain old telephone service) — what you are using if your have a telephone modem to connect to the Net. It is the basic voice phone service you get from your phone company and the term is used to differentiate this type of connection from ISDN or a leased line like T1. *Related Terms:* ISDN, T1.

power encode — any MICR line-encoding process that is automatically generated by an imaging or other automatic check-processing system information density. The. *Related Term:* MICR.

ppi — (1) in printing bindery and production, an abbreviation for pages per inch. One method used to measure the thickness of paper. (2) in computing and computer imaging, an abbreviation for pixels per inch, or a measure of the degree of detail of an image. The finer the optics of the scanner, the higher the scan resolution. Its used interchangeably, although not exactly correctly, with the more common term dpi (dots per inch). *Related Terms:* dpi; resolution.

pplate, rubber — in flexography, printing plates made by molding and curing rubber in a matrix produced from a relief printing form, i.e., a form with image areas raised above the nonprint areas.

PPP (point-to-point protocol) —the Internet standard for serial communications. Newer and better than its predecessor, SLIP, it defines how your modem connection exchanges data packets with other systems on the Internet. *Related Term:* SLIP.

PPTP (point-to-point tunneling protocol) — a protocol that allows secure transmission of data in TCP/IP packets. It is used to carry secure communications over Virtual Private Networks (VPN) that use public phone lines. *Related Terms:* TCP/IP; VPN.

practicality — the measure of whether a typeface is available and economically suitable for a job.

pre-cut — printed materials that have been previously cut to size.

preemptive multitasking — the ability of a computing device to handle more than a single task at a time, interrupting one to perform another and the ability to return to the spot of departure on the first program without rebooting or other interruption.

preface — the introductory remarks that may provide the reason for the book along with the goals and scope of the book, written by the author as part of the front matter.

pre-layout planning — all of the thinking involved in answering important production issues and questions before preparing a set of graphic layouts; usually summarized on a prelayout planning form.

premask — an auxiliary mask used in the two-stage masking system to obtain color-correcting masks without their normal contrast-reducing aspects. *Related Term:* two-stage masking.

premastering — the creation of a recorded file that contains the exact "image" or file layout of a CD-ROM, with error correction and timing information, ready for mastering.

premium — any merchandise or service given away as part of a promotion.

premium paper — paper considered by its manufacturers to be better than #1 paper.

prep – *Related Term:* prepress.

prepack — a point-of-purchase display, often made of cardboard, for holding books or other merchandise. *Related Terms:* POP; dump.

preparation — *Related Term:* prepress.

prepress — a term to describe all printing operations prior to presswork, including design and layout, typesetting, computer applications of full page composition, graphic arts photography, color separation, image assembly, color proofing and platemaking. *Related Term:* preparation.

prepress proof — a color proof for checking made directly from electronic data or film images, prior to platemaking or printing. They are a simulation of the final printed piece by photochemical methods such as a cromalin or matchprint (which have dye or pigment images on transparent film base) instead of photomechanical methods such as ink on paper. *Related Terms:* off-press proofing; color keying.

preprint — any printed work done in advance for later inserting into another publication or for imprinting.

preprinted paste-up sheets — paste-up sheets commonly prepared for jobs of a common size and layout specifications, printed with nonreproducing blue lines to help the paste-up artist position composition and artwork.

preprinted stock — paper, cover stock, envelopes or business cards that have already been printed with a design or some other repeating element, usually in color, and are designed for imprinting or specific personal or company information.

preprinted symbol — a symbol which is printed in advance of application either on a label or on the article to be identified.

preproofing, photo contact — the making of full-color proofs from final negatives or positives by using dyes or pigments transferred to a white base material, or by using dyed or pigmented transparent films that,

as an overlay, provide a colored transparency of the copy. *Related Terms:* laminated proof; cromalin.

prepublication copies — the copies of a book that are sent out before a book's publication date in order to generate interest and sales.

prepublication price — special discount prices offered to specific individuals or organizations before a book's publication date.

prepunched tab strip method — strips of film punched with holes and used to hold flats in place over pins attached to light table.

pre-scored — flat printed material that has received one or more scored impressions to facilitate folding.

prescreen — a halftone positive print that can be combined with line copy in paste-up, thus eliminating the need to strip in a screened negative with a line copy negative. *Related Terms:* Velox; dot-for-dot reproduction.

prescreened film — a high-contrast orthochromatic film that has been pretreated to produce a halftone negative by direct exposure to continuous-tone copy without the use of a separate halftone screen.

presensitized indirect film — factory-manufactured, light-sensitive film that is exposed, developed and washed out before attaching to the fabric mesh of a stretched screen printing frame. Also referred to as indirect photo screen and indirect photo stencil. *Related Terms:* direct photo screen; indirect film; hand-cut stencil.

presensitized plate — an image carrier, usually a sheet of metal or paper supplied to the user with a light-sensitive material already coated on the surface and ready for exposure to a negative or positive.

presentation folder — a (often) high-quality printed cover piece, usually with internal pockets to hold additional printed or photographic data. Used widely for conventions and sales meetings, etc. Normally made of high-grade cover papers, and often foil stamped or embossed.

presentation graphics — all of the charts, graphs, transparencies, posters, videos and related supplemental materials used to reinforce the spoken work at meetings and business gatherings.

presentation software — software designed to assist in the preparation of audio/visual media presentations. They can make 35mm slides, overhead transparencies, etc.

preservatives — chemicals used to lengthen the usefulness of photographic developers.

presort — to categorize mail by postal code and class prior to mailing. It reduces postage costs, as the postal service gives incentives for this operation. It is designed to speed the delivery process.

press — a precision piece of equipment that includes all controls, elements and other parts to produce high-quality printed products on a substrate.

press check — one method of quality assurance in which the customer visits the printing plant as the first few copies of a print job come off the printing press. The customer, assisted by the press operators, examines the sheets from the press and provides authorization to begin production.

press date — the date a publication or job goes on press.

press kit — a collection of publicity materials used to promote a book or a product and often enclosed in a pocket folder. *Related Term:* media kit.

press operator — a person who attaches the plate to the printing press, adjusts the ink, water and paper feeding systems and does the production run. *Related Term:* pressman.

press proof — a proof pulled on the actual production press using the plates, paper, and ink specified for the job. Usually incurs a surcharge for the process and possibly for the waiting time until the OK for production is issued.

press release — the announcement of a new product sent to a news organization for publication. *Related Terms:* press kit; media kit.

press run — (1) the total of acceptable copies required to be run during a single printing. (2) the actual operation of the press during the printing process. *Related Term:* print run.

press sheet — (1) during the production operation, a single printed sheet of paper as it comes off the press for the purpose of checking color, registration, and other reproduction elements. (2) in bindery and finishing operations, one sheet of paper, regardless of size, delivered from a specific printing press, but not yet folded or trimmed.

press, rotary — a classification of printing press on which the substrate (paper) and image carrier (plate or stencil) are both wrapped around cylinders during the image transfer (printing) operation.

press, rotary screen — a screen printing press design with the fabric and stencil attached around a cylinder and the squeegee positioned in the center.

pressboard — a hard, heavy sheet used as packing material when dressing a platen press.

pressing boards — the flat wooden boards with a metal flange edge used in pressing handmade case-bound books in the final binding stage.

press-on type — transparent or translucent sheets of letters, that can be rubbed off onto another surface such as paper by use of a burnishing tool. *Related Terms:* transfer type; dry transfer type.

pressure checks — tests made at the contact points between the rollers and cylinders of a printing press before a run; after a blanket, rollers or roller coverings are replaced; when troubles develop with the plate image; and when a plate or printing paper of different thicknesses is to be used.

pressure-sensitive — a substrate material that has an adhesive coating that will stick to a surface when pressed lightly. It is protected by a waxed carrier sheet which holds and protects the pressure-sensitive material until needed.

pressure-sensitive paper — substrates with adhesive coatings that are covered with a backing sheet until ready to use. Needs no moisture to adhere. *Related Terms:* label paper; pressure sensitive.

presswork — all operations performed on or by a printing press that lead to the transfer of inked images from the image carrier to the paper or other substrate. Presswork includes make ready and any in-line finishing operations specific to the press (folding, perforating, embossing, etc.).

Pretty Good Privacy (PGP) — a commonly used encryption system, invented by Phillip Zimmerman. PGP is based on public key encryption, and is one of the authentication schemes available through Network Solutions' Guardian plan. *Related Terms:* encryption.

Preucil ink evaluation system — a color evaluation system that utilizes a reflection densitometer to measure a printed ink film through Wratten #25, #58, and #47 filters. These measurements are converted into hue error and grayness parameters and plotted as color diagrams. *Related Terms:* hue error; grayness; color circle; color triangle.

preventive maintenance — the scheduled routine of minor repair, lubrications, etc., performed on operating equipment to increase productivity, lengthen the equipment life span, and to prevent emergency and/ or major repair procedures.

preview mode — (1) originally a mode in word processing or desktop publishing software that wasn't WYSIWYG could (usually inaccurately) show a representation of the output as it thought it would look when printed. The quality ranges from acceptable to worse than useless. (2) in modern systems it has been replaced by print preview modes which provides a miniature preview of the page(s) to be printed, including margins, column spaces, etc., and are generally very accurate reproductions of the final printed product.

PRI (Primary Rate Interface) — an ISDN service that is used mainly by Internet service providers (ISPs) and businesses because it provides a lot of bandwidth.

price break — the quantity level at which unit cost of paper or printing drops.

price estimate — in printing sales, a preliminary cost analysis provided by a printer on how much a job will cost, broken down by price per piece and total cost per thousand. Different from a quote, which is a legally binding, signed agreement between a printer and a publisher in which the cost is guaranteed not to fluctuate for a specified period of time. *Related Terms:* estimate; quote.

price quote — in printing, a legally binding agreement between a printer and client that lists the costs of a particular job, in which the price does not fluctuate for a specified period of time. More solid than a printing estimate, which is a preliminary report on how much a print job is expected to cost but is not legally binding.

primaries — colors that can be used to generate secondary colors. In the additive system, these colors are red, green, and blue. In the subtractive system the colors are yellow, magenta, and cyan. The printing process employs the subtractive color system.

primary cache — a method of boosting system performance. It is a solid-state storage site for some instructions and is usually integrated into the CPU chip, and typically limited to 16-32K. It usually relies on a secondary (L2) cache for the all-around boost of system performance. *Related Terms:* cache; L1 cache; L2 cache; secondary cache.

primary colors — *Related Term:* primaries.

primary leading — the interline spacing that is used in the general body of text or in a particular set of lines, as opposed to secondary leading, which might typically be limited to between paragraphs or after headlines.

primary mailbox — the mailbox set up automatically when you set up your Internet account with a Internet service provider (ISP).

primary mailbox password — an account holder's services password to log in to the primary mailbox.

primary mailbox username — the primary mailbox uses the account holder's username.

primary optical center — the usual point at which a reader's eye first sights on a page, toward the upper left quadrant, approximately 5/8 of the distance from the bottom of the page. *Related Term:* optical center.

primary plates — relief plates intended for producing duplicate plates.

primary server — on Network Solutions' Domain Name Registration Agreement, the section where the registrant indicates the host name and Internet protocol (IP) number of the name server that will contain authoritative information for the domain name and will be used to resolve that domain name to its corresponding IP number(s). The designation of "primary" means that this name server will be used first and will be relied upon before any of the other name servers that may be listed on the Domain Name Registration Agreement. *Related Terms:* secondary server; additional servers; host; name server; domain name registration agreement.

primitive — in graphics, a primitive is one of a basic group of shapes, such as circles, polygons, and squares.

print — (1) to reproduce materials using any of the five classifications of printing and their respective processes. (2) a single copy. (3) the collective term applied to printed works and their human components, e.g., out of print, printed works, the print media, etc. (4) the output of a photographic negative when printed in an enlarger or contact printing frame.

print dryer — a machine on which photographic prints are dried by using heat and a ferrotype plate and/or high-pressure air knives.

print engine — the parts of a page printer which perform the print imaging fixing and paper transport, i.e., everything but the controller.

print quality — (1) in optical scanning, the measure

of compliance of a bar code symbol to the requirements of dimensional tolerance, edge roughness, spots, voids, reflectance, PCS, quiet zone, and encoding. (2) in paper, the properties of the paper that affect its ability to reproduce well. (3) in photography and printing, the relative level of the reproduction when compared to others or the original. (4) in code and symbol work, the measure of compliance of a bar code symbol to the requirements of dimensional tolerance, edge roughness, spots, voids, reflectance, PCS, quiet zone, and encoding.

print run — the total number of copies of a publication to be printed. *Related Term:* press run.

print server — a special server to handle the printing of images. Often used so that compressed images can be sent across a network in order to avoid creating unnecessary network traffic.

print spooler — a buffering device which holds files in queue until the printer can handle them.

print wheel — *Related Term:* daisy wheel.

printability — (1) a very subjective term referring to how well paper accepts ink. (2) one of the major factors in type selection, along with legibility, availability and readability. Relates to how well the selected face will reproduce on the substrate selected. For example, small type in screen printing rarely works.

printer — (1) in computing, any computer device that produces images in readable form on paper. (2) in separation and camera work, a color-separated halftone film that will transfer the characteristics of a specific process color in a given job to the corresponding printing plate prior to presswork. (3) in printing, a person who owns or manages a print shop or runs a printing press.

Printer Command Language (PCL) — a language developed by Hewlett-Packard for use with its own range of printers. Essentially a text-oriented language, it has been expanded to give graphics capability. *Related Terms:* PostScript; imPRESS; Printer Control Language.

Printer Control Language — *Related Term:* Printer Command Language.

printer control software — software designed to drive label printers that typically run on PCs or Macintosh computers. A new feature sometimes found is the ability to drive a desk-top laser printer in addition to the label printer.

printer font — in desktop publishing, a font a printer can use; often downloaded into the printer memory prior to imaging.

printer, dye sublimation — a computer printing system which utilizes successive layers of sublimate inks or transfer color mounted on a carrier sheet. The imaging head is a heated device which, when brought in touch with the inks and polymer substrate, causes the ink to change from its solid state to a gas. Simultaneously, the polymer substrate expands. The molecules of the ink commingle with the expanded substrate. As the material passes the head it cools and locks the ink into the substrate. Successive passes for CMYK produce a full color print. A very slowly created, but generally high photographic quality, print results.

printer, impact — any device that uses pressure from a type bar, type head, or matrix pin and inked ribbon to strike a direct impression on a substrate.

printer, line — an impact-type computer output device that simultaneously prints all of the characters in a line of type from single or double tiered segmented type bars repeatedly containing all the letters and numerals, etc. The several segments traverse laterally across the substrate and as respective letters arrive over their designated spot, an electromechanical hammer "types" the character in place. Much faster than a single element typing ball or daisy wheel. Used for high-volume printing applications associated with data processing operations.

printer's error — a nonchargeable mistake made by the printer in preparation for printing, as opposed to an author's alteration. *Related Term:* PE.

printer's ornament — *Related Term:* ornament.

printer's spread — the imposition of pages as they will be assembled and reproduced on press, as opposed to a reader's spread, which is how the pages will appear in the final bound publication. *Related Terms:* imposition; stripping.

printing — all of the art and methods by which an original is reproduced in quantity. Any process that repeatedly transfers to paper or other substrate an image from an original such as a mechanical, die, negative, stencil, electronic memory or printing plate.

printing belt — a strong but flexible belt used for mounting and thin photopolymer relief plates on the Cameron Book Production Press System.

printing classifications — there are five basic and

physically different reproduction classes: relief, intaglio, planographic, porous, and nonimpact. Within each class is a multiple of processes which share all or most of the physical characteristics of the class., i.e., all relief processes utilize carriers where the printing areas are raised above the surface; intaglio has the image areas below the surface; planographic image carriers have image and nonimage areas on the same plane. Porous processes all transfer ink through the carrier from top to bottom to create images; nonimpact systems tend to be electronic in nature and do not utilize physical pressure as much as electron transfer to accomplish imaging, etc.

printing couple — the portion of a printing press that applies a printed image to one side of a press sheet. The inking and dampening systems and plate, blanket, and impression cylinder on an offset press are considered parts of a printing couple.

printing cylinder — *Related Term:* cylinder.

printing frame — *Related Term:* contact printing frame.

printing plate — the surface carrying the image to be printed. *Related Term:* image carrier.

printing process — the selected technique within one of the five specific printing classifications that is used to reproduce given written and pictorial matter in quantity. The major conventional printing processes are lithography, letterpress, gravure, flexography, and screen printing. The major nonimpact printing processes are ink jet, electrophotography, ionography, magnetography, and thermal transfer printing.

printing side — the bottom side of a prepared screen printing frame; the side which touches the product being printed.

printing trade customs — the business terms and policies codified by the Printing Industries of America and selectively followed by printers. *Related Term:* trade customs.

printing unit — the sections on printing presses that house the components for reproducing an image on the substrate. In lithography, a printing unit includes the inking and dampening systems and the plate, blanket, and impression cylinders. *Related Terms:* printing press; printing couple.

printing, multicolor — printing in more than one specific ink color in a given work, but not including 4-color process.

printing, nonimpact — the fifth classification of printing processes especially including most electronically driven devices that reproduce an image without striking the substrate. Some examples include xerography or laser printing, in which the image is created by fused toner particles, and ink jet priming, in which a stream of ink propelled from printing heads forms the image. In this classification are thermal heat transfer, dye sublimation, ionography, electrophotography, etc.

print-through — the amount of visible image or printing on one side of a printed sheet which is discernible from the other. *Related Term:* show- through; opacity.

prism — a transmissive body with two nonparallel planes, often used to disperse a light beam into its component colors.

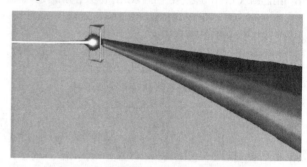

GLASS PRISM
Used to separate a light beam into its spectral components through refraction.

private brand — *Related Term:* merchant brand.

private key encryption — an encryption method in which both the sender and recipient of a message share a single, common key that is used to both encrypt and decrypt the message. *Related Terms:* encryption; Pretty Good Privacy.

privately printed — *Related Terms:* self published; vanity press.

process blue — *Related Term:* cyan.

process camera — *Related Term:* camera, process; camera, horizontal.

process color — reproducing full color art or photos by a halftone process and using three transparent subtractive primary color printing inks: yellow (Y), magenta (M), cyan (C). The black (K) information is interpreted from the other three. Each color is separated from the original by using filters or electronic circuits to obtain on the film only a single primary color's information. When the four different halftone colors are overprinted, they optically combine to produce the visual effect of the full color original. When blended, CYMK reproduce only a small portion of all the colors found in nature, but they can reproduce the widest range with the fewest inks when printing. *Related Terms:* CMYK; process color; photography; full color.

process color reproduction — a printed color reproduction of an original using the three process inks or the three process inks and black.

process inks — a set of transparent yellow, magenta, and cyan inks used for full color printing. A black ink is also included in a four-color process ink set.

process photography — (1) creating line and halftone images for photomechanical reproduction. (2) the equipment, materials, and methods used in preparing color-separated printers for color reproduction. *Related Term:* graphic arts photography.

process printing — *Related Term:* four-color process printing; process color.

process printing, three color — a somewhat inferior process printing technique utilizing only cyan, magenta and yellow inks and no black printing plate. While it is theoretically possible to reproduce all of the hues found in an original by using only these three printing pates, the dark (black) areas usually are a muddy brown. The most serious fault lies in the impurity of the pigments and the fact that each one either reflects or absorbs outside of its theoretical range. *Related Term:* four-color process.

process red — *Related Term:* magenta.

processing — chemically treating photographic papers, films, and plates after exposure, usually to produce a visible image. *Related Terms:* developing; developer.

processing tray — a pan used to hold chemicals for developing, stopping, fixing and washing photographic film; made of plastic, hard rubber, fiberglass or stainless steel.

process-ink gamut chart — a color chart for comparing the gamut or color limits that can be produced from any given ink set and substrate combination or an imaging system such as a computer printer or a monitor.

processor — an automatic machine that feeds exposed photo paper (such as typeset galleys) or film over rollers and through chemical development baths to develop and dry them before they reach the delivery area. *Related Term:* film processor.

producer — the person who is responsible for the project, including time, budget, manpower, etc.

product ID — in the U.P.C. code, the 5-digit number assigned by a manufacturer to every consumer unit in its product catalog. The product ID is different for every standard package (consumer unit) of the same product.

production artist — (1) a person who does pasteup. (2) a person who utilizes computerized systems to create mechanical art and layouts.

production control — (1) a department or management section of most printing establishments that directs the movement of every printing job throughout the facility. (2) applications where bar codes and other forms of AIDC are used to reduce data entry errors and slow response time. *Related Term:* scheduling.

production department — the mechanical manufacturing departments of a printing plant.

production manager — the person who coordinates designers, printer and others who work together to produce printing.

production workflow — a type of workflow that exists when there is a high volume of consistent work that must be accomplished through a standard set of tasks and rules to arrive at a predetermined outcome. Typical of the assembly lines usually found in larger organizations that have broken down the work into a series of basic operations. The successful productivity of each basic operation is vital to ensure the overall success of the process.

prog — *Related Term:* progressive proof.

program — (1) generally, a collection of instructions and operational routines, necessary to activate and operate a computer system. (2) graphically, a set of instructions for a typesetting machine, that is stored in the computer and used to direct its operation. The more sophisticated the programming, the more versatile and the higher the quality of composition the system will provide. *Related Term:* application.

programmable function keys — a keyboard key which can be programmed to perform specific tasks within a software program..

programmer — any person who writes the codes or instructions to enable a computer program to function.

programming — the output of a programmer.

progressive JPEG — a variant of JPEG images, which are displayed one line at a time from top to bottom, they are displayed in alternating lines, then filled in on a second pass. Depending on which graphics viewer or Web browser is being used, progressive JPEGs may produce a "venetian blind" effect or simply a blurry or blocky image that gradually sharpens. Pages using progressive JPEGs let people see at least the outline of an image sooner, and often appear to load faster than pages using normal JPEGs. Although progressive JPEGs are fairly new, most modern browsers support them. If a browser does not support progressive JPEGs, it will display the image as a normal JPEG. *Related Terms:* interlaced GIF; JPEG.

progressive margins — *Related Term:* shingling.

progressive proof — a set of press proofs that includes the individual colors, interspersed with overprints of the two, three, and four color combinations in their order of printing. A typical set of progressive proofs would include yellow; magenta; yellow and magenta; cyan; yellow, magenta and cyan; black, yellow magenta, cyan and black. The proofs are used by printers on single-color presses as an aid when printing process-color work. Often called "progs." *Related Term:* progressives.

progressives — color proofs taken at each stage of printing showing each color printed singly and then superimposed on the preceding color. *Related Term:* progressive proof.

projection print — an enlarged photograph made from the images on a negative using an enlarger.

prolly — Internet chat room shortcut for probably.

prologue — the text that appears at the beginning of a story which sets the stage or introduces the story, as opposed to an epilogue, which appears at the end of a story and offers parting comments.

promo copy — free copy of a work given for promotional purposes.

promotional material — all of the media and/or other merchandise designed to publicize and sell a product.

prompt — the flashing screen symbol where you type or place your mouse on the screen. This is when the host system asks you to do something and waits for you to respond. For example, if you see "login:" it means type your user name.

proof — (1) a prototype of the printed job; a reason-ably accurate representation of how a printed job is intended to look. It can be made photomechanically from plates (a press proof), photochemically from film and dyes, or digitally from electronic data. Pre-press proofs serve as samples for the customer and guides for the preps operators. Press proofs are approved by the customer and/or plant supervisor before the actual press run. (2) in editorial work, short for proofreading copy or checking the color and position of text and images on a page layout. *Related Terms:* press proof; prepress proof; brownline; Vandyke; silver print; blueline; blueprint; color key, etc.

proof OK — a customer signature approving a proof and authorizing the job to advance to the next stage.

proof correction marks — *Related Term:* proofreader's marks.

proof press — a machine used to make copies (proofs), generally of relief or intaglio plates or typeset material so the type can be proofread in advance of further production.

proof sheet — in photography, a term for a sheet of images made by contact printing multiple negatives onto a single sheet of photographic paper.

proofing — the process of testing the final stripped images from a flat on inexpensive photosensitive material to check image position and quality. The act of making a proof.

proofread — to examine a manuscript for errors in writing or composition; the act of closely reading a proof of some typeset material to detect errors-spelling, spacing, positioning, etc.

proofreader — the person who carefully reads and reviews typeset copy for errors, spelling, spacing, positioning, etc.

proofreader's marks — shorthand symbols employed by copyeditors and proofreaders to signify alterations and corrections in the copy. *Related Term:* proofing marks.

properties — (1) the settings and values that characterize an item in a Web page, such as the title and URL, the file name and path of a file, or the name and initial value of a form field. (2) the characteristics particularly ascribed to files in a Windows operating system, such as color depth, graphic format, etc.

property release — *Related Term:* photo release.

proportion — a design characteristic concerned with size relationships of both length and width of sheet size and image sizes in relation to one another and the page.

proportion scale — in reproduction photography and platemaking, a circular slide rule used to establish enlargement and reduction percentage and the ratio of the reproduction size to the original size. *Related Terms:* proportion rule; proportion wheel; percentage wheel; scale.

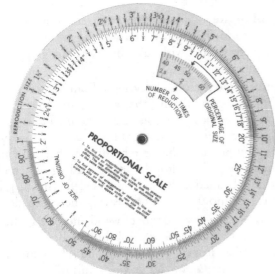

PROPORTION SCALE OR WHEEL
A circular slide rule type of device used to determine proportional enlargements and reduction percentages by matching the existing size and the intended reproduction sizes on the two circular scales.

proportion wheel — *Related Term:* proportion scale.

proportional spacing — a method of letter escapement and spacing whereby each character is spaced to accommodate its respective real width. From character to character, widths of letters or figures will be different; an "i" gets less space than "p," "W" gets more space than "a," etc., increasing readability. Books and magazines are set proportionally spaced; standard typewritten documents are generally monospaced. *Related Terms:* fixed spacing; monospacing.

proportionality failure — common condition in halftone color printing where the ratio of red-to green-to blue-light reflectance in halftone tints is not the same as that in continuous ink solids.

proportionally spaced composition — *Related Term:* proportional spacing.

proposal — a detailed plan for a research project, book or article, including an outline, author bio, marketing strategy (if necessary) , and any other information used to persuade a contract offer for the work.

protected mode — an operating mode of x86 chips that lets the PC access the largest possible amount of memory. In protected mode, different parts of memory are assigned to different programs. This way, memory is "protected" in the sense that only the assigned program can access it. Often, device drivers operate in this mode. *Related Terms:* device driver; real mode.

protective paper — *Related Term:* safety paper.

protocol — a formal description of formats and the rules two computers must follow to exchange messages. It is a specific method of accessing a document or service over the Internet, such as File Transfer Protocol (FTP); HyperText Transfer Protocol (HTTP), PPP, TCP/IP, SLIP, etc. *Related Terms:* HTTP; FTP; PPP; SLIP; TCP/IP.

prototype — a mock-up; a model that is the pattern for the final product. *Related Terms:* dummy; mock-up.

proving — a process of testing final stripped images from a flat on inexpensive photosensitive material to check image position and quality. *Related Term:* proofing.

proximity searching — finding search terms that occur within a specified number of words in the text. Proximity searching is not as stringent as phrase searching, which requires that the terms occur in the text exactly as entered for the search.

proxy or proxy server — a technique used to cache information on a Web server which acts as an intermediary between a client and that server. It holds the most recent, commonly used user content from the Web in order to provide quicker access and to increase server security. Proxy servers are particularly common for ISPs that have a slow Internet link. They can also be configured to allow direct Internet access from behind a firewall. They open a socket on the server, and allow communication via that socket to the Internet. *Related Term:* server.

pseudonym — an assumed name used to conceal an author's true identity. *Related Terms:* pen name; non de plume.

public domain — (1) in media, a channel of communication such as radio, television, newspapers, and magazines. (2) in computing, downloadable software with very few conditions of use. Public domain downloads have no copyright restrictions or conditions of use. Shareware, generally expects a fee. Freeware, usually has copyrights and other restrictions. *Related Term:* mass media; shareware; freeware.

public key encryption — an encryption method using two keys, a public key that is made available to everyone and a secret key known only by the private key holder. The two keys are designed to work together. Anyone can use the public key to encrypt data, but to protect the content of messages, only the person with the corresponding private key can decrypt it. In a similar manner, anyone can use the public key to decrypt data, but to ensure the identity of the person sending the data, it must have been encrypted with the corresponding private key. PGP is a type of public key encryption. *Related Terms:* encryption; keyring; pretty good privacy; PGP.

Public Law 94-553 — the current copyright law, adopted in 1978.

public method — one of several modifiers you can use to control access to a method in object-oriented languages such as Java. If you use no such modifier, only classes declared in the same file have access to the method. Public method is available to any class in any file. *Related Terms:* Java; object.

publication — the final output of a joint effort. Originally assigned only to printers, it has grown to include the Internet, etc.

publication date — not as often thought. It is the actual date on which a book becomes available for purchase, and on which the promotion is slated to peak.

publicist — a person who prepares promotional materials and schedules media appearances for their client(s) such as a book signing tours, personal appearances, engagements, etc..

publish — to pay for producing and distributing or marketing a publication or other images such as a book, newspaper, compact disk, tape, etc.

publisher — the owner and/or primary producer of a periodical or line of books, electronic or electronically distributed media, etc.

publishing — a printing industry category describing those who prepare and distribute materials such as books, magazines and newspapers, although they may not physically print the materials.

publishing paper — a paper made in weights, col-

ors, and surfaces suited to books, magazines, catalogs, and freestanding inserts.

puck — a mouse-type device generally associated with electronic digitizing tablets. It may have a crosshair or other elements to assist in directing it over a flat original for the purpose of drawing an image into a computer.

pull — (1) an alternative term for proof. (2) to develop film for less than normal time to compensate for over- exposure.

pull quote — a quotation extracted from the main text of an article and printed in large type on the page, frequently offset with ruled lines or other graphic elements.

pull-down menus — developed from Xerox research for desktop publishing, they are a method of providing user control over software without cluttering up the screen with text. Using the mouse or cursor keys, the user points to the main heading of the menu and the menu pulls or drops from the heading. When the required function has been selected the menu rolls back up into the menu bar, leaving the screen clear.

pulp — the raw mixture of wood and/or rag fibers, chemicals, and water which is broken down by mechanical and/or chemical means and from which paper is made. *Related Terms:* stock; furnish.

pulp substitutes — an intermediate grade between pulp and wastepaper that does not require de-inking; sometimes referred to as white waste.

pulping — processing the wood chips and logs to release the wood fibers from their lignin bonds during the papermaking process.

pulpwood — logs, chips, sawdust, shavings, slabs, and edgings that have been ground or shredded in preparation for papermaking.

pulse code modulation (PCM) — a not universal, but most common, method of encoding analog signals into digital bit streams.

pulsed xenon lamp — the primary light source in graphic arts photography, rich in ultraviolet and low-frequency blue radiation. It provides a very constant output with a spectral composition resembling sunlight. Used in photography and platemaking. Brighter than quartz light, pulsed xenon lights are used for both black-and-white and color work.

punch and register pin method — a method of multiflat registration in which holes are punched in the tail of the flat or in strips of scrap film taped to the flat; the punched holes fit over metal pins taped to the light table surface and cause the flat to fall in correct position.

punch through — a condition, in relief printing, caused by too much press packing, in which the relief type presses too hard on the paper and makes a raised image on the back side of the substrate.

punching — bindery operations performed by punching holes into printed sheets in preparation for plastic or mechanical binding. It is also done to make holes for looseleaf ring and post binders. *Related Term:* drilling.

punctuation marks — typographic symbols used in text to clarify, increase readability, and direct the reader's attention. Typical are , (comma), . (period), ? (question mark or interrogatory), : (colon) ; (semicolon) and ! (exclamation point or bang).

purge — the purposeful removal of all unwanted, outdated or old data from a memory core, hard disk, network, etc.

purity — *Related Term:* saturation.

push — (1) on the Internet, a system where users don't hunt their way through pages to see what the Web has to offer. Instead, they simply enter a variety of interests and needs into a server-side database. The server collects all the information it feels is relevant and sends ("pushes") the data to the user's desktop, sometimes through a standard browser but often via a proprietary interface like PointCast. (2) in photography, to develop film longer than normal to compensate for under-exposure. *Related Term:* active channels.

pushbutton — a form field that lets the user react to questions. It may be to submit a completed form or to reset the form to its initial state. Similarly, it may be to indicate selections of answers to presented questions. Basically it is nothing more than an active area of the screen which reacts in a preprogrammed manner to activation.

put under — a reference to coating an offset plate with asphaltum to keep the image ink-receptive when the plate is to be stored for a long period of time. *Related Term:* gumming.

pyramid — an arrangement of advertisements on a page to form a stepped half pyramid.

pyroxylin — a resin coating used to imbue paper. It is very printable, is available in metallics and as embossed finishes. In modern production, it is often used to protect casebound books.

Qa

QA (quality assurance) —the process of checking randomly selected printed pieces as they come off the printing press or out of the bindery to make sure they meet quality standards.

QPSK (quadrature phase shift keying) — a digital frequency modulation technique used for sending data over cable netmodems. It is easy to implement and fairly resistant to noise. Primarily used for sending data upstream from the cable subscriber to the Internet.

quad — a unit of space in setting type. (1) originally, in hand composition, a piece of type metal used to fill out lines and spacing between words. Usually two (2-em quad) or three (3-em quad) times wider than the em, where an em quad is the square of the particular type size, i.e., 10-point em-quad is 10 points x 10 points, etc. A half em, called an "en," is half the width of an em quad. (2) the term also refers to horizontal spacing and alignment in phototypesetting, such as "QC" = quad center, "QL" = quad left, "QR" = quad right.

quad center — to set copy in the middle of a line, with all excess space equally placed at each end of the line.

quad left — a command to a phototypesetter or imagesetter that instructs the machine to position all text to the left end of the line. A minimum of interword spacing and letterspacing is used in the portion of the line containing characters and the right portion contains only space. *Related Terms:* flush left; quad right; tab.

quad middle — a command to a phototypesetter or imagesetter that instructs the machine to insert all justifying space at both ends of the line, pushing text material equally toward the center of the line. A minimum of interword spacing and letterspacing is used in the text portion of the line. *Related Term:* quad center.

quad right — a command code to a phototypesetter or imagesetter that instructs the machine to position all text to the right end of the line. A minimum of interword spacing and letterspacing is used in the portion of the line containing characters and the left portion contains only space. *Related Terms:* flush right; quad left; tab.

quadding — filling a composed line of type with spacing material after all characters have been set in such a manner as to position the type properly in relation to the right- and left-hand margins, dependent upon the compositor's instructions.

quadrant makeup — a plan for a page in which each quarter is given a strong design element.

quadratic texture mapping — this technique, used with Nvidia-based 3D graphics boards, speeds up texture mapping and redrawing by reducing the amount of work required. Nvidia chips use fewer polygons to render an acceptable-looking rounded object. Filling the screen, therefore, takes less time and CPU horsepower than it would using another rendering chip. *Related Term:* Nvidia.

quadratone — a black and white photo printed with four halftone plates. May reproduce a multitone effect or extend the tonal range by having each plate concentrate on a particular area of the image, i.e., highlight, midtones or shadows.

quadrille paper — paper printed with a non-reproducing blue ink square grid pattern that is used as a guide for mechanical paste-up operations.

QuakeSpy — a Windows Internet program that retrieves lists of active servers for gamers to play the game Quake.

quality — when applied to printed images, it could mean (1) the aesthetic aspect, influenced largely by the creativity and knowledge of the designer; (2) the technical aspect, the way the original is processed through the photomechanical system; (3) the consistency of the printed image.

quality control — applications that use automatic identification to make sure the right material is in stock so it can be delivered for the right cost, to the right user, at the right time.

quality paperback — in publishing, a trade paperback book, usually the softcover version of the hardcover book, as opposed to a mass market paperback book, which is printed on less expensive paper and may have a smaller trim size than the hardcover edition.

quality printing — a subjective evaluation of how

well a product was produced; has different meaning for every printing job. Judgment of quality depends upon context, time, and the intended usage of work being evaluated or compared.

quantizing — the second of three steps in the process of converting an analog signal into a digital data stream. The others are sampling and encoding.

quarter bound — a casebound book whose spine is covered in one material and its corners and sides are covered in another, as compared to full and half bound.

quarter tone — picture tonal values in range of a 25 percent dot. *Related Term:* midtone.

quarto — (1) a book which is 9 X 12 inches, and made from twice folded or quarto signatures. (2) a sheet folded twice at opposite angles, producing a final folded size which is one-quarter that of the parent sheet.

quartz light — a type of bright light that does not dim with age, used in black and white photography and platemaking.

query — generally, a short term for a query letter. In Web communications, a question, usually used in connection with a search engine or database, to find a particular file, Web site, record or set of records in a database.

query letter — an introductory letter that a potential author sends to a book or magazine publisher describing an idea for a book or article outlining the author's qualifications.

queue (pronounced "Q") — a waiting area for e-mail messages, files, print jobs, or anything else that is being sent from one device to another, or a stream of tasks waiting in line to be executed. Related Topic: buffer.

Quick – (1) an early term for a skilled a compositor (especially foundry type composition). (2) an early second generation phototypesetter from the Linotype (now Heidelberg) Corporation.

quick printer — (1) a printer whose business emphasizes basic quality, small presses and fast service and who normally uses paper or plastic plates. Usually specializing in simple print runs of 10,000 or less. (2) incorrectly used to describe a convenience copy center utilizing xerographic reproduction in color or black and white. This is a copy shop; not a print shop.

quick printing — reproduction centers that provide a fast turnaround of short-run (1000-15,000 impressions) monochrome jobs, generally no larger than

11x17 in., produced on high speed photo duplicators and photocopiers or presses that are 14x20 in. (366x508 mm) or smaller. These printers, many of whom own franchises of larger quick-printing concerns, produce some color work, but seldom accept four-color process jobs. The introduction of computers, color copiers and direct digital press systems capable of this work has begun to significantly change the amount of spot and full-color work performed by these concerns.

QuickDraw — a native graphics language on the Apple Macintosh.

quickset ink — ink that dries rapidly on paper.

QuickTime — a software-based file format for integrating and storing sound, graphics, and movie files and other multimedia data between a wide range of applications. A MOV file on the Web or on a CD-ROM, it is a QuickTime file. It was originally developed by Apple Computer for the Macintosh, but player software is now available for Windows and other platforms. QuickTime player can be downloaded in versions for either Mac or PC from Apple's Web site. *Related Terms:* .AVI; .MOV; .MPEG.

QuickTime VR — a software program which brings virtual reality to the desktop without any special equipment. It enables the user to experience a 3D photographic or rendered representation of any person, place or thing. With mouse and keyboard instructions, participants can rotate objects, zoom in or out of a scene, look around 360 degrees, and navigate from one scene to another.

quiet zone — a clear space, containing no dark marks, which precedes the start character of a symbol and follows the stop characters. Used to avoid reading confusion during scanning.

quire — nominally, 1/20th of a ream (25 sheets). In fine papers the quire is twenty-five sheets, but in coarse papers it is twenty-four sheets.

quoins (pronounced coins) — expanding block-like or wedge-shaped devices operated by the use of a special quoin key. They are used as part of the lockup process in letterpress forms to tighten furniture against a type form in the press chase prior to putting it on press.

quotation — in printing management and sales, a legally binding agreement between a printer and publisher that lists the costs of a particular print job, in which the price does not fluctuate for a specified period of time. More solid than a printing estimate, which is a preliminary report on how much a print job is expected to cost but is not legally binding. An endorsement of a book.

quote — *Related Term:* quotation.

QWERTY — generally, the top row or keys on the standard keyboard layout found on computers, phototypesetters, and typewriters used to compose in the English language. *Related Term:* ETAOIN; ASDF layout.

R print — a color photographic print made directly from transparency without using an internegative.

race — *Related Term:* type classifications.

RAD (rapid application development) — the generic name for tools and techniques designed to build the basics of an application quickly, especially the user interface. RAD tool examples are programs such as Visual Basic and PowerBuilder, which provide tools for easily creating and linking dialog boxes.

radio button — a form field that presents the user with a selection that can be chosen by clicking on a button. Radio buttons are presented in a list, one of which is selected by default. Selecting a new member of the list deselects the currently selected item.

radio frequency — nonoptical automatic identification devices that use radio waves to transmit data.

radio frequency data communication (RFDC) — a hand-held or vehicle-mounted unit that sends and receives messages by radio frequency, displaying information on a screen for workers. It allows real-time, two-way exchanges of data between terminals, one often mobile and the other a host computer at a distant location.

radio frequency identification (RFID) — the use of small radio transponders which are activated by a reading transmitter. The transponder can carry a unique ID code or other information in its memory, and can be read at a distance without line of sight.

radio frequency tag — an electronic tag capable of receiving/storing and/or transmitting digital information.

radio frequency terminals — a portable device used to interact with a remote host computer.

rag content — the percentage of cotton or other cloth fiber in paper. The higher the percentage, the higher quality the sheet.

rag left — the description given to type that is justified (aligned) on the right margin and appears ragged on the left. *Related Term:* ragged left.

rag paper — paper pulp made by disintegrating new or old cotton or linen rags, and cleaning and bleaching the fibers. Quality papers generally contain a minimum of 25 percent rag or cotton fiber and can be made up in the following grades: 25, 50, 75, and 100 percent. Rag pulp is used principally for making premium bond, ledger, writing papers, and other papers requiring permanence. *Related Term:* cotton content paper.

rag pulp — paper pulp made by disintegrating old or new cotton or other cloth fiber rags, and cleaning and bleaching the fibers. Rag pulp is used principally for making premium bond, ledger, writing papers and papers requiring permanence, such as paper money.

rag right — the description of type that is justified (aligned) on the left margin and appears ragged on the right. *Related Term:* ragged right.

ragged — lines of type that do not start or end at the same position.

ragged center — any type set with each line centered; ragged margins on left and right.

ragged left — a term pertaining to type which is justified on the right margin and ragged on the left. *Related Term:* quad right.

ragged left/right — successive lines of type which are of unequal length and which are aligned at either (but not both) the right- or left-hand margin.

ragged right — lines of text which are flush against the left-hand margin but end at various points on the right margin. *Related Term:* quad left.

RAID (redundant array of independent [or inexpensive] disks) — a convenient, low-cost, and highly reliable storage system which saves data on more than one disk simultaneously. Designed on several levels, at its simplest (RAID-1) it consists of two drives that store identical information. If one drive goes down, the other continues to work, resulting in no downtime for users. It isn't a very efficient way to store data, however. High-numbered RAID configurations attempt to address efficiency and cost issues. Unfortunately, there is almost always a performance trade-off: depending on the type used, it will be slower than a single drive at either reading or writing data.

railroad board — a thick, coated, often waterproof stock for products such as signs and cards.

railroading – the unwise but sometimes necessary practice of setting type and sending it on for final presswork without proofing or reproofing. While not a recommended procedure, it is often practiced in some deadline-intensive applications such as magazines and newspapers. It particularly applies to original or final correction composition.

raised cap — an initial capital letter, used as a design element, at the beginning of a chapter or book and which is raised above the first line of type in the text. *Related Term:* drop cap.

raised printing — *Related Term:* thermography; Virkotype.

RAM — an acronym for "Random Access Memory." Memory chips that store information only when the computer is on; when switched off, the information is lost forever. It can contain both application programs and the user information, but to save information in RAM, it must be transferred to a disk. *Related Terms:* DRAM; EDO RAM; SRAM; virtual memory; von Neumann architecture.

RAMDAC (random access memory digital-to-analog converter) — this microchip sits on a VGA card or other graphics display board and translates the digital representation of a screen full of information into an analog signal that the monitor can display. The faster the RAMDAC (measured in MHz), the higher the screen refresh rates that the card will support at each given resolution. *Related Term:* refresh rate.

ramped screen — *Related Terms:* graduated screen tint; gradient screen.

random proof — *Related Term:* scatter proof.

range finder — a part of a camera used to help the photographer focus a camera and frame the subject.

rapid access — a process of preparing a film image using an emulsion and developer that allows for wide latitude in skill to produce an acceptable film image.

raster — an image composed of a set of horizontal scan lines that are formed sequentially by writing each line following the previous line, particularly on a television screen or computer monitor.

raster count — (1) the number of addressable coordinate points on a video screen. (2) in computer graphics, the number of lines in one dimension within a display space.

raster file — an electronic file in which the pixels are scanned and input in rows.

raster graphics — images defined as a set of pixels or dots in a column and row format. *Related Term:* bitmapped graphics.

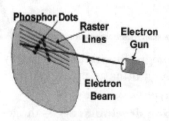

RASTER DISPLAY
Electron beam excites phosphor dots and cause them to emit visible light. The phosphors are arranged as tiny dots and the electron gun sweeps across them in a horizontal line at 30 or more times per second. The speed is high enough that the viewer doesn't see the action.

raster image display — the image on a CRT, TV screen or computer monitor, composed of a pattern of horizontal lines made of small dots or pixels.

raster image processor (RIP) — the hardware engine which calculates the bitmapped image of text and graphics from a series of instructions. It interprets all of the page layout information for the marking engine of the image-setter. PostScript or another PDL (page description language) serves as an interface between the page layout workstation and the RIP. *Related Terms:* imagesetter; page description language; PostScript; rasterization.

raster scanning — a raster is the set of parallel horizontal scan lines that form a television or video display image. Each scan line is reproduced many times per second. Raster scanning is the generation of a beam of light or electrons according to a predetermined rather than a random access pattern.

rasterization — the function of the raster image processor. Converting mathematical and digital information into a series of dots by an imagesetter for the production of negative or positive film.

raw stock — a substrate which has had neither coating or ink applied.

RC photographic paper (RC) — (1) light-sensitive photo print paper with a special resin-coated surface. (2) paper for typesetting and PMTs that, when properly processed, does not yellow with age.

RCA connectors — a cable-connecting system originally developed to connect turntables to stereo systems. With the advancement of audio-video technology, they are now used mainly for carrying video and line-level audio on stereos, TVs, and VCRs. They have a center contact (a pin or a hole) and an outer contact (a flower or a cylinder).

RD (receive data) — a modem light that flashes off and on during data transfers, telling you that the modem is receiving signals from a remote computer.

read — the activity of a computer when the heads are collecting and inputting data from storage into memory.

readability — one of the major factors in type selection, along with legibility, printability and availability. Pertains to the perception of the type as easy or hard to read, etc.

reader service card — usually a business reply card either bound or loosely inserted into periodicals, magazines, newspapers, etc., with boxes or check lines or numbers to circle to request response from advertisers or on specific subject matter in the publication. Sometimes referred to as a bingo card.

reader's proof — a galley proof, usually the specific proof read by the printer's proofreader, which will contain queries and corrections to be checked by the client. *Related Term:* printer's proof.

reader's spread — how pages will appear in the final, bound publication, as opposed to a printer's spread, which is how the pages are arranged to be printed on press. *Related Term:* imposition.

readout — (1) the headline or part of a headline between a banner head and the story; (2) devices for breaking out selected body matter such as quotes (pull quotes) to set them in display type and be embellished with typographic devices.

real mode — an operating mode of x86 chips that replicates the memory management used by 8086 or 8088 chips. Real mode limits the processor to IMB of memory and provides no memory management or memory protection features. The phrase is often used to describe device drivers that operate in this mode. MS-DOS runs in real mode. *Related Terms:* device driver; protected mode.

real time — (1) computing at a speed that can produce results without a noticeable delay. (2) in video switching, the ability to alternate between live and canned images.

ream — 500 sheets of paper of any size paper; a standard measurement.

ream marked — unwrapped sheets of paper, in a carton or on a skid with markers placed every 500th sheet.

ream weight — *Related Term:* basis weight.

ream wrapped — wrapped paper packaged in bundles containing 500 sheets. Abbreviated RW.

reboot — to turn on or power up a computer; the process of loading the boot program and setting up the computer to perform useful tasks. A warm boot is when the reset button is depressed and the power remains on. A cold boot is when the power is completely turned off and the power switch is activated. Warm boots are most common, but some application problems can be cleared only with a cold boot.

receiver sheet — in the diffusion transfer process, the piece of the two-part material that is not sensitive to light. Sensitive to chemicals instead, it accepts the images from "donor" materials during processing.

receptors — the material on which the printing is done. Some receptors are paper, glass, wood, metal, plastic, cloth and ceramic. *Related Term:* substrate.

recessed — a lowering or indentation of the image area from the non-image area. The intaglio and gravure image areas are recessed by mechanical or chemical means.

reciprocity — the use of set combinations of apertures and shutter speeds, designed and paired to obtain the same correct exposure, regardless of which combination is used.

reciprocity law — a photographic rule stating that exposure is a function of the product of both exposure time and light intensity, but not of either variable alone.

reciprocity law failure — the failure of the effect of exposure to be the same for a given value of the product of illumination intensity and time (when the factors making up the product are varied), which may happen if one of these factors is extremely small.

record — in a database, one complete entry consisting of one or more fields of data.

recto — the right-hand page of a book, magazine, or other publication. Page 1 is always a recto page and rectos always bear the odd-numbered folios. *Related Term:* verso.

recto–verso — two-sided printing. *Related Terms:* duplex; backup.

recycle — to return once-used materials for reprocessing and/or reuse.

recycled fibers — fibers recovered from wastepaper, printing and converting waste, forest and lumber mill residues, etc., to be used for the manufacture of paper or paperboard. *Related Terms:* recovered fibers; secondary fibers.

recycled paper — (1) new paper manufactured from de-inked, used paper and bleached pulp or from printing and converting waste. (2) wastepaper or stock separated from other solid waste and designated for reuse as a raw material. Papers that are heavily contaminated (with color or coating) may not be recyclable. *Related Terms:* wastepaper; waste.

Red Book — another name for the CD-DA audio CD format introduced by Sony and Philips, the Red Book standard defines the number of tracks on the disc that contain digital audio data and the error correction routines that save sound from minor data loss. *Related Terms:* CD-DA; CD; CD-I; CD-ROM; Green Book; Orange Book; White Book; Yellow Book.

reduced instruction set computer — *Related Term:* RISC.

reducers — (1) in printing, compounds such as varnishes and solvents used to reduce the consistency of printing inks. (2) in photography, a term sometimes applied to certain film-processing chemicals.

reducing back — a camera back that takes smaller film holders than those for which the camera was designed.

reduction — making the size of a copy smaller than the original by mechanical or electronic means.

reel — (1) in papermaking, a roll of paper wound directly off the papermaking machine, thus measuring the full width of the Fourdrinier wire. This becomes the master roll from which smaller rolls are cut. (2) in publishing, a roll of paper for installation on a web printing press.

reel-fed press — *Related Term:* web press.

re-etching — a hand color correction technique done primarily on color separation relief plates.

reference marks — symbols used in text to direct the reader to a footnote. E.g., asterisk (*), dagger (†), double dagger (‡), section mark (§), paragraph mark (pillcrow) (¶).

refining — the papermaking step in which the wood fibers are cut and shortened to produce a more uniform pulp. This is accomplished by grinding them between revolving serrated metal disks prior to bleaching. Performed in pulp refiners, it causes the fibers to shred and collapse, opening up additional fiber-to-fiber bonding surfaces. Without it, the bleached fibers are hollow and somewhat rigid, and have insufficient bonding surfaces to form paper.

reflectance — generally, the ability of a surface or material to reflect light. Specifically, the ratio of the amount of light of a specified wavelength or series of wavelengths reflected from a test surface to the amount of light reflected from a barium oxide or magnesium oxide standard. *Related Term:* brightness.

reflection copy — any copy which must be properly viewed by light reflecting from its surface as opposed to transillumination. A photographic print, painting, or other opaque copy used as original art for reproduction. *Related Terms:* reflective art; transmission copy; transillumination.

reflection densitometer — an instrument that measures the amount of incident light that is reflected from the surface of a substrate, such as ink on paper. *Related Term:* densitometer.

reflective art — *Related Term:* reflection copy.

reflective media — normal copy. Any artwork or media which is viewed by reflected light as opposed to transillumination such as used for transparencies.

reflector — the antenna's main curved "dish," which collects and focuses signals onto the secondary reflector or low-noise amplifier.

reflex blue — a toner used in black inks to neutralize the brownish tinge of carbon black pigments.

reflow — the process of reworking typography when undesirable line breaks occur in digitally typeset copy change due to alterations in the layout.

refresh rate — the frequency at the electron guns in a CRT redraw the image on the screen. It is an important measure of how steady the image will appear. Refresh rates of 60 Hz (in which the screen is redrawn 60 times per second) such as a normal television almost always flickers, especially under fluorescent lights, which refresh at about the same rate, and causes eye fatigue. Refresh rates above 70 Hz are preferable. Refresh rates vary depending on the screen resolution. It is important that both the monitor and display adapter run at a 70-Hz or higher refresh rate. *Related Terms:* color balance; color tracking; CRT; electron gun; pixel; plug and play; RAMDAC.

register — (1) to place an image where it should be, particularly the proper alignment of two or more images in relationship with each other. (2) generally, the positioning of printing in proper relation to edges of paper and other printing on the same sheet. Such printing is said to be in register.

register bond — *Related Term:* form bond.

register marks — small reference patterns, guides, or crosses placed on originals before reproduction to aid

REGISTER MARK
One common register mark which is placed on overlays to aid in proper registration of colors, multiple flats, etc. Sometimes called a "bullseye."

in color separation and positioning negatives for stripping. They are also placed along the margins of negative film flats to aid in multicolor registration and the correct alignment of overprinted colors on press sheets.

register pins — *Related Terms:* registration pins; gauge pins.

register punch — a punching machine used to make accurately spaced round and oblong holes in masking sheets and plates for positioning copy during the prepress stage and, sometimes, printing. *Related Terms:* pin register; register.

registered user — a user of a software or a Web site with a recorded name and password.

registrant — the individual or organization that registers. (1) on the Internet, it applies to a specific domain name with Network Solutions. This individual or organization holds the right to use that specific domain name for a specified period of time, etc. (2) generally applied to anyone who acknowledges the purchase and responsibility for software usage and reports it to the manufacturer. *Related Terms:* domain name registration agreement; registration fee.

registration — (1) the overall agreement in the position of printing detail on a press sheet, especially the alignment of two or more overprinted colors in multicolor presswork. Register may be observed by agreement of overprinted register marks on a press sheet. In stripping, film flats are usually punched and held together with pins to ensure register. The punched holes on the film flat match those on the plate and press specified for the job. (2) the process through which individuals and organizations obtain a domain name. Registration of a domain name enables the individual or organization to use that particular domain name for a specified period of time, provided certain conditions are met and payment for services is made. *Related Terms:* registrant; registration fee; domain name registration agreement.

registration fee — the charge for registering a domain name. It covers the cost of processing the initial registration and maintaining the domain name record for two years. For domain names with registration dates on or before March 31, 1998, the fee was $100.00. For domain names with registration dates on or after April 1,

1998, the fee is $70.00. After a two-year period, the domain name is subject to renewal (reregistration) and the renewal (reregistration) fees will be due on an annual basis if the registrant wishes to renew the domain name's registration. *Related Terms:* registration; maintenance; renewal (reregistration); domain name.

registration forms — forms that are used to submit and process registration requests, weather for Internet domains or for software, hardware, etc. Any input matrix which ultimately relays appropriate requested data. *Related Terms:* host form; contact form.

registration mark — crosshairs or other graphic devices applied to originals prior to reproduction or by imagesetters for positioning multiple films in proper register (alignment).

registration pins — a system for flat and film registration. The most common type is 1/4 in. in diameter and attached to a flat plate which is attached to the film back of a process camera. Film which has been punched with holes along an edge is then positioned on the pins and an exposure made, using the pins for each different filtered exposure. The camera person will end up with four different negatives that, when processed and put back on pins, will be in perfect register without having to use registration marks. Also used to register film in the stripping, proofing and platemaking processes.

registration unit — the mechanism(s) on a printing press that ensures that successive sheets are held in the same position each time an impression is made.

registry — a registry is responsible for delegating Internet addresses such as Internet protocol (IP) numbers and domain names, and keeping a record of those addresses and the information associated with their delegation. (2) a Windows system configuration file that details how Windows looks and behaves. It stores user profile information such as wallpaper, color schemes, and desktop arrangements in a file called user.dat. It also stores hardware and software-specific details, such as device management and file extension associations, in a file called system.dat. Registry details can be edited using a program called Regedit (which ships with Windows) and exported to text format as a file with the extension REG. The registry can also be modified by control panel settings.

reglets — the thin pieces of furniture used to fill small spaces and protect larger furniture from quoin damage during a letterpress lockup. *Related Term:* furniture.

regular sizes — sizes other than basic sheet size in which paper is commonly cut and stocked by paper suppliers.

reject — in automated document handling, a character, field, or document that the recognition system is unable to recognize.

reject rate — the percentage of the total items to be recognized that the recognition system knows it cannot identify.

rejustify — the process of making a line of type the correct length after corrections have been made.

related industries — a category of industries allied with printing and its services that includes raw material manufacturers (ink, paper, plates, chemicals), manufacturers of equipment, and suppliers that distribute goods or services to printers. *Related Terms:* half duplex; full duplex; simplex.

relational data base — a record-oriented collection of information stored in as many segmented files as there are defined fields. If a record consists of a person's name, street address, city, state, zip code and phone number, the data may be stored in segmented files keyed to each other by the record number. This file arrangement permits extremely flexible search capabilities but also is slower because of the extra time needed to search and verify data presented. *Related Term:* data base.

relative humidity — a calculation of the ratio of moisture present in air to the amount required for saturation. The number is represented as a percentage, i.e., 29.2 percent, etc.

relative unit (RU) — a measure of the width of a character, especially in proportionally spaced fonts.

relative URL — the Internet address of a page or other Web resource with respect to the address of the current page. It gives the path from the current location to the location of the destination page or resource, and may include a protocol.

release paper — specially coated paper used as a carrier for pressure-sensitive label material. It is designed to be a support for the label until it is used, but the coating prevents the label glue from adhering to it.

relief — *Related Term:* printing classifications.

relief printing — a printing process which uses, as an image carrier, a metal, rubber, or photopolymer printing plate on which the image areas are raised above the nonimage areas. *Related Terms:* flexography; letterpress, block printing, rubber stamps.

remaindering —selling off of remaining stocks of a book, usually at a huge discount, after sales decrease or cease.

remnant space — any advertising space in a magazine or newspaper that has not been sold at the regular rate and is available at a discount.

remote login — accessing a remote computer, using a protocol over a dedicated line or computer network, as though it were the host machine. Commonly used protocols include telnet for PCs and rlogin for UNIX environments.

render — (1) traditionally, to apply materials to a canvas or board for the purpose of creating an image. (2) in digital applications, using a computer processor to scale, adjust color, compress or process video, audio or still images.

renewal (Reregistration) — on the Internet, the process of reinitiating a domain name's registration for a specified period of time.

renewal fee — the maintenance charge made for renewing an existing domain name registration. Domain names with anniversary dates on or before March 31, 1998, cost $50.00 . Domain names with anniversary dates on or after April 1, 1998, cost $35.00. *Related Terms:* renewal (reregistration); registration maintenance.

renewal notice — the notice provided to registrants 60 days before the anniversary of their registration date to let them know that their domain name will be due for renewal and that an invoice will be sent for the fees. *Related Terms:* renewal (reregistration); renewal (reregistration) fees; invoice.

repeat length — the length of image a web press can print before it begins printing the same image a second time. Generally determined by the circumference of the image carrier cylinder.

repeatability — the ability to make successive replications of a step, process or technique with apparent identical characteristics, indistinguishable one from the other.

repeater — a bidirectional device used in channels to amplify or regenerate signals to be forwarded over a distance. Frequently utilized in microwave and other line of sight transmission systems.

repp finish — a paper finish that simulates coarsely woven fabric.

reprint — printing a book again using the original

materials, or to print a book in another version, such as the paperback version of a previously published hardcover book.

repro — *Related Term:* reproduction proof.

reproducibility — one factor in making a typeface or art selection for reproduction. It is the measure of how well a typeface or other original art can be reproduced by using one or more of the printing processes.

reproduction — (1) duplicating an original by an photographic or photomechanical process. (2) a copy of an original work.

reproduction percentage — a percentage calculation made with a proportion scale to guide the camera operator in setting up the camera for correct sizing of artwork. *Related Terms:* proportion scale; percentage wheel.

reproduction proof — (1) in relief composition, an exceptional quality proof made for the expressed purpose of being photographed for reproduction. (2) in photocomposition, the continued use of the term to indicate type in which corrections have been made and all elements are in position and ready to have negatives and plates made for printing. (3) generally, camera-ready pages ready for platemaking. *Related Term:* reproduction proof. Repros are often pulled on a special coated paper to improve their reproduction quality.

reprographics — an internationally agreed to term used to describe the overall areas of xerography, diazo, and other methods of copying used by designers, engineers, and architects. Not regular commercial printing, but sometimes includes small in-plant operations which use duplicator-class presses.

reprography — a general term applied to any type of duplication.

request for comments (RFCs) — in advertising and design, a formal review or polling process to evaluate concepts prior to execution. Usually conducted on several levels, ranging from management to final consumer. (2) in computing, the official document series of the Internet Engineering Task Force (IETF) that discusses aspects of computing and computer communication with a focus on networking protocols, procedures, programs, and concepts. *Related Terms:* Internet Engineering Task Force (IETF).

rescreening — the process which occurs when a previously halftoned original is reproduced after the introduction of a new screen and new screen specifications. Usually performed whenever resizing is to be

done to eliminate enlarged or compressed halftone dots in the reproduction.

resident fonts — fonts which are stored in the output device as opposed to being stored as part of the operating sytem or computer unit.

resin — one type of binding material found in both inks and varnishes.

resin coated paper — *Related Term:* RC; RC photographic paper.

resist — any material that blocks or retards the action of some chemical. A number of different types are extensively used in the printing industry. They are light-hardened chemical stencils used in photomechanics to prevent etching of nonprinting areas on a printing plate. *Related Term:* photo resist.

resize — to change output size from that of the original. To enlarge or reduce in dimension.

resolution — (1) the narrowest element dimension which can be distinguished by a particular reading device or printed with a particular device or method, often measured in dots per inch. The precision with which a system can render detail in visual image and separate adjacent small details either visually, photographically, or photomechanically. The higher the resolution (the greater the number of dpi), the sharper the image. If a resolution is agreed upon as a standard, it is called a graphics standard. (2) generally, a measure used to describe what a printer can print, a scanner can scan, and a monitor can display. Printer and scanner resolution is measured in dots per inch (dpi), the number of pixels a device can fit in an inch of space. (3) in computer monitors, resolution refers to the number of pixels in the whole image, because the number of dots per inch varies depending on the screen's dimensions. A resolution of 1,280 by 1,024 means that 1,024 lines are drawn from the top to the bottom of the screen, and each of these lines is made up of 1,280 separate pixels where each may have any number of combinations of red, green, and blue intensities. Monitor resolutions in the PC world include VGA (640 by 480), SVGA (800 by 600, 1,024 by 768, and 1,280 by 1,024). (4) in typography, the measurement used to express the quality of output in dots per inch. The greater the number of dots, the smoother and cleaner an appearance the character/image will have. Currently page (laser) printers print at 300 - 800 dpi. Typesetting machines print at 1,600 - 4,000 dpi or more. (5) in film recording, resolution usually refers to the number of lines that make up the entire screen on a display or on film. Their resolution ranges from the lowest PC standard (200 - 350 lines) up to 10,000 lines. (6) on the Internet, the process by which domain names are matched with corresponding Internet protocol (IP) numbers. *Related Terms:* dpi; pixel.

resolution enhancement technology — one type of printer technique which varies the toner dot size and position to provide smoother edges on raster images, particularly type.

resolve — the matching of domain names with their Internet protocol (IP) numbers. "Resolution" is accomplished by name servers, a combination of computers and software, which use the data in the domain name system to determine which IP numbers correspond to a particular domain name. *Related Terms:* domain name system (DNS), name server.

respi screen — a type of contact screen with a 110-line screen ruling in the highlights and a 220-line screen in the midtones and shadows to produce smoother gradations.

rest in proportion (RIP) — an instruction when giving size to artwork or photographs that other parts of the artwork are to be enlarged or reduced in direct proportion to the marked piece.

restrainers — usually bromides, benzotriazole and chlorides used to retard (reduce the speed of chemical activity) the developer action and reduce fog.

résumé — a brief one- or two-page written record of prior work experience, highlighting the personal strengths, experience, education, etc., of the applicant.

retouching — to enhance or alter a photographic print, artwork or color separations or correct its flaws, either manually or digitally using techniques such as spotting and air brushing or computer programs for photo manipulation. *Related Terms:* retouching.

retouching pencil — a special type of pencil used by retouch artists to add detail or repair continuous tone images.

retrofit — backward integration of advanced capabilities into an older or obsolete device or program that was not originally intended for that purpose.

return key — an alternate name for the enter key on a computer keyboard. It developed from the carriage return feature of the conventional typewriter and the name persists.

returns — unsold product that is returned from the retailer to the manufacturer for credit.

reveal — what appears in a single viewing of a piece, without opening or unfolding. One panel. Each unfolding or opening produces another reveal.

reverse — (1) a copy in which areas that normally would be white are made black; e.g., white lettering on a black background. (2) an infrequently used term that refers to a technique of creating an image by using an open area in the midst of another ink image; for example, type or a graphic reproduced by printing an ink background to establish its outline and allowing underlying color or paper to show through and form the image. *Related Terms:* reverse; tonal reverse; lateral reverse.

reverse image — *Related Term:* reverse.

reverse kicker — a headline form in which the kicker is larger than the primary headline. *Related Term:* kicker.

reverse leading — the action of a phototypesetter in reversing the imaging film to cause the next line of type to print higher on the page than the previous line. On some machines it is employed as a technique used so superior figures or mixed display type can be added to the line. *Related Term:* negative leading.

reverse out — to reproduce as a white image out of a solid background. *Related Terms:* reverse; dropout.

reverse type — *Related Term:* reverse.

reverse-reading — a photographic negative or positive that reads from right to left when viewed from the emulsion side. *Related Terms:* wrong-reading; right-reading; reverse; lateral reverse.

review — a critical evaluation of an article, book, or other written work.

review copy — a complimentary copy of a work sent to reviewers, potential customers, and any other person who could conceivably help promote the work.

revise — (1) in production, a specific stage in the correction process, e.g., first revise, second revise. *Related Term:* revision number. (2) editorially, to change the original materials as previously printed.

revised edition — a new printing of a publication in which changes to the original materials have been made. If a book, it requires a new ISBN.

revision number — indicates the stages at which corrections have been incorporated from earlier proofs and new proofs submitted such as first revise, second revise. *Related Term:* revise.

rewriteable disk — any computer data storage disk

that may be written to and erased multiple times. *Related Terms:* magnetoptical; CD-ROM-RW, etc.

RFI — abbreviation for request for information.

RFP — abbreviation for request for proposal.

RFQ — abbreviation for request for quotation.

RGB (red, green, blue) — (1) the transmissive (additive) primary colors. Sometimes call the scientific hues, that, when mixed together in equal amounts, create white light. (2) a method of displaying color video by transmitting the three primary colors, each as a separate signal using one of two methods: TTE RGB, which limits the color signal to a few discrete values; and analog RGB, in which color signals take on any values between their upper and lower limits, providing a wider range of colors. *Related Terms:* subtractive primaries; CYMK; reflective primaries, hexadecimal.

Rhodamine — a bluish red pigment used for making magenta ink. It has the best blue light reflectance of the commonly used magenta pigments.

rhythm — a design principle that gives a sense of order and movement in the page or visual content; an important factor in layout planning.

ribbon — a cloth or plastic tape coated with several layers of material, one of which is ink-like, that produces the visible marks on a substrate. Used on formed font impact, dot matrix, thermal transfer and hot stamp printers. *Related Term:* foil.

RIFF (resource interchange file format) — a 1990 platform-independent multimedia specification that allows audio, image animation and other multimedia elements to be stored in a common format. *Related Term:* MCI.

right-angle fold — two or more folds that are at right angles to each other; for example, a French fold.

right-reading — a positive or negative with orientation of its image so it can be read in normal fashion, with the top up, and read from left to right. *Related Terms:* lateral reverse; wrong-reading; reverse-reading.

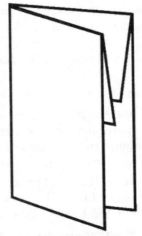

RIGHT ANGLE FOLD
This is a modification of the right angle fold. Most often the panels are of equal length.

right-reading film positive — a high-contrast photographic film positive with the emulsion on the right-reading side. For use when making photo stencils in screen printing.

rights — the conditions and terms of a licensing agreement between copyright owner and client, vendor and customer, to print, publish, and sell, etc. They can include rights for domestic, foreign, multimedia, electronic media, broadcast, translation, film, television, serialization, etc.

ring binder — a binder with a spine which has openable "O" or "D" shaped rings, designed to hold drilled paper or other materials. Typical is the 3-ring notebook.

RIP — (1) in digital composition, the raster image processor, a device that converts the description of a page from high-level PDL coding to low-level scanning instructions for an imagesetter. The part of an output device that rasterizes information so that it may be imaged to film or paper by a laser printer or photographic imagesetter. (2) in camera work and digital prepress, rest in proportion. An instruction indicating that one measure is given and all remaining are to be proportionally sized if not marked.

ripple finish — a finish on paper that looks like small waves in a pool of water.

RISC (reduced instruction set chip) — an approach in computer chip design that dramatically reduces the number of instructions a chip must perform when compared to conventional chips. They are usually faster than conventional instruction set chips (CISC), easier and less expensive to manufacture and their increased speeds made them ideal for everything from multimedia applications to personal digital assistants. The Motorola's PowerPC chip is a RISC chip. Intel also used RISC techniques in 486 and Pentium processors. *Related Term:* microprocessor.

river — the optical path of white space that sometimes occurs when word spaces in successive lines of type happen to end up immediately below each other for some distance. This accident is considered undesirable in typography and may be minimized by moving words from one line to another to reposition word spacing.

RJ-11 — the standard telephone connector; a tab snaps into the socket and has to be pressed to remove the connector from the wall. An ordinary phone circuit uses two wires. The RJ-11 jack has room for up to four wires, but at a glance it's easy to confuse with the larger RJ-45 jack, which can house up to eight wires. *Related Term:* RJ-45.

RJ-45 — similar to a standard phone connector but twice as wide (with eight wires). RJ-45s are used for hooking up computers to local area networks (LANs) or for phones with lots of lines. *Related Terms:* LAN; RJ-11.

RNA — ring no answer. This is the symptom used to describe a modem at a local PoP that rings, but does not pick up the incoming call.

robotics — the general technology of producing robots.

robots — manipulative devices which are programmable and designed to perform mechanical functions.

ROFL or ROTFL (rolling on the floor laughing) — a shorthand term used in postings and online chat to show enthusiastic appreciation of a witticism in a previous posting. *Related Term:* LOL (laughing out loud).

role account — a group name under which more than one individual may perform a specific function. Typically used to prevent a critical function, for example, handling incoming registrations, from being tied to a particular individual.

roll fed — a printing press or converting machine that receives paper as continuous webs from rolls, instead of as sheets. *Related Term:* web press.

roll set — a permanent curl in paper that lies closest to the core (center) of a roll. *Related Term:* core curl.

roller — (1) on a printing presses, cylindrical drums on steel shafts, their outside composed of either metal or a rubber compound. They are used to distribute ink or dampening solution throughout the ink or dampening train and to apply it to the printed form. (2) the cylinders used to convey the paper web through the press or papermaking machine. *Related Terms:* rollers; oscillator.

roller stripping — the action that occurs when the ink does not adhere to the metal ink rollers on an offset press; often the result of poor maintenance and cleaning.

roller, intermediate — a cylinder driven by friction that transfers and conditions ink on press. Located between the ductor and form rollers, it is referred to as a distributor if it contacts two other rollers and as a rider if it contacts a single oscillating drum.

roll-fed press — *Related Term:* web press.

roll-to-roll printing — printing from a web of substrate and then rewinding the printed material back to a roll.

rollup — a technique used to cover an area on a gravure cylinder for re-etching.

Rolodex card — a small diecut card with address, phone, and other contact information printed to fit into a specially constructed holder.

ROM (read-only memory) — a memory storage area that typically contains hard-wired instructions for use when a computer starts (boots up) and which are not lost when power is turned off. The instructions are contained in a small program called the BIOS (basic input/output system) which loads from ROM. The instructions are specific for starting up the hard disk so that the operating system (OS) can be loaded and the remaining start sequence completed. Some newer ROM chips can be updated with new BIOS instructions. They are called EEPROMs or flash BIOS. *Related Terms:* BIOS; EEPROM; RAM.

roman — one of the six basic classes of type in which the letter forms are upright and have vertical stems, like the type you are now reading. More specifically, an upright letter form with serifs derived from Roman stone-cut letter forms and characterized by variation in stroke weights; based on fifteenth century humanist manuscripts. To some, the unmodified version of a roman serif typeface, with no bold or italics applied. *Related Term:* roman type.

roman numerals — selected roman letters used as numerals until the tenth century AD. Examples: I = 1, V = 5, X = 10, L= 50, C = 100, D =500, and M = 1,000. (As opposed to arabic numerals, 1, 2, 3, 4, 5, 6, 7, 8, 9, 0)

roman type — *Related Term:* roman.

room integration — a reference to the specifications for the equipment, associated electronics and specially designed environment and the construction of a total teleconferencing room.

root — The top of the domain name system (DNS) hierarchy. Often referred to as the "dot." *Related Terms:* dot or "."; domain name system (DNS).

root server — name servers that contain authoritative data for the very top of the domain name system (DNS) hierarchy and have the software and data needed to locate name servers that contain authoritative data for the top- level domains (e.g., root servers know which name servers contain authoritative data for com, net, fr, uk, etc.) *Related Terms:* root; top level domain; name server; domain name system (DNS).

ROP (run of press) — color or other matter that isn't given a specific special position in the publication. A type of "as available" opportunity. Used extensively in newspapers for late advertising matter that cannot be scheduled in advance.

rosette — the minimum accepted moiré which produces the flower-like pattern when the four CMYK color halftone screens are printed at the traditional angles. The rosette pattern is noticeable only under magnification.

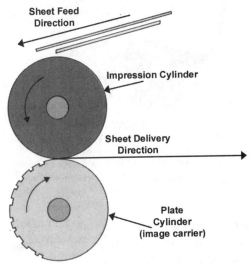

ROTARY PRESS
Two cylinders and a curved image carrier make this configuration the most popular in all classifications and processes where speed and maximum productivity are important.

rotary press — *Related Term:* press, rotary.

rotary screen press — *Related Term:* press, rotary screen.

rotating distribution rollers — rollers in ink or water systems that distribute by rolling against each other. *Related Term:* oscillating distribution rollers.

rotation error — the condition when not all screen images appear squarely in a monitor's display. When the entire image tilts from the vertical position. Some monitors include an adjustment to correct this error so that the image is oriented correctly. Sometimes called tilt error. *Related Term:* screen geometry.

ROTFL — chat room shorthand: Rolling On The Floor Laughing.

rotogravure — an intaglio-class printing process that uses a cylinder as an image carrier. Image areas are etched below nonimage surface areas in the form of tiny wells. The cylinder is immersed in a fairly liquid ink which covers the surface and fills the wells. Excess ink is scraped off by a stainless steel or polymer straight-edge called a doctor blade, leaving ink only in the wells. When the substrate contacts the printing cylinder under great pressure, ink transfers, forming the image. A big advantage of this method of printing is its high speed and long run capability. Used mainly for mail order and magazine supplement work, wrapping paper printing and publications. *Related Terms:* gravure; photogravure; intaglio; halftone gravure.

rotopaper — a supercalendered newsprint made specifically to print on a gravure press.

rough — *Related Term:* rough layout.

rough layout — a usually full-sized layout that is a refinement of the image element arrangement in a thumbnail sketch. A preliminary sketch of a proposed design giving a general idea of the size and position of the various elements to be included. *Related Terms:* dummy; comp; keyline; thumbnail.

round-back spine — in case binding, the rounding of the spine to make the face of the book convex and offering a degree of protection. *Related Term:* flat spine.

rounding — a step in the case binding method in which the book body is given a convex shape on the backbone and a concave shape on the front page.

router — hardware that routes data from a local area network (LAN) to a phone line's long distance line. They can also act as usage monitors, allowing only authorized machines to transmit data into the local network to help keep private information secure. Routers may also handle errors, keep network usage statistics, and handle security issues. *Related Term:* LAN.

routing — (1) in the digital world, the process of directing the movement of information from point to point with minimum delay (and/or minimum cost path) in a network for a message or packet to reach its destination. (2) in traditional printing, the process of cutting away the nonprinting areas of a plate as in letterpress printing. *Related Term:* router.

row — the horizontal orientation of typographic elements in a table or on a page. The opposite of columns.

royal — the size of printing paper 20 in. x 25 in. (508 x 635 mm).

royalty — the percentage of a publication's gross or net sales paid to the author as specified in the author/publisher agreement.

RS-232 (recommended standard 232) — the original nine-wire interface standard for teletype machines. Its third revision (RS-232-C) is the standard for computer serial-port transfers. It is probably the only computer component that's 40 years old and still working. One wire is used as the ground; the rest are dedicated to detecting carrier signals, managing the timing of data transfer, and sending and receiving data.

RSA — a public key encryption algorithm. It has been adopted for many different types of encryption, from digital signatures to the SSL (secure socket layer) used to protect transmissions between Web browsers and servers. Even several commercial encryption packages, such as PGP, use the RSA algorithm. *Related Terms:* encryption; SSL; PGP.

RTF (rich text format) — an original Microsoft file format that enables the saving of word processor text files with formatting, font, text, and much page layout information intact. While saving in the software's own format has similar results, this protocol was developed for exchanging files between all varieties of word processors. *Related Term:* ASCII.

RTFM (read the f - - - ing manual) — a classic acronym response of overworked technical support consultants to stupid user questions.

RTP — (1) in electronic communications, routing table protocol. A communications protocol that uses a list of steps or instructions or routing table for incoming calls. The tables are used for directing outgoing calls across long-distance networks. (2) in publishing presswork, reel-tension-paster, a device for automatically replacing expended rolls of paper on a web press.

RTS (ready to send) — one of the nine wires in a serial port used in modem communications, it carries a signal from the computer to the modem indicating ready to start or when the other is ready. *Related Term:* CTS; RS-232.

RTV (real time video) — an on-line, symmetrical, DVI video compression algorithm.

rub — an element that prints a continuous line or lines using friction. Usually used to frame type, art, or a layout.

rub fastness — *Related Term:* abrasion resistance.

rubber plate — in flexography, printing plates made by molding and curing rubber in a matrix produced from a relief printing form, i.e., a form with image areas raised above the nonimage areas.

rubber stamp — a vulcanized rubber surface containing a relief image, generally mounted on a holder with a handle. When holding the handle, the rubber

RUBBER STAMP AND PAD
As a member of the relief classification of imaging, rubber stamps continue to be invaluable in both small and large office and clerical applications throughout the world.

plate is first impressed upon a saturated inking pad; a subsequent impression on a substrate will create an image of the rubber form.

rubbing up — the name given to the procedure used to process a presensitized lithographic plate.

rub-off and marking — the characteristic of ink to not dry completely or to be subject to smear or abrasions and scuff marks when touched.

rub-on lettering — *Related Terms:* dry transfer lettering; Letraset.

rub-proof — a condition in postpress where an ink has reached maximum dryness and will not scratch without excessive abrasion.

ruby window — a window on mechanical made with red film acetate such as Rubylith.

Rubylith — the Ulano Company's trademark name for clear acetate sheeting with an affixed thin red plastic film which can be lightly cut and peeled away, leaving the remaining layer to form a "window" for the illustration which will appear as a black "mask" on Orthochromatic photographic materials. Used by graphic artists, designers, and strippers to indicate placement and size of illustrations. *Related Terms:* masking film; peelable window method.

rule — (1) in printing, a printed line, or the brass material which makes it, usually specified by its arrangement and thickness or "weight" as hairline, 2-point, 6-point, parallel, etc. (2) in bookbinding, a stamping die used to form borders, panels, etc. (3) in general industry, a metal or other stable-based calibrating measuring device; a ruler. (4) in computing, rulers displayed on the screen of a layout or word processing program that show measurements in inches, picas or millimeters.

ruler — (1) a metal or other stable-based calibrated measuring device. (2) rulers displayed on the screen that show measurement in inches, picas, millimeters or other units appropriate to the software application. *Related Term:* line gauge.

ruling — (1) general term applied to the dpi rating of halftone and mechanical screen tints. (2) in layout and pasteup the process of placing dimension and activity lines on a sheet of layout paper or board, masking sheet or other substrate. *Related Term:* screen ruling.

ruling pen — a technical pen made especially for drawing precise straight or curved lines. *Related Term:* technical pen.

run — (1) in computing, the complete execution of one computer program, routine, or several routines linked to form an automatic operation. (2) in print press operations, the total activity to complete a printing order is called a press "run"; the total number of copies ordered or printed. *Related Terms:* machine run; press run.

run down — a reference to feeding a number of waste sheets through a press to remove the ink on the plate after all of the sheets for the run have been imaged.

runability — a statement of how well (or poorly) a given sheet behaves on press. The properties of paper that affect and describe its ability to run on press.

runaround — (1) generally, the ability within a program to run text around a graphic image within a document, without the need to adjust each line manually. (2) a term used to describe such type. *Related Term:* text wrap.

run-in — (1) in composition, emphasizing the first word or words in a paragraph by setting them in a different style, case and/or size of type. (2) in design, to set type with no paragraph breaks or to insert new copy without making a new paragraph. (3) in proofreading, a notation to indicate that the text should not be broken, as with a new paragraph.

run-in head — *Related Term:* run-in.

run-length encoding — a compression technique that reduces file sizes, especially for black-and-white or cartoon-style line art graphics. It works by replacing "runs" of the same color with a single character. The more runs there are, and the longer the run sequence, the greater the compression. *Related Term:* .rle.

running — (1) in production, a printing problem that occurs when the ink penetrates the paper and can be seen on the reverse side of the sheet. (2) generally, the condition of equipment while it is in operation, i.e. the press is running, etc. *Related Term:* show-through.

running copy — an infrequently used term describing the body text of an article or story.

running foot — information or a title that is repeated at the bottom of each page or a series of pages. Running heads or feet may or may not include folios.

running head — information or a title that is repeated at the top of each page or a series of pages. Running heads or feet may or may not include folios.

running sides — information or a title that is repeated on the outside margins of each page or a series of pages. Running sides may or may not include folios.

running text — *Related Terms:* running foot; running head; running sides.

running water bath — a tray bath used in tray film processing that washes away all traces of the chemicals used during development. Designed to prevent later darkening of the clear areas of the film. *Related Term:* wash.

RW — the abbreviation for ream wrapped.

RWhois — a relatively new protocol which extends and enhances the older Whois protocol in an effort to provide a scalable, decentralized, and efficient means of storing and retrieving information related to hosts, network information systems, and the individuals associated with those systems. RWhois uses

RWHOIS

the hierarchical nature of the information related to the network to provide the shortest and most efficient path between network data and the person who needs it. *Related Terms:* Whois; domain name system (DNS).

Ss

S/MIME (secure multipurpose Internet mail extensions) — a means to make e-mail messages more secure by adding both digital signatures and encryption. Using compliant e-mail packages, users can make sure that a message really came from the supposed sender (thanks to the signature), and that no one else could have read the message before it arrived (thanks to encryption).

S/N or S/NR (signal to noise ratio) — the ratio of the signal power to the noise power in a specified bandwidth, expressed in dbw.

S/N ratio — a db rating assigned to a particular audio or video signal indicating the signal strength expressed against the noise and interference in it. The higher the S/N, the more clear and better quality the signal.

S/S (same size) — an instruction to reproduce to the same size as the original.

S/T interface — the part on the ISDN modem to which all other ISDN devices connect in order to communicate. *Related Term:* U interface.

S-100 bus — the circuits that move information from a computer's CPU to the RS-232 port. The standard bus used in many computers.

saddle — the center of a printed and folded signature where it is stitched, stapled or bound.

saddle binding — *Related Term:* saddle stitching.

saddle sewing — the process of creating thread stitches along the entire length of the spine to secure signatures in a book.

saddle stitching — the insertion of two or three wires, or staples, inserted through the crease of the spine to hold the sheets together. The folded sheets rest on supports called saddles as they are transported through the stitching machine. Saddle-wire is the com-

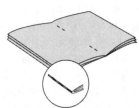

SADDLE STITCHING OR BINDING
A wire stitch or staple from the outside cover through to the saddle or center spread holds all elements securely. It is one of the most economical methods of binding for both large and small runs, but is somewhat limited in the number of pages that can be practically bound.

mon binding method for small booklets, brochures, and pamphlets and will accommodate 64 pages, and sometimes 96, if the paper is light enough. Usually limited to about 64 pages in size due to image creep. The cost is low and the pages will lie flat when the booklet is used. *Related Terms:* saddle wire; wire stitch; saddle-wired.

saddle wire — the wire used in saddle stitching.

saddle-wired — a form of binding that uses staple-shaped wires through the gutter fold. *Related Term:* saddle-stitched.

safelight — a darkroom light with a limited spectral output that inhibits it from exposing or fogging specific light-sensitive materials.

safety paper — paper designed to plainly show evidence of erasure, ink eradication or other tampering. Used for checks, stamps, tickets, stocks and other articles with fiduciary value. *Related Terms:* check paper; protective paper; security paper.

sales rep — a shortened version of sales representative. Any person who represents a firm and who is specifically charged with taking orders. *Related Term:* account representative.

same size (SS) — making a reproduction of size equal to the original copy. *Related Terms:* SS; enlargement; reduction.

sample book — *Related Term:* swatch books.

sampling — the first step in the process of converting from analog to digital. It involves measuring the value of the analog signal at predetermined intervals. The results are referred to as "samples." The sample values are then encoded in such a manner as to provide a digital representation of the original signal.

sandbox — in computer memory, a protected, limited area where Java-based are allowed to "play" without risking damage to the entire system.

sandwich — a short notice placed within the body of an article; often, but not always, with header and footer rules.

sans serif — literally, "without

SANS SERIF
Unadorned strokes, with without terminal adornments.

serifs." One of the six type classifications; is characterized by vertical letter stress, uniform strokes, and absence of serifs. For example, Helvetica, which lacks the small extensions on the ascenders and descenders. *Related Term:* serif.

SASE (self addressed, stamped envelope) — usually not pronounced as a word, but spoken as four letters. The descriptive term for a prestamped and addressed envelope supplied by a customer into which a vendor places information, or merchandise.

satellite relay — an active or passive satellite repeater that relays signals between two earth stations.

satin finish — (1) a slightly embossed finish on text paper. *Related Term:* dull finish.

saturated colors — strong, bright colors, especially reds and oranges, that do not reproduce well in video. They tend to saturate the screen with color or bleed around the edges, producing a garish, unclear image.

saturation — the attribute of color perception that expresses the amount of gray or purity of a color. The higher the gray content, the lower the saturation.

SC paper — an abbreviation for supercalendered paper.

SCA-FM (subsidiary communication authorization) — the electronic technique that places the radio signal on the FM spectrum. These signals can only be picked up with special tuners that distinguish the SCA from the FM signals. Often the method used for broadcasting background music used in restaurants, stores, or reader news for the sight-impaired, etc.

scale – to reduce or enlarge an image or a page. *Related Term:* scaling.

scaling — (1) to identify the percent at which artwork should be enlarged or reduced. The process is called "scaling" and can be accomplished by using the geometry of proportions, the use of a desk or pocket calculator, logarithmic scale, or a disc calculator called a proportion scale or wheel. (2) the act of using a proportion wheel to determine enlargement or reduction percentages. (3) the means within a program to reduce or enlarge the amount of space an image will occupy. Some programs maintain the aspect ratio between width and height while scaling, thereby avoiding anamorphic distortion. *Related Terms:* proportion scale; proportion wheel; scale.

scaling wheel — *Related Term:* proportional scale.

scallop — a page layout in which columns of unequal length are aligned at the top so their bottoms vary.

scamp — (1) in design, a sketch of a design showing the basic concept. (2) in production, the name for a once popular direct plate system.

scan — (1) in optical reading, the process of capturing an image of some object, usually a paper document for purposes of data storage. (2) the term applied to graphics when an electronic sensing device is utilized to digitize photographs of artwork for inclusion into pages being prepared for reproduction using digital make-up.

scan converter — a device that converts video frequency signals to audio frequencies and vice versa; used in freeze-frame video to transmit video signals over telephone lines.

scan pitch — in array scanning, the number of samplings or exposure lines per inch or per centimeter of copy.

scan rate — a measurement of scanning speed. Usually it is listed with two numbers separated by a slash, such as 100/1,200. The first number is the number of seconds needed to scan one inch. The second is the scan pitch. Thus 100/1,200 means that it will take 100 seconds to travel one inch along the original copy and it will sample the copy 1,200 times, making 1,200 lines of resolution. In the case of electronic dot generation, the rate depends on the speed of the output and the selected halftone ruling being used, i.e., 65 lpi ruling will output faster than a 200 line per inch scan.

scanner — (1) an electronic digitizing device that uses a light beam to examine color transparencies or reflective art and isolate each process color on an individual piece of film, or photographic separation, to be used in the reproduction process. (2) device used in conjunction with desktop publishing systems to scan line art, logos, photographs and convert them into bitmap graphics, or with optical character recognition (OCR) capabilities, typewritten or printed materials into word processing files. After the artwork, photographs, and text have been scanned into the system and stored on disk, all can be called up on the computer screen and manipulated using appropriate software and, finally, output to a desktop printer, or on film or paper from a high-resolution imagesetter. *Related Terms:* imagesetter; OCR; optical character recognition; flat bed scanner; drum scanner; electronic scanner.

scanner, matrix symbol — used in optical reading, they are designed to evaluate data from a two- dimensional area, locating each printed element in both *x* and *y* coordinates simultaneously. Because of this, CCD camera technology is currently employed in some of the available scanners.

scanning — the active use of a scanner for digital graphic creation.

S-CDMA (Synchronous Code Division Multiple Access) — a proprietary version of CDMA, it was developed for data transmission across coaxial cable networks. It scatters digital data up and down a wide frequency band and allows multiple subscribers connected to the network to transmit and receive concurrently. This method of data transmission was developed to be secure and extremely resistant to noise. *Related Term:* CDMA; wide-band broadcasting.

scene graph — the data structure defined in VRML. Developers can link external media such as graphic textures or sound files, and like HTML, the scene graph files are simple text files. *Related Term:* VRML.

schoolbook perforations — a perforation placed right next to the spine to facilitate easy page removal from a book.

score — (1) as a verb, the compression of paper along a straight line to weaken its fibers to facilitate accurate folding. It also helps prevent cracking of paper coatings at the fold point. *Related Terms:* crease; creasing; perforating. (2) as a noun, the indented line in paper that makes it easier to fold; the result of scoring.

scramble — the deliberate distortion of information to permit only authorized reception. *Related Term:* encryption.

scraper board — a board with a China clay surface, covered with India ink. Tonally reversed drawings are produced by scraping away the ink to expose the China clay surface. *Related Term:* scratch board.

scratch board — *Related Term:* scraper board.

screamer — (1) in publishing, a banner headline. (2) also, a slang term applied to the exclamation point punctuation mark. *Related Terms:* streamer; banger; exclamation point.

screen — (1) an alternate name for a halftone or tint screen. (2) the fabric used in screen printing. (3) the conversion of continuous tone images such as photographs into halftones or solids into tinted

values by use of either a halftone or mechanical screen. *Related Terms:* halftone screen; mechanical screen.

screen angle — the angle at which the halftone screen or halftoned copy are placed to avoid displeasing dot patterns (moiré) in reproduction. Black halftones are usually shot at 45 degrees from horizontal to assure regular dot patterns on the edges of a square-cut halftone. Deviation from this angle causes the saw-tooth effect sometimes seen on the edges of a halftone reproduction. In full color work, each individual color halftone has a specifically assigned angle for positioning the rows of dots in relation to a reference grid with horizontal and vertical lines. The most dominant color screened is positioned at a 45° angle to the reference grid to avoid moiré patterns.

screen base — serves as the printing surface and the frame hinging support for a hand-operated screen printing unit.

screen density — the amount of ink expressed as a percent of coverage that a specific screen allows to print.

screen determiner — a device for accurately determining the screen ruling of a previously printed piece, a screened print, or other fixed-frequency materials.

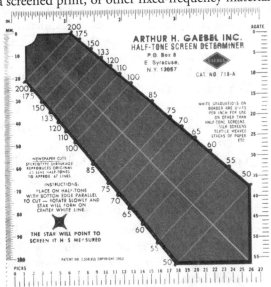

SCREEN DETERMINER
A scale device which, when laid over a halftone reproduction, will develop a controlled moiré pattern which cooincides with the ruling value on the determiner. Where the pattern developed intersects the nmber of the edge of the ruled lines, it reveals the true screen ruling or lpi of the reproduction.

screen fabric — the porous material that is tightly stretched over a screen printing frame and which provides the base to hold the stencil.

screen frame — the rigid device, wood or metal, that holds the fabric tight in screen printing operations.

screen frequency — the number of lines or dots per linear inch, horizontally and vertically, on a halftone screen or mechanical tint screen. *Related Terms:* screen ruling; halftone; vignette; dpi; lpi.

screen geometry — a general term describing a monitor's ability to reproduce various shapes accurately and without distortion. Monitors are subject to numerous problems with geometry, including pincushioning, trapezoid error, rotation (or tilt) error, and inadequate linearity. *Related Terms:* linearity; pincushioning; rotation error; trapezoid error.

screen indicator — a ruled or lined device used to determine the ruling of any screen used in halftone or mechanical screen reproduction. *Related Term:* screen determiner.

screen mesh count — the measure of the number of threads or openings per linear inch of a screen printing image carrier.

screen pattern — the shape of dots produced by a halftone screen. Some common shapes are square, round, elliptical and linear. Different shapes cause different effects in the final ouput. Some computer manipulation software, as well as drum scanners, allows specification of screen types. Adobe PhotoShop allows a diamond dot which is now supported on some imagesetters.

screen percentage — the dot area of a screen. A 20 percent screen will produce a 20 percent ink coverage on the press, etc.

screen press — the equipment that holds individual screen frames, one for each color; usually designed to rotate around a central axis and move each screen into position sequentially over a stationary object.

screen printing — a member of the porous classification of printing, it is a method of producing printed images by forcing ink through a hand or photographically prepared stencil that is attached to screen fabric.

screen ruling — *Related Term:* screen frequency.

screen tint — an image area printed with dots so ink coverage is less than 100 percent and simulates shading or a lighter color. Tints are films with percentage openings rated by their approximate printing dot size value, such as 20-, 50-percent, etc. When placed in direct contact with a plate or other receiver and covered with a normal lithographic negative, the openings in the negative will take on the characteristics of the screen, i.e., a 20 percent shading, etc. *Related Terms:* screen tone; mechanical screen tint.

screen translation/screen reformatting — in optical mark reading, making an individual screen of an existing software package look different, usually for the smaller display of a hand-held device.

screened print — an positive image with a halftone screen made from a halftone negative or by diffusion transfer, usually associated with mechanical paste-up. It will be photographed as a line shot along with the type and line art on the page. *Related Term:* Velox.

screening — (1) in process camera and prepress, the process of converting a continuous-tone photograph to a matrix of dots proportional in size to the highlights and shadows of the continuous-tone image. Screening is accomplished photographically by imposing a halftone screen directly in front of the photographic emulsion that will receive the screened image. A similar effect can be accomplished by electronic scanning and computer manipulation. (2) in gravure printing, the objectionable screen pattern that appears in the solids if cylinder or plate cell walls are excessively shallow or wide.

screw and post bind — a bindery system using screws which screw into threaded steel posts which pass through both the front and back cover. All included materials must be punched or drilled with holes which coordinate with the position of the posts.

scribe — the process of scratching the emulsion side of a negative or positive to make it transparent and able to pass light. The effective opposite of opaquing. The process is sometimes used to correct broken type or rules.

script — (1) in computing, a type or computer code string that can be directly executed by a program that understands the language in which the script is written. Scripts do not need to be compiled into object code to be executed. (2) in typography, one of the six major typeface classifi-

SCRIPT TYPE
The letter forms are called cursive, but whenever the individual letters are connected, as in handwriting, the process itself is called incursive writing.

cations. One in which the letters are modeled to resemble handwriting; non-joining letters are called cursive script. Scripts come in a variety of weights of formal and informal styles.

scroll — (1) historically, an early book form made from a strip or roll of animal skins, bark, papyrus, etc., often fastened at each end to a wooden stick onto which the material was rolled for storage or transport. (2) in computing, to move a story up or down on a video display terminal screen, usually for editing purposes. (3) on the Internet, the act of moving a browser's scroll bar to see what else is on a Web page other than what appears in the initial screen.

scroll bar — a device that helps the operator navigate through a document on a computer or workstation screen by selectively revealing certain vertical and horizontal sections of the much larger image. *Related Term:* scroll.

scrolling text box — a labeled, multiple-line form field in which users can type one or more lines of text.

SCSI (Small Computer System Interface) — an industry-standard interface between computers and peripheral device controllers. It provides high-speed access to peripheral devices and has gone through several iterations, i.e., fast SCSI, wide SCSI, etc. Pronounced "scuzzy."

scuff resistance — *Related Term:* abrasion resistance.

scuffing — undesirable print abrasions caused by surface wear or rough handling. Particularly problematic in packaging, scuffing may be minimized with scuff-proof inks, varnishes, and/or other coatings.

scum — nonimage areas on a lithographic plate that, during the production run, become sensitized and begin to take ink.

SD (send data) — a modem light that flashes off and on during data transfers, telling you that the modem is sending signals to a remote computer.

SDK — a technology that significantly improves the performance of removable media storage peripherals such as CD-ROM drives. The traditional system of connecting directly to the PC via an EIDE interface is replaced by connecting peripherals directly to the hard drive through the SDK interface.

SDRAM (synchronous dynamic RAM) — a new feature set for DRAM designed to keep up with the 66-100 MHz bus speeds used on newer Pentium II systems. It incorporates new features that allow two sets of memory addresses to be opened simultaneously. Data is retrieved alternately from each set, thus eliminating delays that normally occur when one bank of addresses must be shut down and another prepared for reading during each request. *Related Terms:* DRAM; EDO RAM; RAM.

search and replace — the process of automatically locating a specific word or symbol in a word processing or page layout file and replacing it with another word or symbol.

search engine — on the Internet, a mechanism for finding things. Popular search engines include Yahoo, Lycos, Alta Vista, and many others.

seasoning — acclimation of paper or other substrate for a period of time, to adjust humidity and temperature, prior to presswork. *Related Term:* condition.

SECAM (Sequential Couleur avec Mémoire) — the television broadcast standard in France, the Middle East, and most of Eastern Europe, it broadcasts 819 lines of resolution per second and is one of three main television standards throughout the world. *Related Terms:* NTSC; PAL.

second coming type — newspaper talk for the largest type which can fit on the front page. *Related Terms:* screamer headline; banner headline.

second generation plate — a duplicate printing plate made from a mold or pattern taken from an original plate. *Related Terms:* stereotype; electrotype; duplicate plate.

second level domain — that portion of the domain name that appears immediately to the left of the top level domain. For example, the prof-lyons in prof-lyons.com. Second level domain names are often descriptive and have come to be used increasingly to represent businesses and other commercial concerns on the Internet. *Related Terms:* domain name system (DNS); top level domain.

second pass/dry pass — the extra passage of a sheet through the press for additional color impressions or coating applications.

second serial rights —rights granted to publish a serialized version of a work after it has already been published.

secondary cache — a technique of caching data in fast, pricey memory, to speed up system performance. It is bigger than the primary cache which is usually in the same chip as the central processing unit, and fits between it and main memory (RAM). It's faster than main memory, but slower than primary cache memory. Typically it is around 256K in size.

secondary color — a hue (color) produced by mixing two primary colors. *Related Term:* complementary colors.

secondary leading — a separate leading parameter that may be specified by a typesetting operator to generate a particular amount of leading between paragraphs, as opposed to primary leading, which appears between lines of the main text.

secondary mailbox — an additional mailbox customers can add to their accounts.

secondary waste — fragments of finished products from a manufacturing process. For example, secondary waste includes printer's trimmings.

seconds merchant — *Related Term:* job lot merchant.

section — a printed sheet folded to make a multiple of pages. *Related Term:* signature.

section mark (§) — a character used at the beginning of a new section. Sometimes also used as a footnote or corollary symbol.

secure online payment system — a system that allows Network Solutions customers to pay for their domain name registration and renewal (re-registration) fees 24 hours a day, seven days a week, by entering their credit card information directly via the Web. The system provides a fast and secure method of payment and requires that the customer have a browser that will support secure socket layer (SSL). *Related Terms:* registration fee; renewal fee.

secure socket layer (SSL) — a low-level protocol that enables secure communications between a server and a browser.

security certificate — digital data, often in the form of a text file, used by the secure socket layer protocol to establish a secure connection. Contains information about who it belongs to, a unique serial number or other identification, who issued it, the term of validity and some form of encrypted identifier that can be used to verify contents of the certificate and the transaction.

security paper — paper incorporating special features (dyes, watermarks, etc.) for use on checks to prevent forgery or alteration. *Related Term:* safety paper.

segmented file — a file in a record-oriented database that is stored in as many segmented areas as there are fields. For example, if a record consists of company information, the company name, street address, city, zip code, officers, products, each of these, and many other fields, may be stored in segmented files and keyed to each other by a record number. This is a classical file structure of a relational data base which allows more flexible searching and output, but requires more time to prepare and verify.

selection bar — an unmarked column along the left edge of the FrontPage Editor window that is used to select text with the mouse.

selector — the string used to identify a declared style value in cascading file sheets. It is attached to objects on the page (either as an HTML tag or as an attribute of a tag), and it determines what style the HTML elements receive. *Related Terms:* cascading style sheets; contextual selector; HTML.

selenium cell — an electronic element which varies its electrical voltage output in proportion to its intensity of illumination.

self cover — a cover made of the same stock or paper used in the rest of the book. Used for booklets or pamphlets when the cover stock does not have to be particularly strong, or to save the cost of the extra materials and operations required to produce and bind a cover.

self mailer — a printed piece designed to be mailed as a stand-alone without an envelope.

self publisher — a publisher who is also the writer of the publication. *Related Term:* vanity press.

self-checking — a bar code or symbol using a checking algorithm which can be independently applied to each character to guard against undetected errors. Commonly used in bar coding, but also a technique used to verify or authenticate such numbers as ISBN numbers, etc.

self-copy paper — *Related Term:* carbonless paper.

semichemical pulp — the combination of both chemical and mechanical pulp in the papermaking process.

semicoated carbon paper — a carbon paper coated one side.

semiconductor — a class of materials that allow electrical current to flow through them under certain conditions. Semiconductors are used to create common electronic components, such as diodes and transistors. *Related Terms:* diode; transistor.

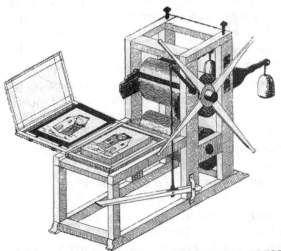

SENEFELDER'S ORIGINAL LITHOGRAPHIC PRESS

Senefelder, Alois — inventor, in 1798, of lithography and a number of related parts, including inks and image transfer systems. He said that the most important invention, to him, was his method of transferring existent images to a stone for reproduction. He had a long career as chief lithographer to the court of Austria.

sensitivity guide — a small, calibrated, continuous-tone quality control device used for establishing and maintaining proper exposure and processing conditions in cam-   era, proofing and

12-STEP CAMERAMAN'S SENSITIVITY GUIDE
A precise photographic control device with steps of increasing density. When photographed with the original copy, it provides a known point for determining exposure and development times.

platemaking operations; also used in one method of direct color separation. *Related Terms:* step guide; cameraman's sensitivity guide; photographer's sensitivity guide.

sensitivity guide method — a method of controlling the development stage of film processing through use of a cameraman's sensitivity guide, which is photographed along with the original copy. As development progresses, steps of the guide turn progressively black. The negative is removed from the developer when the desired step is totally developed. *Related Terms:* step guide; step wedge;

cameraman's sensitivity guide; gray scale; inspection development.

sensitization — making the sensitive layer of a photopolymer plate very sensitive to ultraviolet light. *Related Term:* carbon dioxide conditioning.

sensitize — (1) to make the image areas of a printing plate more ink receptive. (2) the application of a diazo coating to a wipe-on offset plate. *Related Term:* diazo process.

sensitizer — the light-sensitive liquid powder or crystal concentrate material that is mixed with the base material to form the ready-to-use emulsion for direct photo screen printing.

sensitometer — a calibrated instrument that created precisely controlled sensitivity guide exposures onto light-sensitive materials.

separate entry typesetters — a somewhat obsolete system where the typesetter is operated primarily by punched paper tape, magnetic tape or electronic signal. Operators on separate keyboard units produce the tapes and signals that direct the typesetter to compose the phototype on paper or film. After developing, the type is ready for paste-up as camera copy. *Related Term:* front-end phototypesetter.

separation — the film negative or positive to be used for making a plate for a single color of ink from among several colors to be printed. *Related Term:* color separation.

separation filters — red, green, and blue filters used on camera to create individual color printers during color separation. Each one transmits about one-third of the spectrum.

sepia —a pale reddish-brown toner used in photography to lend an antique flavor to a print.

serial — a publication, such as a series of books, or single article parts issued at regular intervals.

serial line Internet protocol (SLIP or CSLIP) — a protocol used to run IP over serial lines, such as telephone circuits or RS-232 cables, interconnecting two systems. *Related Terms:* point-to-point protocol.

serial port — the communications port on a computer, also called the COM or RS-232 port. It's called serial because, although it has nine pins, the computer sends data on only one wire and receives data on one other wire. All the data bits have to follow one another on the single wire, as opposed to the parallel port, where eight separate wires transfer each bit of a byte. *Related Terms:* COM port; parallel port.

serial transmission — data transmission that sends one data-bit at a time.

series — the face variations within a family of type; commonly bold, extra bold, condensed, thin, expanded, and italic.

serif – (1) one of three variables in alphabet design; refers to small strokes that are, usually, a perpendicular line found at the end of the unconnected or finishing stroke. Used to provide visual balance to the character shape. Serifs may vary in weight, length, and shape, and contribute greatly to the style of the typeface. (2) one of the six classes of type, the others being sans-serif, Gothic, square serif, script, and novelty or decorative. *Related Term:* sans serif.

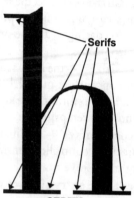

SERIFS
The extensions at the ends of the main strokes of letters. Often resembling "feet," they can be of any weight and may or may not have connecting brackets.

serif type — any type with serifs.

serif, square — one of six basic classifications of type wherein the serifs are often the same weight, or heavier, than the main strokes. *Related Term:* square serif.

serigraph printing — printing done by any porous process which utilizes a woven fabric, plastic or metal that allows ink to pass through. Typified by screen printing and mimeography.

serography — *Related Term:* screen printing.

server — (1) the major part of a client/server setup, it is usually a computer that provides the information, files, Web pages, archives and other services to clients as they log in. (2) also the software and operating system designed to run server hardware. *Related Term:* client; host.

server-side image map — an image map that passes the coordinates of the cursor to a CGI handler routine on the server. Server-side image maps require your server to compute the target URL of the hyperlink based on the cursor coordinates.

server-side include — a feature provided by some Web servers that automatically inserts text onto pages when they are given to the browser.

service bureau — a business that specializes in outputting high resolution paper and film (from computer flies) on laser imagesetters as well as scanning and related services. It often caters exclusively to the graphics communication trade and does not solicit retail clients.

session — a complete connection between a user agent and a server where information is exchanged between the two computers. *Related Terms:* cookie; HTTP.

set – (1) the ability of ink to stick to paper; properly set ink can be handled without smearing. (2) an inappropriate alternate name for set width.

set size — the width of the type body of a given point size.

set solid — type set without extra leading (line spacing) between the lines. Most often type is set with extra space, e.g., 9 point set on 10 point (the type + 1 point of extra leading).

set width — also called set. (1) in metal type, the width of the body upon which the type character is cast; the distance across nick or belly side of foundry type. (2) in digital typesetting, the width of the individual character, including a normal amount of space on either side.

set-off — the condition that results when wet ink on the surface of the press sheets transfers or sticks to the back of successive sheets in the delivery pile. Sometimes inaccurately referred to as "offset."

sew — to use thread to fasten signatures together at the spine of a book.

sewing frame — a device used when sewing signatures together in hand case binding.

sewing tape — a nonadhesive strong fabric tape used to help hold sewn signatures together in hand case binding.

sewn binding — a bookbinding method that uses thread to fasten sheets and signatures of paper together.

SGML (standardized general markup language) — an ISO markup language for representing documents on computers. HTML used on the Internet is based on SGML concepts.

SGRAM (synchronous graphics RAM) — the speed-enhancing features of SDRAM plus graphics capabilities that enhance 3D graphics performance. It can work in sync with system bus speeds up to 100 MHz. *Related Terms:* DRAM; EDO RAM; RAM; SRAM; SDRAM; WRAM.

shade — a darker hue obtained by adding black to a color.

shaded — a typeface that gives a gray instead of solid black imprint by having a textured face.

shading — (1) in general printing, the introduction of a screen tint into art to create ink values of less than 100 percent. (2) in 3D graphics, the process of coloring the polygons. The techniques for doing it are often more complicated than just flooding a shape with a single color. *Related Terms:* screen tint; flat shading; Gouraud shading; Phong shading.

shadow — (1) a general term customarily applied to the darker areas of a photograph, halftone negative or a printed halftone in which densities are above 76-80 percent full value. (2) a dark outline on one or more sides of an illustration or type which create a dimensional effect. *Related Term:* drop shadow.

shadow area — *Related Term:* shadow.

shadow dot — the larger dots which appear in shadows and denser areas of halftone reproductions.

shadow mask — in monitors, a metal plate with holes in it that focuses the beams from the electron guns at the back of the CRT. The distance between these holes is called the dot pitch. *Related Terms:* CRT; dot pitch; electron gun.

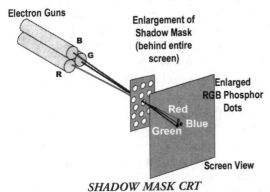

SHADOW MASK CRT
A picture tube with a perforated grill against the inside face of the screen. The mask is used to control precisely which phosphors will be struck by a specific electron gun. In the case of color, it prevents one color electron gun from activating adjacent phosphors and causing untrue color.

shaft — the steel core or center support of a press cylinder or roller.

shareware — copyrighted software that is tried and used on the honor system. A wonderful alternative to commercial software, it is available from centralized archives on the Internet and local bulletin board systems. Usually, trials are free for a specified time period; if you continue to use it, you're expected to register the program

and pay a fee to its developer. Some programs are partially disabled, stop working after a set period of time, or contain "nag screens" that pop up frequently to encourage you to register, but fees are usually no more than $50, and some selfless developers ask only that you send a postcard letting them know you like their product. Registering often gets you full documentation or free software updates. Software that doesn't involve a fee is called freeware. *Related Terms:* freeware; public domain.

sharp — the defining characteristic of an image at optimum focus.

sharpen — in electronic imaging, the heightening of the contrast between the dark and light tones of an image to give the appearance of greater resolution. *Related Terms:* dithering; focus.

sheet — (1) a single piece of paper. (2) in poster work refers to the number of double-crown sets in a full size poster, i.e., 8, 16, 32 sheet, etc. (3) in production, a reference to the paper being printed, i.e., custom sheet, house sheet, etc.

sheet off — the process of removing excess ink from ink rollers by carefully hand rolling a sheet of paper through the ink system and then removing it.

sheet separators — elements used in the press feeder unit to ensure that only one sheet is fed into the registration unit at a time.

sheeter — a (usually) inline device on a web press which cuts printed produce into flat sheets at press speed.

sheet-fed — *Related Term:* sheet-fed press.

sheet-fed press — printing press that feeds and prints on individual sheets of paper (or other substrate). Some sheet-fed presses employ a roll feed system in which rolls of paper are cut into sheets before they enter the feeder; however most sheeted presses forward individual sheets directly to the feeder.

sheetwise — a method of printing a two-sided signature where half of the required number of copies is printed on the first side. The pile is turned upside down and the remainder is printed on the second side. Sometimes called work and back. *Related Terms:* work and back; work and tumble; work and turn.

shell — the basic user interface to an operating environment.

shift key — (1) originally, a function key on manual typewriters that shifted the position of the carriage, causing the upper portion of each key's hammer, the part that contains the capital letter, to strike the paper through the ribbon. (2) in computing and digital imagesetting, a key on the keyboard which performs a similar function, that of creating capital characters, but also can be used in combination with other auxiliary keys to perform tasks such as some formatting functions.

shingling — (1) an arrangement where a series of sequenced sheets is jogged even at one edge and the (usually) opposite edges are trimmed in such a manner as to have each successive sheet slightly longer than the previous one. Often done to expose different subject titles appearing on each sheet. (2) the allowance made during paste-up or stripping to compensate for creep. *Related Terms:* image creep; image shift; saddle binding.

shipping and receiving — applications that keep track of all goods that enter or leave your facility. *Related Term:* inventory control.

shipping container symbol — the 14-digit number applied to intermediate packs and shipping containers containing UPC marked items. It is always encoded in the interleaved 2 of 5 symbology.

shore durometer — a device designed to measure rubber hardness in units called "durometers"; the lower the rating, the softer the rubber.

short grain — *Related Term:* grain short paper.

short ink — an ink that does not flow freely; a thick ink such as used in offset lithography. Related Term long ink.

short lens – *Related Term:* wide angle lens.

short rate — in commerce, a reduction in the suggested retail price of less than 40 percent.

short run — generally, a print run of less than 10,000.

short stop – a mildly acetic solution designed to neutralize residual alkaline developing agents on film or paper, at the conclusion of the developing step. Generally made of Glacial Acetic acid and water. Commercial preparations have color indicators and other chemical enhancements. *Related Term:* stop bath.

shot — a single exposure made by a camera.

shoulder — (1) in photography, the area on a characteristic (DlogE) curve which indicates that further exposure will make no difference in density. The area of maximum density. (2) in composition, particularly hot metal, the area around a character which gives it distance from an adjacent letter.

shouting — chat and Usenet posters will often tell other chatters and users to "stop shouting." It's another way of saying, "TURN OFF YOUR CAPS LOCK!" In general, chatting and posting in uppercase is considered rude.

showcard — *Related Term:* poster board.

show-through — (1) an undesirable effect in general printing, when printing on one side of the paper can been seen from the other side under normal lighting conditions. (2) in optical reading, the generally undesirable property of a substrate that permits underlying markings to be seen. *Related Terms:* opacity; print-through.

shrink wrap — a method of tightly wrapping a package in heat-reactive plastic film, and then passing it through a warm air tunnel which causes the plastic to shrink to tightly fit the wrapped product.

SHTTP (secure hypertext transfer protocol) — a protocol developed to keep your money safe when used in commercial transactions on the Internet.

shutter — the mechanism that controls the duration of the passage of light through the camera lens and aperture by activating one or more overlapping blades which inhibit the light path when not activated, i.e., the positions are normally closed, and open only during exposure.

shutter priority camera — a type of camera where the photographer sets the shutter speed or duration and the camera computer automatically adjusts the aperture, based on the set film speed. *Related Term:* camera, shutter priority.

shutter speed — the length of time the shutter is in the open position.

side bearing — the nonprinting area around a character, particularly before and after. The space which prevents leading and following characters from touching. Incursive script faces have nonexistent or very small side bearings.

side binding — the technique of fastening pages or signatures by passing wire stitches or staples through the product edge (usually on the left or top margin). *Related Term:* saddle binding; saddle stitching.

side guide — a device that serves as the third point of a three-point register system (including the front guides) on the feed board. Side guides move the sheet sideways just prior to entering the front guide and grippers to facilitate consistent register.

side heading — subheading set flush into the text at the left or right margin.

side sewing — a method of sewing used in case binding. All folded signatures for one book are sewn at once through the whole book, parallel to the folds. It produces a stronger binding than does Smyth sewing.

sideband — on Internet web pages, a vertical bar positioned on the side of the screen. May contain instructions, additional menu items, advertise-ments, etc. *Related Term:* sidebar.

sideband addressing — a feature of AGP. Sideband addressing provides additional channels for transmitting data requests between the graphics processor and the system. *Related Term:* AGP.

sidebar — (1) a block of information related to and placed near an article, generally on the outer columns. (2) a vertical bar usually positioned on the right-hand side of a Web page screen.

side-stabbed or -stitched — the folded sections of a book are stabbed through with wire staples at the binding edge, prior to the covers being drawn on.

side-wire stitching — a method of binding that is sometimes used when the number of pages is too great for saddle wiring or stitching. Staples are placed through the side or edge of the lifts of paper. The advantage is low cost as well as great strength, but since the wires are placed about 1/4 in. in from the spine, the binding is "tight," and pages will not lie completely flat. The page layout must be planned in advance to provide a generous inside margin. *Related Terms:* side-stabbed; side binding; saddle stitching.

SIDE WIRE STITCHING
A wire stitch, similiar to that used in saddle stitching, is used to pierce flat sheet piles through the cover, all sheets and the back. It can be used for very large page count documents.

SIG (special interest group) — people with a common interest who meet or exchange e-mail messages on a particular topic in an organized way.

sign paper — a rigid, water-resistant paper for outdoor signs.

signature — (1) a letter or figure printed on the first page of each section of a book and used as a guide when collating and binding. (2) the reference to all of the pages on both sides of a given press sheet after folding for inclusion in the binding operation. (3) one or more printed sheets folded to form a section of a book or pamphlet. (4) the personal mark made by clients on

contracts to indicate mutual agreement. (5) the three- or four-line message at the bottom of an e-mail message or Usenet news article identifying the sender.

signature imposition — the process of passing a single sheet through the press and then folding and trimming it to form a portion of a book or magazine.

signed applet — an applet whose source and integrity are guaranteed by its author. This is done by attaching a digital signature to the applet that indicates who developed the applet, when, and whether it has been tampered with since that time.

silhouette — (1) originally a 100 percent black subject with no midtone or highlight detail against a 100 percent white background. (2) now, any art in which the background has been removed. *Related Terms:* outline halftone; silhouette halftone.

silhouette method — one technique of preparing artwork for photosensitive screen printing.

silhouette/outline halftone — a halftone in which the original background has been removed so that the subjects may be printed on another background or plain paper.

silicon — the semiconductor material used in the manufacture of integrated circuits and microchips.

silicon chip — the microchip.

silk screen printing — an outdated and obsolete term. *Related Term:* screen printing.

Silurian paper — another name for tinted fiber paper such as that used in U.S. Treasury notes. *Related Term:* tinted fiber paper

silver bromide — the most common light-sensitive salt used in emulsions, either alone or with iodides and chlorides.

silver halide — a silver salt such as silver chloride, silver bromide, and silver iodide suspended in gelatin which serves as the light-sensitive emulsion of photographic film.

silverless film — a type of film that uses an emulsion containing light-sensitive compounds other than silver. The most common alternate sensitizer is diazo compounds.

silverpoint drawing — a drawing done with a metal pencil on special paper.

SIMM (single in-line memory module) — the most widespread form of RAM available. Generally available in 1 MB to 32 MB configurations.

simple mail transfer protocol (SMTP) — a protocol used to transfer electronic mail between computers. It is a server to server protocol, so other protocols are used to access the messages. *Related Term:* post office protocol.

simplex — a communications system capable of sending information in only one direction.

simulation — to artificially induce activity, motion, presence, etc.

simultaneous edition — two or more different versions of a book published at the same time, such as foreign language editions, etc.

simultaneous submission — a manuscript or query letter that is sent to more than one publisher for consideration at the same time.

single color press — a printing machine which is capable of printing only a single color of ink for each sheet pass. Multiple colors can be printed by reprinting the sheets after inks and plates have been changed.

single copy order — the description of a bookseller order for only one copy of a book.

single-edge razor blade — a tool used to cut and trim materials during paste-up. *Related Term:* X-acto knife.

sinkage — (1) a point below the top margin of a page where chapter openings or other materials are set. (2) the extra white space at the top of a chapter opener.

SIT — files created by the StuffIt compression and decompression software. Files with the .sit extension require a decompression program to open. Although such files are typically compressed using Macintosh software for other Mac users, they can also be opened using some PC programs. *Related Term:* zip.

site license — the license to use and possibly copy software which is granted to an organization, as opposed to an individual. It presupposes multiple users, machines, etc., and is structured accordingly regarding charges, terms, etc.

site or Web site — a specific location or place on the Internet or World Wide Web and the all-encompassing body of information as a whole, for a particular domain name. A site is a place made up of all of these Web pages. Sites can also refer to an FTP or archive locations which are directories on remote computers which have been set up to allow users to log in and retrieve or upload files.

six-color — a printing process that uses six different colors, for example, the standard four-color process

inks (cyan, magenta, yellow, and black) plus two spot colors and/or metallics or fluorescents.

sixteen-sheet (16-sheet) — a poster set of 16 pieces whose individual sheets measure 120 in. x 80 in. *Related Terms:* 8-sheet; 32-sheet.

six-up — The imposition of six items to be printed on the same sheet in order to take advantage of full press capacity and minimize paper consumption. *Related Terms:* four-up; eight-up, etc.

size — a solution based on starch, alginate, casein or other materials which is applied during paper manufacture to reduce ink absorbency, lessen the ability to absorb moisture, and therefore is better able to hold its shape.

size handle — the black rectangle displayed on a selected form field or hotspot. When you select a size handle, the cursor becomes a bidirectional arrow. Click and drag a size handle to reshape the field or hotspot.

size, basic — the sheet size, in inches, used to define basis weight for a particular paper. *Related Term:* basis weight.

sizing — (1) in papermaking, the process of treating paper with chemicals to impart resistance to water, oils, and other fluids; seal down its surface fibers; and improve its surface strength. (2) in bookbinding, an adhesive used to apply gold leaf or another color to book covers. (3) in graphic arts photography, scaling a photograph for reproduction. *Related Terms:* scale; size.

skeleton black — the black printer in full color printing, used to enhance detail, particularly in the shadow areas.

skew — rotation of a bar code symbol about an axis parallel to the symbol's length.

skewed indent — an indent whose value is changed for each line, giving the margin a "slanted" appearance.

skid — a shipping and selling unit of approximately 2,500 pounds of paper, in sheets. *Related Term:* pallet.

SKU (stock-keeping unit) — a reference to the number associated with a particular item included in a data base.

skyline — a headline or story at the top of the first page of a publication, above the nameplate. *Related Term:* banner headline.

slack sized — a paper specially treated with relatively small amount of size because it doesn't need to be especially water resistant, as compared to hard sized. *Related Term:* soft sized.

sleeve — the wrapper placed around publications to protect them during delivery. Very often made of kraft paper.

slick paper — *Related Term:* gloss paper.

slide — a projection transparency in its mount.

slide scanner — a scanner that can only utilize transmission or transparency images, as opposed to reflective or opaque images. *Related Terms:* flat bed scanner; drum scanner.

SLIP (serial line Internet protocol) — a standard for connecting to the Internet with a modem over a phone line. Higher level protocols like PPP provide vastly superior error correction. *Related Term:* PPP.

slip sheet — a blank sheet placed between newly printed product sheets to prevent setoff or scuffing between sheets during handling and shipping.

slipcase — an open-sided box designed to hold a single book or a series of periodicals, which are placed with the spine toward the outside. They serve as protective covers for their contents as well as facilitate handling in the case of periodicals.

slipsnake — a fine, abrasive stone or hard rubber eraser used to remove images from (usually) planographic plates.

slit —(1) a straight cut in paper made with a slitting disk or wheel. Often done at the delivery end of a printing press to cut larger sheets into smaller ones prior to reaching the delivery pile; also, at the paper mill when log rolls are cut to press size and rewound. (2) the cutting of small slots within a sheet to facilitate insertion of business cards, brochures or other collateral matter.

slow film — film that requires a relatively large amount of light (long exposure or large f/stop aperture) to record image. Generally characterized by having very small "grain" and yielding very sharp images at optimum focus. Most process camera films are of this type. A film with a relatively low ISO or ASA number.

slow scan video — (1) a device that transmits and/or receives still video pictures over a narrowband telecommunications channel; (2) may refer specifically to a still frame video unit that accepts an image from a camera or other video source one line at a time, necessitating that the subject in front of the camera remain stationary for a number of seconds.

SLR (single lens reflex) — a camera configuration that uses a system of mirrors to allow viewing of the subject through same lens that focuses the image on film. *Related Term:* camera, single lens reflex.

slug — (1) in hot metal composition, a complete line of type cast in a single piece of metal. (2) in publishing, the single tag line that refers a newspaper reader to a story that is continued on another page. (3) in gravure, the term used to describe cylinder cells that are printing blurred or unclear. (4) in hand composition, strips of metal, generally 6 or more points thick, used to space and secure lines of foundry type. (5) in general, a short phrase or word that identifies an article as it goes through the production process; usually placed at the top corner of submitted copy.

slug caster – slang term for machines used for casting complete lines of relief type. *Related Terms:* Intertype; Linotype; Elrod.

slur — Undesirable phenomenon that may occur during printing when halftone dots become slightly elongated. Usually a mechanical adjustment problem. Caused by paper, plate or blanket slipping during the impression of the image.

slurry — *Related Term:* furnish.

slush pile — the unsolicited manuscripts and query letters accumulated by a publisher.

small caps — alphabet sets in which a smaller version of a typeface's uppercase letters are cut to approximately the x-height of the lowercase letters and used in their place. Traditionally used for letters after drop caps, and for display type purposes. These are small caps. Abbreviated sc.

small format camera — any camera making negatives 35 mm or smaller. *Related Terms:* camera, large format; camera, medium format; camera, small format.

smashing — a step in the case binding method in which the entire group of sewn signatures is pressed together to ensure a consistent book thickness.

SMATV (satellite master antenna television) — a distribution system that feeds satellite signals to a hotel, motel, apartment complex, etc.

SMIL (synchronized multimedia integration language) — a June 1998 standard for delivering multimedia including synchronized animation, graphics, text, video, and audio content over the Internet. Using the standard's specifications, developers can use audio and video together as enhancements to online presentations.

smiley :-) — the official term for this is "emoticon." They are used to convey emotion and bring an extra nuance to an ASCII world. The idea is to tilt your head to the left 90 degrees and it looks like a smiley face. There are an endless number of variations on the theme. *Related Terms:* emoticon; ASCII art.

smooth finish — the most level finish on offset paper.

SMTP (simple mail transfer protocol) — a protocol that regulates what goes on between the mail servers. When mail is exchanged on the Internet, it is what keeps the process orderly. *Related Terms:* IP: NNTP: TCP/IP.

Smyth sewn — a method of sewing in case binding in which signatures are sewn through the center fold and then all the sewn signatures are sewn together to form a single book. *Related Term:* side sewing.

snail mail — a term used by e-mail zealots to describe the regular paper-based mail service; a reference to the differences in speed between e-mail and regular mail, which seems a lot slower by comparison. *Related Term:* e-mail.

snap fitter and dowel method — *Related Term:* pin register.

snap set — bound multipart forms which are glued on one edge, with a perforation. Individual parts can be removed by the perforation, or the entire set can be disassembled in the same manner.

snap to (guide or rules) — a WYSIWYG program feature for accurately aligning text or graphics. The effect is exercised by various nonprinting guidelines such as column guides and margin guides, which automatically place the text or graphics in the correct position when activated by the mouse. The feature is optional and can be turned off.

sodium thiosulfate — the prime chemical used to fix the image on a photographic film after developing. *Related Terms:* fixer; hypo.

soft bind — *Related Term:* perfect bind.

soft copy — copy in computer memory or on disk, not on paper ad a hard copy. The digital version of a manuscript or typeset copy before it is printed to paper.

soft cover — a book cover made from heavier paper or paper fiber material with greater substance than that used for the body of the book but with much less substance than binder's board. Bound without a case; usually perfect bound, but also sewn and bound with a paper cover. *Related Terms:* paperback; paperbound.

soft dots — a type of dot in a line screen negative whose edge is not smooth or circular due to excessive edge halation. This can create a fuzzy image. By contrast a hard dot has a very smooth edge.

soft ink — ink that has the consistency of lithographic inks.

soft or discretionary hyphen — a specially coded hyphen which is only displayed when formatting of the hyphenated word puts it at the end of a line, and that will not remain if the text happens to reflow. *Related Term:* word wrap.

soft proof — a proof of an image or page layout on a computer monitor, as opposed to a hard proof, which is on paper of some other tangible substrate.

soft return – a return inserted in a word processing text by the software, as opposed to hard returns, which are inserted by the operator when using the carriage return key. Soft returns are usually associated with a "word wrap" feature of the word processing software and will not remain if the text happens to reflow.

soft sized — *Related Term:* slack sized.

soften — decreasing the background areas around an image to enhance its presence. Usually makes the image appear to be closer into the foreground than the original.

software — the digitized instructions (or program) and/or related documentation associated with a system; the controls that cause a computer to operate and perform desired functions. *Related Term:* hardware.

solarization — a photographic technique utilizing both under and overexposure to create an aesthetic effect.

solid — (1) in printing, any area of the sheet receiving 100 percent ink coverage. (2) in composition, a reference to type set with no additional leading between the lines.

solid state laser scanner — in optical mark scanning, a type of laser that has become quite successful. It emits light at a wavelength of 670 nm and also at 780 nm.

solidus (/) — the mark used to separate numbers in a fraction or to unify words (e.g., and/or); a slash.

sort — the process of arranging elements, items, data and other materials into categories, i.e., alphabetic, numeric, color, size, etc., where all similar items are grouped together. Often the sorting process has a chronological aspect where the oldest, darkest, etc., appear first and then newer, lighter, etc., appear in descending or decreasing order thereafter.

sorts — pi characters that are obtainable but not ordinarily included in a font of type, such as mathematical signs and special punctuation and accent marks. (1) in hot metal composition they were manually inserted by the operator from a sorts bin. (2) in modern composition, sorts are normally an additional font which is addressed directly from the keyboard. Most systems have a number of these special fonts for special job requirements. (3) in hand composition, the term was sometimes used to specify individual replacement letters for a font of type and were used for replenishing used-up type in a case. *Related Terms:* dingbats; wingdings; webthings.

sound — the computer's audio; sounds you can hear. Computers can generate sound a number of ways; (1) the internal speaker that beeps when the machine is booted up. (2) CD audio, by putting an audio CD into your CD-ROM drive. (3) the use of waveform sounds, which are digital recordings of any kind of sound or music (for example, Windows' warning-sound file ding.wav). (4) the use of MIDI, a kind of digital sheet music that instructs synthesizer chips on the notes, tempo, and instrumentation of a musical passage. *Related Terms:* AIFF; AU; MIDI; VOC; WAV; waveform.

source image — any reflective or transmissive copy to be scanned.

source marking — the process of labeling an item with a bar code at the point of its initial production.

sp — an abbreviation for "spelling." A proofreader's mark used to signify that the spelling of a particular word should be checked.

space — (1) in typography, the nonprinting areas between words, lines and art. (2) in code and symbol scanning, the lighter element of a bar code usually formed by the background between bars.

space width — in optical mark reading, the thickness of a space measured from the edge closest to the symbol start character to the trailing edge of the same space.

space, two-dimensional — area where words, artwork and photographs are to be located and printed;

a design element in layouts.

spaces — in handset type, fine pieces of metal, less than type high, inserted between words or letters to adjust and ensure proper spacing.

spadia — an advertising sheet designed to half wrap a publication. Often seen as a wrap on Sunday comics and advertiser inserts.

spam — computer junk e-mail. It can be an unsolicited mass mailing to bulletin boards, newsgroups, or lists of people. Spam is usually intrusive and unwelcome. If you spam or get spammed and respond, it precipitates a flame war. *Related Terms:* flame; flame war.

SPARC — a scalar processing architecture developed and marketed by Sun Microsystems as a work terminal or workstation.

spatial resolution — horizontal and vertical resolutions as a result of subdividing and quanitizing a screen into individual parts.

spec — (1) an abbreviation for specification. the process of establishing production specifications such as types, papers, inks, outside purchases, etc. in advance of estimating and production, or the characteristics of a process color expressed in various percentages, the specific instructions for reproducing an image or a page layout, etc. (2) in markup, to determine and communicate the characteristics of typeset copy, color, or other aspects of page layout.

spec sheet — the instrument sheet on which specifications are drawn.

special — (1) in Internet administration, a status code generally used when there is a problem applying a payment to a domain name, e.g., a wrong invoice number, a misspelled name, or other problem that delays processing payment. (2) in production, a job, supplement or other production work which must be worked into an existing schedule for early finish and delivery, and which has priority over all previously planned work.

special character — a character not in the standard 7-bit ASCII character set, such as the copyright mark (©).

special effects — the general term for reproduction of photographs using techniques such as line conversion and posterization.

special purpose printing — any printing operation that accepts orders for only one type of product, such as forms work, legal printing, or labels. *Related Term:* specialty printer.

specialty advertising — the printed advertising on products such as mugs, matchbooks, jewelry, and pencils.

specialty paper — any other than normal paper. A distributor term for any of the nonwood substrates which substitute for paper. Typical are synthetics, pressure-sensitive, carbonless, remoistenable gummed, etc.

specialty printer — a printer specializing in making a particular product, usually novelty or advertising items such as key tags, napkins, etc. *Related Term:* special purpose printing.

species — an often used basic division of type; a face of type. *Related Terms:* class; family; series; font.

specification sheet — *Related Term:* spec sheet.

specifications — in printing production, the complete and precise descriptions of paper, ink, binding, quantity and other features of an anticipated printing job. For type, instructions about type face and size, line measure, indentations, headlines, etc. Often the basis for bidding or quotation. *Related Term:* specs.

specifications, film — film micrometer thickness, screen rulings if required, lateral reverses, right-reading, emulsion up or down, coverage area, if film is to be loose or stripped, if the projected work will run on a web or sheet fed press, and the printing method, i.e., litho, flexo, gravure, etc.

specifications, finishing — final fold and trim size, type of binding, inclusion of slip cases, post-press operations such as foil stamping or embossing, die cutting, etc.

specifications, paper — a description of all paper required, including the number of sheets, the parent size, the weight, the grain direction, color, finish, grade and manufacturer.

specifications, printing — specifics on quantities, ink colors, allowable dot gain, special ink mixing, varnish or aqueous coating, UV coating, substrate name, press sheet size, running instructions such as work and turn, etc.

specifications, type — specifics about type face, size, measure, indents, headlines, interline spacing, etc.

specking — (1) small dots of ink that can be seen between halftone dots and the clear areas of printed material. (2) referring to the act of creating job specifications.

specs — *Related Term:* specifications.

spectral highlight — the white portion of a photograph with no detail, such as bright, shiny reflection from a metal object. Quite often such areas are represented in the printed product as white paper without any dot structure.

spectral reflection — *Related Term:* spectral highlight.

spectral response — the variation in sensitivity of a reading device to light of different wavelengths.

spectral sensitivity — response to lightwave energy. While the human eye responds only to the visible spectrum, some films and electronic components respond to infrared and ultraviolet energy as well.

spectrophotometer — an instrument that measures light at many points on the visual spectrum, with results that can be plotted on a graphic curve.

spectrum — any spatial arrangement of electromagnetic energy in order of wavelength. For instance, the visible spectrum has the complete range of colors, from blue (short wavelengths) to red (long wavelengths).

spectrum, visible — the electromagnetic spectrum from about 400 to 700 nanometers that causes the sensation of a visible rainbow-like band of colors. *Related Terms:* prism; white light.

specula highlight — the lightest highlight area that does not carry any detail, such as reflections from glass or polished metal. Normally, these areas are reproduced as unprinted or nearly unprinted white paper.

speed — (1) a film's sensitivity to light. Rated by ISO numbers, e.g., ISO 1600, 1000, 800, 400, 200, 100, etc. The higher the number, the faster the film, and the less light required to record an image. (2) the relative production cycles of a piece of equipment when measured against time, i.e., 10,000 impressions per hour, etc.

spellcheck software — the facility of certain word processing and page makeup programs to enable a spelling error check to be carried out by individually comparing words in a new computer file with correctly spelled words in the computer dictionary, noting exceptions in spelling, capitalization, etc. and in some instances, recommending word spelling replacements. Since dictionaries of American origin may not conform to European English standards, the option should be available within the program to modify the contents. Dictionaries can contain any number of words, but 60,000-200,000 words is not uncommon, and can be even larger on more sophisticated machines. Nearly all newer word processing packages provide the ability to add custom words in either a separate custom dictionary or to the standard dictionary. The additions can be made separately or as they are encountered during the spelling check process.

SPID (service profile identifiers) — numbers that identify the services and features the telephone central office switch provides to ISDN devices. *Related Terms:* DN, ISDN

spider — a class of robot software that explores the Web by retrieving a document and following all the hyperlinks in it. Web sites tend to be so well linked that a spider can cover vast amounts of the Internet by starting from just a few sites. After following the links, spiders generate catalogs that can be accessed by search engines. Popular search sites like Alta Vista, Excite, and Lycos use this method. Also known as a Web spider. *Related Terms:* bot; crawler.

spine — the binding edge at the back of a book (hard- or softcover), i.e., the midpoint area between the front and back covers of a book or magazine; the center point of the outside cover. *Related Term:* backbone.

spine out — the displaying of books on a shelf so that their spines are showing, as opposed to face out.

spiral, plastic and coil binding — three methods of "mechanical" binding that are used for calendars, notebooks, etc. Holes or slots are cut along the edge of the spine or the cover and all content pages. A spiraled wire or plastic coil or comb device is inserted. All of these bindings will allow the pages to lie flat. In fact, the spiral and coil bindings allow pages to rotate 360 degrees; some plastic binding is almost as flexible. Plastic binding allows the individual user to insert or remove pages. To do this requires the purchase of a device called a punching and closing machine. Spiral bindings will not allow for changing pages once the sheets are bound.

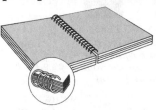

SPIRAL COIL BINDING
Individual pages are drilled along the binding edge and a continuous metal or plastic coil is threaded through them, sequentially. The ends are bent or other wise locked to prevent the uncoiling and removal of the binding.

spirit duplication — sometimes generically called "ditto." An imaging process in the planographic classification that makes a limited number of copies from an aniline dye impregnated carbon master (image carrier). The dye image is dissolved with an alcohol base solution so that it transfers when pressed against a special grade of imaging paper.

splash page — an extra "first" or "front" page of a Web site, with a logo or message, announcing that you have arrived. It usually provides links to succeeding sections or pages, because the real information and navigation for the site lies on the homepage or welcome page. *Related Term:* buffer page.

splice — joining of the end of one roll of paper to the beginning of another.

split fountain — a technique using an ink fountain divider, either chemical or mechanical, to enable a fountain to handle two or more colors at one time. Oscillating rollers tend to blend the two colors in the ink train and on the image carrier, creating a multihued reproduction. It is an inexpensive way to put some color on a page without the attendant charges for multicolor plates, press, etc.

split run — dividing a print run into two related jobs with minor variations, such as printing a portion of books in soft cover and the other quantity of the same books in hardcover, or printing half the brochures in one color and the other half in another color.

spoilage — paper used during setup, make-ready, printing, or bindery operations.

spoilage allowance — planned extra sheets delivered to the press to allow for inevitable waste and to ensure that required number of quality products are delivered to the customer.

spooler — a computer memory utility that manages printers on a network by storing data and passing it along as it can be processed, avoiding the need to interrupt normal CPU functions and inhibit continued production. Another type intercepts print files until the printer is ready to print them, thus enabling the computer screen and keyboard to be freed up for additional input while the printer processes and prints work which has been sent.

spot — (1) in reproduction, the undesirable presence of ink or dirt in a space. (2) in amplitude-modulated imagesetting called a halftone spot. The precise exposures (usually less than 14 microns in diameter) created by mathematical algorithm to give the appearance of random spacing. A single laser pulse image.

spot color — the selective addition of a single solid (or screened) nonprocess color printed using one separation plate, as opposed to a process color printed using two or more separation plates.

spot illustration — a small nondescript drawing, used to provide graphic interest to an article or story that may not lend itself any regular type of illustration.

spot meter — a reflective light luminance meter capable of measuring as small as 1 degree of luminance on the subject. Usually coupled with some sort of magnifying viewfinder. Used to measure precisely specific areas of a scene with minimum influence from areas surrounding it.

spot plater — a machine used to build up the metal in small nonimage areas on a gravure cylinder.

spot varnish — a clear coating applied to a particular area of a printed piece that provides protection as well as a dull or glossy appearance, depending on the type of varnish.

spotting — a studio photography term sometimes used incorrectly in place of prepress opaquing.

spread — (1) in design, two pages that face each other. Also the layout, especially of photos, on such pages. (2) in prepress, a photographic technique used to increase the size of an image during contact printing. Often used in conjunction with another trapping technique called a "choke," in which another color area is made slightly smaller. *Related Terms:* double truck; choke.

spread sheet — a mathematical program which is organized into columns and rows. Intersecting points are called cells and data entered into them can be formulated based upon contents of other cells or groups of cells.

spread spectrum — spread spectrum is a type of modulation that scatters data transmissions across the available frequency band in a pseudo-random pattern. Spreading the data across the frequency spectrum makes the signal resistant to noise, interference, and snooping. Spread-spectrum modulation schemes are commonly used with digital cellular phones, as well as with wireless local area networks (LANs) and cable modems.

SPRR (second pass read rate) — the ratio of the number of successful reads to the number of attempts on the second effort.

spur — the little stick-out portion of some font letters like G or t.

SQL (structured query language) — a type of pro-

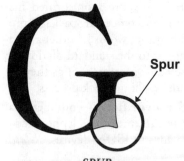

SPUR
Strictly a design appendage which occurs only on some letters and only in some type designs, particularly the majuscule G.

gramming language used to construct database queries and per-form updates and other maintenance of relational data-bases, it is not a full-fledged lan-guage that can create stand-alone applications, but it is strong enough to create interactive routines in other database pro-grams.

square dot — one type of halftone dot shape capable of carrying exceptional detail onto the printed sheet.

square halftone — a square-finish halftone. A rectangular, not necessarily square, halftone, i.e., one with all four sides straight and perpendicular to one another.

SQUARE SERIF CLASS LETTERS
One of the six classifications of type, it is typified by strong, angular serifs, usually without brackets and which are the same or nearly the same weight as the main letter strokes.

square serif – (1) one of the six classes of type. (2) a typeface characterized by monotone strokes and equally weighted, squared serifs without fillets or rounds. Sometimes called slab serif. *Related Term:* Egyptian.

squeegee — in screen print-ing, an imple-ment that con-tains a pli-able rubber, syn-thetic or other flexible blade that is used to force ink through the stenciled image carrier.

squeegee side — the side of an attached screen printing stencil which the squeegee is rubbed against when making a printed image.

SRA — a paper size in the series of ISO international paper sizes slightly larger than the A series, allowing extra space to for work that bleeds.

SRAM (static RAM) — a dynamic memory that stores its data in capacitors that don't require constant recharging to retain their data, it performs better than DRAM. *Related Terms:* EDO RAM; DRAM; RAM.

SRDS (Standard Rate and Data Service) — a reference used by advertising agencies to determine advertising rates and other statistics of various print and broadcast media.

S-registers — hardware settings that can be programmed using AT command strings. They vary from modem to modem, which is why some software and some operating systems insist on knowing the manufacturer and model of your modem. *Related Terms:* Hayes-compatible; AT commands.

SS — Abbreviation for "same size."

SSI (server side included) — a process by which HTML authors can include content on Web pages without actually coding the properties of the content in the HTML document, thus allowing great flexibility in changing/adding complex portions of information that change often on a Web page without actually having to edit the syntax of a specific HTML document that contains the simple SSI statement. It's called "server side" because the execution of this program takes place on a server.

SSL (secure socket layer) — a transaction security standard developed by Netscape Communications to enable secure commercial transactions to take place over the Internet. It's one of several competing security standards. *Related Term:* SHTTP.

stab waste — *Related Term:* butt waste.

stabilization — a photographic process that utilizes a two-chemical bath for processing. Once very popular, especially for phototypesetting paper and rapid photographic printing.

stable base jig — an exposed piece of stable-base film containing register marks used as a carrier for color transparencies during the color separation process.

stable base sheet — a substrate sheet that does not change size with changes in temperature.

STAC (Symbol Technical Advisory Committee) — *Related Term:* Uniform Code Council.

STAC LZS compression (STAC Lempel-Ziv standard compression) — a data compression standard developed and marketed by STAC Inc. to be used over Internet connections. It is widely supported by many types of equipment and can triple data rates with highly compressible files. *Related Terms:* LZW compression; PPP; ISDN.

stacking flexographic press — the original flexo-

graphic press design, similar to rotary letterpress and containing up to six color printing stations. *Related Terms:* in-line flexographic press; central impression cylinder flexographic press.

staging – (1) the process of applying a protective lacquer or varnish to a negative or positive image in preparation for selective dot etching, especially in areas of soft focus detail. (2) covering the bare metal of a gravure cylinder in preparation for image etching. *Related Term:* stopping out.

staging solution (stage) — the solution used as a resist in the etching process; prevents dots or tones from being etched.

stain — (1) in photography, a dark liquid material used by retouch artists to add detail on continuous tone images. (2) in printing, the placing of ink on the edges of books to achieve a marbling effect.

stamping — using a die and often colored or metallic foil to press a design into a book cover, a sheet of paper, or another substrate. The die may be used alone (in blank stamping) if no color or other ornamentation is necessary. Special presses fitted with heating devices can stamp designs into book covers. *Related Terms:* foil stamping; embossing.

stamping die — a deeply etched or engraved brass or zinc relief plate used to impress designs on book covers. Brass plates are used when the stamping process requires heat.

standard — (1) a verified calibration reference against which instruments are compared and adjusted for accuracy. (2) a set of rules, specifications, instructions and directions used as a regular guide to conduct.

standard generalized markup language (SGML) — (1) a language for marking text for typesetting and disk publishing to allow sophisticated formatting and searching. SGML enables the publisher to mark text just once for multiple uses. (2) the basis for HTML, used for web page description.

standard inks — a set of process inks made to the color specifications of a standards-setting organization. Common in the publications industry for proofing advertising reproductions, but otherwise uncommon.

standard viewing conditions — a prescribed set of conditions under which the viewing of originals and reproductions is to take place, defining both the geometry of the illumination and the spectral power distribution of the illuminant. For the graphic arts, the standard specifies 5,000 K color temperature, 90 color-rendering index, transparent gray surround with transparencies, and viewing at an angle to reduce glare.

standing form — a type form that is never destroyed, melted, or distributed, but is held so that it can be reprinted.

standing head — a headline whose words and (often) position stay the same, issue after issue. An example could be a series of columns heads appearing on a number of pages, with each page having variable data, but retaining a uniform set of column heads, regardless of the page on which the tables or charts appear.

stapling — attachment, together, of loose sheets or folded signatures with individual cinched wire fasteners. The term is sometimes incorrectly used interchangeably with stitching or wire stitching.

star network — a communications system consisting of one central node with point-to-point links to several others.

star target — a resolution test image often containing alternating light and dark wedge shapes that meet at a point in the center of the target. Commonly used for monitoring dot gain, slur, and doubling on press or for evaluating a given photographic or photomechanical system.

start and stop bits — bits at the beginning and end of asynchronous transmissions that notify the receiving computer that a character has begun or has ended.

start bit — original modem communications alternated between two tones, 0 and 1, which covered the entire binary language. There was no silence, as that meant the phone connection had been terminated. Between data transmissions, it continuously transmitted the 1 tone, indicating to the other modem that it was in an idle state. When more data was to be sent, it would send a single 0 bit to say there was real data on the way, and the next eight bits the receiving modem heard were the real data. The line then went back to idle for at least one more bit. A 0 tone just before the data was called the start bit. The 1 at the end of the data was called the stop bit. *Related Term:* asynchronous communication.

start of authority (SOA) resource record — a record type used in the distributed domain name system (DNS) database to indicate that a particular name server contains authoritative data for a particular domain. *Related Terms:* domain name system (DNS); name server.

start/stop character or pattern — a special bar code character that provides the scanner with start and stop

reading instructions as well as scanning direction. The start character is normally at the left-hand end of a horizontally oriented symbol. The stop character is normally at the right-hand end of a horizontally oriented symbol.

stat — (1) a general term for an inexpensive photographic print of line copy or halftone. (2) the trademarked name given to output of the now obsolete photostat process for producing authenticated replicas which were generally tonally reverses. (3) a colloquial term for statistics. *Related Term:* photostat.

stat camera — (1) a common name for any small process camera. (2) (obsolete) the name given to photostat cameras which produced tonally reversed images directly from camera speed print paper. *Related Term:* camera, stat.

state — (1) a term of reference for the condition of a component, i.e., on/off, black/white, 0/1, etc. (2) the state of a Web page, such as its attributes, configuration, or content. For a Web page, state is usually maintained only as long as the page is in the browser. *Related Terms:* cookie, stateless.

stateless — a Web server which considers each page request independently. The request specifies the entire document, without requiring any context or memory of previous requests. Unless a cookie is set, no information is carried across requests. *Related Terms:* cookie, state.

static eliminator — any device designed to reduce or neutralize static electricity in an environment. Most commonly used on presses and duplicators where a piece of copper tinsel, mounted and grounded, is placed in paper path of the delivery unit to remove static electric charge that make it difficult to accurately stack individual sheets of paper.

static imbalance — a defect in cylinder balance on a press that occurs when a cylinder is not perfectly round or has different densities within a cross section.

static IP — an IP address which is the same every time you "log on" to the Internet.

static neutralizer — an attachment for a printing press that removes static electricity from the paper in order to improve sheeting handling and avoid ink setoff and problems with the paper feeder.

station — (1) in communications, the assigned orbital position of a satellite; its location. (2) in graphic communications, the specific assignment as it relates to a multiposition machine such as a carousel screen press or an automated gathering and inserting machine. Each assigned employee has a "station" for which they are responsible.

stationery — a collective term applied to letterheads, envelopes, business cards, and other printed materials for correspondence. *Related Term:* business cabinet.

status bar — an area on the computer screen that displays information about the currently selected command or about an operation in progress.

status T — the accepted standard of spectral response for wideband reflection densitometers.

stay-flat binding — a perfect binding technique used in publications where the cover spine is not glued to the edges of the bound pages, which enables the book to lay flatter when opened. *Related Term:* lay-flat binding.

stem — usually the main vertical stroke of a type character.

stencil — any material that serves as the image carrier or matrix and is mounted on the woven fabric in screen printing or other porous-class imaging systems such as mimeography. It passes ink in the image areas and blocks ink passage in nonimage areas.

step and repeat — the exposing of multiple images onto a single film or a single printing plate from a single negative or positive flat. Special step and repeat contact frames, projection plate makers, and multi-imaging cameras are used to automate this process. *Related Terms:* multi-imaging camera; photo composer.

step tablet — *Related Terms:* gray scale; cameraman's sensitivity guide; step wedge.

step wedge — (1) in process and general photography, a slang term for a test strip, gray scale, sensitivity guide, etc. (2) generally, an exposure measurement and calculation method used to determine the best length of time to expose light-sensitive materials. *Related Terms:* sensitivity guide; photographer's sensitivity guide; gray scale; cameraman's sensitivity guide; step tablet.

stepped head — a typographic arrangement of display type in which the top line is flush left, the middle line (if used) is centered, and the third line is flush right, and all the lines are less than full length to cre-

Stem

ate a stairstep effect.

stereotype — a duplicate relief printing plate made by casting a lead alloy into a papier mâché mold of the original plate.

stereotyping — the process of producing a duplicate relief plate (stereotype) by casting molten metal into a mold (or mat) made from an original lockup.

stet — derived from the Latin for "let it stand". A proofreader's mark that looks like this: *stet.* It is placed in the margin of the copy and indicates copy marked for correction should stand as it was before the correction was made, keeping the original version. Copy to be stetted is always underlined with a row of dots, in addition to the margin notation.

stick — *Related Term:* composing stick.

stick-on letters — alphabet characters printed on paper, usually self-adhesive, for cold-type composition. *Related Terms:* Letraset; dry transfer letters.

stick-up cap — *Related Term:* raised initial letter.

stiffness — the relative ability of paper or other substrate to resist bending and/or support its own weight.

still development method — a method of controlling the development stage of film processing by resting the film on the bottom of the tray when the image starts to appear and leaving the film motionless for the remainder of the developing cycle. It is especially effective for development of fine line negatives and for copydot images.

still image video — a system by which still images are transmitted over standard telephone lines, usually allowing for real-time interaction between locations, but usually at a greatly reduced number of frames/second.

stipple finish — a paper finish whose texture simulates a pattern of dots of various sizes.

stitch bind — to bind with wire staples. *Related Terms:* saddle stitch; side stitch.

stitching — any stapling or sewing technique which uses thread or wire to affix signatures together or attach them to a binding. *Related Terms:* saddle stitch; side wire stitch; stapling; saddle stitch; side sewing; Smyth sewn.

stochastic screening — an electronic screening technique which employs no assigned screen frequency, i.e., the dots are randomly spaced and of equal size rather than geometrically aligned as in halftone dots. Their spacing is closer in dark areas of the reproduction. Typical dot sizes are 14-20 microns. Because of the fineness of the dot sizes, the illusion of the process having no screens, but looking more like a full tone photography results. *Related Terms:* halftone screening; FM screening.

stock – (1) also called substrate. Any material used to receive a printed image: paper, board, foil, etc. (2) in papermaking, pulp which has been beaten and refined, and which after dilution is ready to be sent to the head box and made into paper. *Related Term:* furnish.

stock order — an order for paper that a mill or merchant sends to a printer from inventory at a warehouse, as compared to a mill order.

stock photo — a photograph in a commercial collection. Usually rented for a specific publication, date or time.

stock sheet — (1) the large paper size generally purchased from the paper company and from which press sheets are cut. (2) incorrect term sometimes used to mean house sheet. *Related Term:* parent sheet.

stock solution — a highly concentrated chemical reagent which must be diluted with water prior to use to adjust to proper working strength.

stocking merchant — a paper distributor that maintains an inventory of commonly used papers.

stocking papers — the most popular sizes, weights, and colors of papers that are available for prompt delivery from a merchant's warehouse.

stone lithography — the original form of lithography in a written or drawn image was placed on limestone with wax or grease. The limestone has a naturally hydrophilic property on its surface so that the non-image areas accept water and repel the wax or grease.

stone proof — a proof taken from a form in foundry or hot-type composition without using a machine, often by placing a sheet of paper over the inked form as it lay on the imposition stone. *Related Term:* galley proof.

stop — (1) in camerawork, the aperture or diaphragm setting of a camera lens. (2) in photographic film or print processing, a shortened version of stop bath, the slightly acidic solution used to halt development in film processing, sometimes called "short stop." *Related Terms:* aperture; f/stop; iris; diaphragm.

stop bath — an solution used in film processing to stop the action of the alkaline developer. Commonly composed of acetic acid and water. In addition to stop-

ping development, it makes the hypo (fixer) last longer. *Related Term:* stop.

stop bit — the opposite of the start bit. *Related Term:* start bit.

stop words — words that are common in a full text file but have little value in searching, such as "the" "is" "a," etc. Words in a stop word file will be excluded from the indexes, considerably reducing the size of the indexes and improving search performance.

stopping out — an old term for the process of preparing negatives and printing plates for printing, such as opaquing negatives and protecting certain areas of deeply etched plates. *Related Term:* staging.

storage — a device (a memory tape, disc, or drum) into which data can be entered, in which it can be held, and from which it can be retrieved at a later time.

storyboarding — outlining a video program by producing a series of sequenced illustrations or layouts of scenes.

straight composition — (1) reading copy that contains no charts, tables, formulas or other elements that complicate typesetting. (2) body text copy set in simple rectangular columns with little or no variation, i.e., the first and last characters of the successive lines line up to form vertical columns. (2) a very incorrect term for straight copy. *Related Terms:* justification; justified; straight copy.

straight line — that region of constant slope of the characteristic (DlogE) curve of a photographic emulsion where density is directly proportional to the logarithm of exposure. *Related Terms:* shoulder; shoulder portion; foot; foot portion; density curve.

straight matter — reading matter; body type; the text material in a book. *Related Term:* straight composition.

strap — a subheading used above the main headline in a newspaper article. *Related Term:* kicker.

strawboard — a thicker board made from straw pulp, used in book work and in the making of envelopes and cartons. Not suitable for printing.

streaking — streaks on the printed substrate caused by improper ink and/or water adjustments, poor maintenance, etc.

stream feeder — a press element that overlaps sheets on the registration feed table, allowing the registration unit to operate at a slower rate than the printing unit and producing better sheet-to-sheet registration.

streamer — a banner headline.

streaming — the quick movement of data bits between hardware components. Data doesn't have to be all in one place for the destination device to process it; writing hard disk data is streaming when it's moving data to be written to a tape backup device.

stress — one of the three variables in alphabet design; refers to distribution of visual "heaviness" or "slant" of the character, i.e., the implied angle between the thinnest parts of a curved letter.

stretch ink — ink formulated to retain density as it stretches on a flexible substrate.

strike-on composition — a cold-type composition method of producing images through direct impression such as by striking a carrier sheet with a raised character through an ink or carbon ribbon. A typewriter. *Related Term:* impact composition; daisy wheel printer; line printer; cold type.

strike-through — (1) in printing, a term incorrectly used for print-through, the problem of ink going through the sheet of paper so that ink on the first side can be seen from the back of the sheet. Often called print through. (2) in editorial writing, a typographic feature of some word processors and layout programs whereby each letter is composed with a line through it, i.e., ~~this is strike-through copy.~~

string — a sequence of keyboard character or codes to be processed as a group for the purpose of shortening keyboard operations and ensuring accuracy. Many programs and keyboards enable the operator to create and store strings for reuse as needed. *Related Term:* script.

string and button envelope — the classic interoffice memo envelope with the string that is wound around a riveted button for closure.

strip — short for stripping. Positioning film negatives or positives on a flat (goldenrod) before platemaking. *Related Term:* stripping.

strip in — manually affixing a film negative or positive to another piece of film.

striping — splitting a data stream across two or more drives to increase the possible data transfer rate, such as in RAID level 0.

stripper — the job title of a prepress person who combines negatives into and assembles flats for platemaking.

stripping — the process of assembling pieces of film containing images into flats in preparation for making printing plates. The process of securing them on a masking sheet that will hold them in their appropriate printing positions during the platemaking process. *Related Term:* film assembly.

stripping blade — in porous printing, especially screen printing, a pen-shaped tool with a rounded metal blade used to remove unwanted cut screen printing film and stripping film.

stripping film — a stable-based nonphotographic film made up of two layers, the emulsion, usually red or orange and the base, usually clear. The thin emulsion can be cut and peeled away to reveal clear areas that can be used to form visual images in their own right, or to act as masks for underlying negatives and/or positives during subsequent processing. *Related Terms:* Rubylith; Amberlith; masking film.

stroke — (1) in typography, one of three variables in alphabet design; a line that forms a letter or a part of a letter, e.g., S and 1 have one stroke and B has three (in hand block lettering). (2) in composition, a character image or command generated by the depression of a single key on the keyboard. Short for keystroke. (3) on imagesetting, the generation of a character image on photosensitive material by means of a series of lines (strokes) created by a moving spot of light.

strong text — the HTML character style used for strong emphasis. Most browsers display this style as bold.

style — (1) in copyediting, the rules for treatment of such matters as modes of address, titles, numerals, paragraph indents, pull quotes, etc. (2) in typography, the variations in appearance, such as italic and bold, that make up the faces in a type family.

style guide — (1) in general printing, a collection of tags specifying page layout styles, paragraph settings and type specifications which can be set up by the user and saved for use in other documents. Most page makeup and word processing programs come with a set of editable predefined style sheets. (2) in Web publishing, the set of guidelines written for the purpose of keeping consistent and standardizing the further development of a specific Web site. They include everything from proper HTML usage to acceptable colors and fonts and CGI and JavaScript programming and grammatical specifics. (3) in general and corporate publishing, usually a manual that outlines accepted usage of a corporate identity such as logos and letterheads, as well as the correct spelling of commonly used industry terminology.

style sheet — in computerized composition, a pre-defined script or instruction set for a group of character attributes and paragraph formats that can be applied in one step to a paragraph or range of paragraphs.

stylus — a smooth blunt instrument used to rub dry transfer type from the carrier page and adhere it firmly to the receptor.

sub weight — *Related Term:* substance weight.

subhead — a secondary title or heading that is usually set in smaller type, making it less prominent than a main heading. The subordinate headlines used to mark divisions in a chapter, or a descriptive head, word or phrase that precedes a block of body text in a larger article.

subjective balance — the design characteristic in which images are placed on white space in such a way as to create a feeling of stability.

sublimate — a technology used in some photo-realistic electronic imaging systems and in the heat transfer process. A description of the physical properties of some materials to transfer from a solid state into a gas, without any liquid state.

sublimation ink — ink that is printed on paper and later, through a sublimation process, transferred to cloth with heat and pressure. *Related Terms:* ink, sublimation; heat sublimation.

subsampling — a bandwidth reduction technique which reduces the amount of digital data used to represent an image as part of a data compression process.

subscribe — adding your name and e-mail address to a mailing list, or other discussion group.

subscript — inferior characters as in the symbol for water (H_2O) set in smaller size and below normal x-height. *Related Term:* superscript.

subsidiary rights — any rights which require additional fees for license to a publisher by the copyright owner. Examples are serial rights, broadcast rights, video rights, movie rights, translation rights, television, and other electronic rights.

subsidy press — any publisher who charges the author to print the author's book. *Related Term:* vanity press.

substance — the weight in pounds of a ream (500 sheets) of standard-size business papers. Similar to basis weight of other grades of paper.

substance weight — basis weight in pounds of one ream of the basic sheet size (17 inches x 22 inches).

Used when referring to bond papers. *Related Term:* sub weight.

substitution error — a mis-encoding, misread or human key entry error where a character that was to be entered is substituted with erroneous information. Example: correct information — abcde; substitution —abced.

substrate — any media on which something is printed. The base material used in printing processes to receive an image transferred from an image carrier or plate; common substrates are paper, foil, fabric, and plastic sheeting.

subtitle — (1) in periodical publishing, a phrase that is part of the nameplate and tells why a publication is published, for whom and how often. (2) in book publishing, sentence or phrase that appears after the title of a book.

subtractive color process — a means of producing a color repro-duction or image with combinations of yellow, ma-genta and cyan colorants on a white sub-strate.

subtractive colors — the colors produced by combining cyan, magenta, and yellow inks printed on white paper to absorb, or subtract, the red, green, and blue portions of the spectrum in the printing process. The basis for full color printing. *Related Terms:* cyan; magenta; yellow; CYMK. (See illustration on next page)

subtractive plate — a presensitized lithographic plate with a diazo and lacquer surface applied by the manufacturer. The lacquer is removed from the plate to form the hydrophilic, nonimage areas. Generally utilizes a film negative for exposure.

subtractive primary colors — the colors formed when any two additive primary colors of light are mixed; used in full color process printing. They are yellow, magenta, and cyan (YMC). Yellow is the additive mixture of red and green light; magenta is the additive mixture of red and blue light; cyan is the additive mixture of blue and green light. In inks, they are formulated to be transparent. YMC when combined form a theoretical black. In actual practice, though, they produce a dirty brown. For this reason an artificial black printer is created from information merged from all of them. *Related Terms:* process blue; cyan; process red magenta; yellow. *See illustration on next page.*

successive sheet-feeding system — one of the most common forms of press sheet feeding where the feeding unit picks up one sheet each time the printing unit prints one impression. *Related Term:* stream feeding.

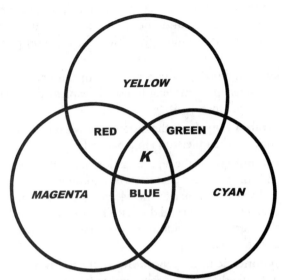

THE SUBTRACTIVE COLOR THEORY
In theory, pure cyan, magenta and yellow absorb all wavelengths of the visible spectrum and should produce the visual sensation of black or the absence of reflected light. In practice, because of dye and pigment impurities, the result is most often a dark muddy brown. For this reason, an artificial fourth printer, black (K), is created to intensify the shadow contrast and detail.

sucker feet — the elements used in the press feeding unit to pick up individual sheets and place them in the registration unit. Usually a small rubber or plastic disk connected to a vacuum system.

suede finish — *Related Term:* dull finish.

sulfate paper — a paper made from pulp cooked in alkaline chemicals. *Related Term:* kraft paper.

sulfate pulp — the paper made from wood pulps cooked in a sulfate solution. Also known as kraft.

sulfite paper — paper made from pulp cooked in acid chemicals.

sulfite pulp — the paper made from wood pulps cooked in a bisulfate solution.

sunken initial — an initial inset in reading matter. *Related Terms:* initial letter; drop capital; hanging initial.

super — the gauzelike fabric that is adhered to the backbone of a book with glue for the purpose of improved strength.

super additivity — the opposite of additivity failure, it occurs when the total density of overprinted ink films exceed the sum of the individual ink densities.

super VGA — a computer monitor standard of reso-

lution, typically 1,280 pixels by 1,024, capable of displaying over 16 million colors. It is a vast improvement over standard VGA.

super VHS — an enhanced VHS video tape format that uses a component video scheme in which the luminance information is separated from the chrominance information, yielding a higher quality picture. It has 400 lines or horizontal resolution; not quite broadcast quality or Hi8 but with much improved color purity when compared to standard VHS.

supercalendered paper — a smooth finished paper with a polished appearance, produced by rolling the paper between calenders (heated, alternating polished chrome and fiber rollers) at the end of the papermaking process; usually separate from the actual papermaking machine. Examples of this are high gloss and art papers. Abbreviated SC paper.

superior characters — the small characters set above the normal letters or figures such as in fraction numerators and math notations such as squared or cubed (x^3; y^3, etc.). Figures set in a smaller size above the normal height. Often called superscripts in computer software. *Related Terms:* superscript; inferior characters; superior characters.

superscript — (1) a term in photocomposition and computing which refers to using a special shift position to multiply the number of levels key positions can address. (2) a burst instruction assigned to and stored as a single keystroke in a script or superscript position of a keyboard, intended to provide complex commands in typesetting, imagesetting, etc., without manual typing.

suppress printout — in electronic publishing, a command in page layout programs that allows the user to skip an image or a page from a series queued for printing while still printing the others.

surface — the relative smoothness or roughness of a paper. A rough or smooth finish, besides influencing the overall bulk, affects the way images print, especially halftones. Rich, dark tones can be printed effectively on a coated sheet or smooth, whereas a rough-surfaced, toothy paper will absorb more ink and leave the dark areas grayer and less intense. *Related Terms:* tooth; paper tooth.

surface plate — A lithography plate to which the image is applied by photochemical means. *Related Term:* deep etch plate.

surface tension — the primary cause of ink to form droplets.

surfing — the informal term used for exploring the Internet (i.e., "surfing the Net"). Most often used in reference to accessing sites on the World Wide Web.

surge supressor — an electronic device designed to trap any power surges or spikes coming through the power or telephone lines. They are designed to protect delicate electrical and electronic equipment which can be severely damaged or destroyed by high voltages.

surprint — the combination of one image over another, such as type across a screen or halftone area or across an illustration. If both images are the same color, it can be achieved by double exposure in the negative stage (double burn) used when it is not possible or practical to accomplish in the artwork. *Related Terms:* knockout; overprint.

S-video — the type of video signal used in Hi-8 and S-VHS. *Related Term:* super VHS.

swap file — an area on the hard disk used as virtual memory. It's called a swap file because virtual memory management software swaps data between it and main memory (RAM). *Related Terms:* virtual memory; RAM.

swash — (1) in composition, a cap letter with an ornamental flourish. (2) in typography, the actual flourish on selected characters of a typeface, separate from basic letter forms.

swatch — a miniature sample of an item or color such as paper, ink, foils, etc. used to determine final specifications.

swatch book — a book with swatch samples. *Related Term:* swatch.

switch — a mechanical or solid-state device that opens or closes circuits, changes operating parameters or selects paths or circuits on a space or time division basis.

swivel blade knife — a cutting tool with a rotating blade attached to a pen-shaped handle for use on screen printing film and stripping film. *Related Term:* bean cutter.

A A

SWASH CHARACTERS
The fancy flourishes which appear as part of certain letters of a standard font or on alternate characters. On the left is the standard majuscule A. On the right is the alternate decorative swash version.

SWOP (Specifications for Web-Offset Publications) — a set of standards for enumerating web-offset specifications for separations, proofing, and printing process color. Used extensively in Europe and as a U.S. standard.

symbol — in bar coding, a combination of char-

acters including start/stop characters, quiet zones, data characters, and check characters required by a particular symbology, which form a complete, scanable entity.

symbol density — in bar coding, the number of data characters per unit length; usually expressed as characters per inch (cpi).

symbol length — the distance between the outside edges of the quiet zone in a bar code.

symbology — the representation of numbers, letters and computerized characters by a combination of bars and spaces. Similar to the Morse Code that encodes characters in dots and dashes. The rules for encoding the characters in wide and narrow bars and spaces are called a "symbology."

symmetric multiprocessing — the ability of a computer to share data processing tasks among and between several on-board processors to increase productivity and decrease throughput time.

symmetrical — the arrangement of elements in formal balance.

symmetrical compression — a compression system which requires equal processing capability for both compression and decompression steps. It is used in applications where both steps will be utilized frequently, such as still image databases, still image (color fax) transmission, video production, videophones, video conferencing, and video mail. *Related Term:* asymmetrical compression.

synchronous — an operation where a series of events takes place under the control of a clocking device; also refers to information that is sent or exchanged at a certain time. *Related Term:* synchronous transmission.

synchronous communication — the technique of choice for ISDN lines. It handles data more efficiently than the typical modem's asynchronous technique which sends small blocks of data with lots of control bits for error correction. Synchronous techniques use big blocks of data with control bits only at the start and end of the entire transmission. Because of the infrequent error correction, it requires a clean, interference-free line. When provided, it can transfer data up to 30 percent faster than ISDN modems. *Related Term:* asynchronous communication.

synchronous transmission — data characters and bits are transmitted at a fixed rate with the transmitter and receiver synchronized. This eliminates the need for start-stop elements, thus providing greater efficiency.

syndicate — a clearinghouse-style organization that licenses and releases articles and stories, under contract to subscribers for simultaneous release.

syndication — the actions of a syndicate when a work is simultaneously licensed by and released to various media, such as newspapers or radio stations.

synthesizer — an electronic device which can be programmed to simulate nearly any audible source, including natural sounds and musical instruments.

synthetic enamel ink — a type of printing ink that gives a glossy surface on decals, outdoor signs and hard surfaces.

synthetic paper — plastic or plasticized materials made into sheets that simulate paper. Synthetic paper resists tearing and does not deteriorate when wet. Typically, Tyvek® and similar.

SysOp — the person responsible for maintenance of a given computer system. Short for "system operator."

system error — any failure on the part of a computer to perform a user command or execute a program function.

Tt

T1 (DS-1) — a leased dedicated line connection capable of carrying data at 1.544 Mbits per second. It theoretically could move a megabyte in less than 10 seconds, but it is still not fast enough for full-screen, full-motion video, which requires at least 10,000,000 bits per second. It is the fastest speed commonly used to connect networks to the Internet for high-volume voice or data traffic and for compressed video. *Related Terms:* ISDN; POTS; T3; T-carrier.

T3 (DS-3) — a leased line connection capable of carrying data at 44.746 megabits per second, more than adequate for full-screen, full-motion video or for major PBX-PBX interconnection. *Related Terms:* ISDN; POTS; T1; T-carrier.

T4S — an abbreviation for trim four sides.

TA (typesetter alteration) — (1) in composition, any change made because of typesetter error. (2) in proofreading, the proofreader mark showing such a change. *Related Term:* AA.

tab — (1) in composition, the presets determining positioning of multicolumn material such as tables and lists. (2) in publication work, half of a broadsheet, an abbreviated form of tabloid. (3) in design and production, the edge of a divider sheet that extends beyond the trim size of the publication. *Related Term:* tabloid.

table — (1) in layout and design, one or more rows of cells on a page used to organize the layout of a page or arrange data systematically. (2) in Web publishing, one type of placement of anything in a table of cells, aligned in columns and rows. Materials placed can include text, images, forms, etc. (3) generally, any column/row format for displaying data, numbers, etc. *Related Term:* tabular material.

table of contents — a page number chart that features the sections of a book, magazine, or other published work, usually in sequential order. *Related Term:* TOC.

table of illustrations — a page number chart that lists the illustrations, credits and page numbers in a published work, usually in order of appearance.

tablet — a graphics computer input device. A pen digitizer.

tabloid — a publication format usually about half the size of a broadsheet.

tabular figures — numerals which all have the same width. This makes it easier to set tabular matter.

tabular material — in typesetting, tables and charts with columns of data that must align, such as statistics arranged in table or columnar form, stock market reports, financial statements, and so on. *Related Term:* tabular setting.

tabular setting — text set in columns, such as timetables.

tack — (1) the characteristic of ink that allows it to stick to the substrate (usually paper). (2) the resistance to splitting of an ink film between two separating surfaces, i.e., stickiness. To improve trapping in wet-on-wet printing, the ink being printed should have a lower tack than the ink that was printed before it.

tackoscope — an inkometer. An instrument for measuring the tack of printing inks

tag — (1) in paper and papermaking, a grade of dense, strong paper used for products such as badges and file folders. (2) in formatting copy, particularly for the Web, a term used to describe the commands or instructions associated with HTML or Web page code. In HTML, tags look like this: <p> , <font size=1>, <head>, <body> or </p>, always with a pair of angle brackets (< >) surrounding the specific instruction. (3) in general markup and composition, the various formats which make up a style sheet, paragraph settings, margins and columns, page layouts, hyphenation and justification, widow and orphan control and automatic section numbering.

tag image file format (TIFF, TIF) — a file format for graphics that is particularly suited for representing scanned images and other large bitmaps. Originally they saved only black and white images in uncompressed form, but newer versions support color and compression. TIFF is a neutral format designed for compatibility with both Macintosh and MS-DOS applications.

tag selection — in some browsers, a method of selecting a group of paragraphs and other objects on a page. It is used to select the members of a list, an entire form, or a WebBot component by moving the cursor to the left of the objects until the cursor becomes the tag selection cursor (an arrow pointing to the upper right), and then double-click.

tagline — the production identifier or line of text that appears at the top or bottom of a printed page that shows the file name, page number, date, and/or time. *Related Term:* slug line.

tail — (1) in typography, a stroke extending below the baseline, like the lowercase j or the uppercase Q. (2) in presswork and imposition, the opposite edge of a press sheet from the gripper edge, the last edge of a press sheet as it goes through the press.

tail-piece — any small ornament at the end of a chapter or story, often a dingbat.

talk — an internet protocol which allows two people on remote computers to communicate in real-time. *Related Terms:* Internet relay chat (IRC).

TANSTAAFL (there ain't no such thing as a free lunch) — a observatory shorthand comment that is sometimes appended to a comment written in an online forum.

tape — usually a punched paper ribbon (between 6 and 31 levels) or magnetic tape (7- or 9-level) produced by a keyboard unit and used as input to activate the photo unit of a type- or imagesetting system or a computer for subsequent processing of the data thereon for typesetting machine output.

tape backup — a file storage and transfer medium. Both nine-track tape and smaller format tape cartridges are commonly used. Tape backup is used for protecting and transferring files, because tape has a high density of data storage.

tape binding — the binding of booklets, catalogs, etc., by use of a tape material with a flexible glue. The glued tape is applied with heat and pressure to the binding edge of the book. *Related Term:* thermal binding.

tare weight — the container weight, i.e., gross weight minus the product. *Related Terms:* gross weight; net weight.

TARGA (TGA) — a file format for exchanging 24-bit color files on PCs.

task — any item on a browser list representing one action you need to perform to complete a Web page. Some programs automatically generate tasks, based upon designer input. Usually the operator can add their own tasks to the list.

T-carrier — a series of transmission systems using pulse code modulation technology at various channel capabilities and bit rates (e.g., T1, T2, T3, T4).

TCP/IP (transmission control protocol/Internet protocol) — a protocol that dictates how computers share information with each other by controlling the transmission of packets of data over the Internet. It checks for lost packets, puts the data from multiple packets into the correct order, and requests a retransmission of missing or damaged packets. It works regardless of computer platform including Macs, PCs, and Unix boxes. *Related Terms:* client; data packet; protocol; IP; TCP; FTP; telnet; SMTP.

tearsheet — a sheet removed from a publication and supplied to advertisers, designers or authors as verification of publication and to show that the work was performed correctly and that a particular ad or article appeared in a particular issue.

technical contact/agent — the person or organization who maintains the primary domain name server and is able to answer technical questions about the domain name's primary domain name server and work with technically oriented people in other domains to solve technical issues that affect the domain name. *Related Terms:* contact/agent; primary server.

technical edit — manuscript review performed by an expert(s) in the field. *Related Terms:* expert reading; peer review.

technical inking pen — a specially designed inking pen used to make precise lines and images. Pen points are available in different line widths.

technological literacy — the ability to absorb and understand, use and exploit technology.

technophile — an ardent supporter and user of technology who, it is claimed, first emerged during the computer revolution of the '70s.

telecommunication — the point-to-point transmission of information through telephone lines, microwave transmission, satellite, etc., over distance using electrical means.

teleconferencing — a two-way electronic communication between two or more groups, or three or more individuals, who are in separate locations, including group communication via audio, white-board, video and computer systems.

telepen — a continuous bar code which encodes the full ASCII character set.

telephone — a two-way device used mainly for voice communications which converts sound into electrical current which can amplified and then be transmitted over communications channels.

telephone company central office switching — all phone calls are routed through the telephone company's central office. These offices form a central location for all calls within a particular area. The switch type refers to the equipment the telephone company uses to receive and transmit data over ISDN. To configure ISDN devices to communicate, knowledge of the switch is essential.

telephone conference bridge — a device that is designed to link three or more telephone channels for a teleconference; usually refers to a bridge that provides only dial-up teleconferencing where an operator calls each participant. Contrast to meet-me bridge. *Related Term:* teleconference.

telephone, electronic — a telephone set which employs electronic circuitry to provide additional features and improved performance, as compared to the original plain electric phone.

telephony — an adjective that covers a multitude of communications issues, telephony has recently permeated the world of small computing, as add-in boards that combine the functions of modems, sound boards, speakerphones, and voice mail systems have begun to proliferate. These DSP telephony boards are relatively inexpensive and are often integrated into new machines targeted for small office and home office users. *Related Term:* DSP.

telephoto lens — any camera lens with a focal length greater than 105 mm and that significantly magnify objects. *Related Term:* lens, telephoto.

Teletext — a broadcasting service using several otherwise unused scanning lines (vertical blanking intervals) between frames of TV pictures to transmit information from a central data base to receiving television sets.

telewriter — an electronic device that produces freehand information that can be sent over a telecommunications channel, usually a telephone line.

Telnet — an Internet program for connecting to a remote host or server. The Telnet interface is text based and a user usually has to enter a login name and password before gaining access to the system. Telnet can check your e-mail, download a program, or chat with other Telnet users. It is one of the oldest Internet ac-

tivities and is primarily used to access online databases or to read articles stored on university servers. It is also possible to Telnet via your Web browser by changing the http:// to telnet:// and entering the site's address. *Related Term:* server.

template — a set of design formats in software packages for text and images and in which print pages and Web pages can be designed and/or developed. Usually they are protected from overwriting and can be used repeatedly in creating new documents.

tensile strength — the maximum stress tolerated by a substrate, particularly paper, before breaking.

terminal — (1) in communications networking, a device in a network or system where information can be entered, removed, or displayed for viewing and arranging. At a minimum, this usually means a keyboard and a display screen and some simple circuitry, and allows you to type commands to a computer located at a remote site. (2) an input/output device designed to receive or send source data. (3) in electrical work and electronics, the point of junction, i.e., the screw under which a wire is secured to complete a circuit.

terminal adapter — hardware that converts the data it receives over ISDN to a form your computer can understand. Sometimes mistakenly called an ISDN modem or a digital modem, a terminal adapter handles data digitally and does not need to modulate or demodulate an analog signal. Terminal adapters can be an internal board or an external box that connects to the computer through the serial port. *Related Terms:* modem; ISDN.

terminal emulation — there are several methods for determining how your keystrokes and screen interact with a public-access site's operating system. Most communications programs offer a choice of "emulations" that let you mimic the keyboard that would normally be attached directly to the host-system computer.

terminal server — a special-purpose computer that has places to plug in many modems on one side, and a connection to a LAN or host machine on the other side. The terminal server then does the work of answering calls and passing the connections on to the appropriate node. Most terminal servers can provide PPP or SLIP services if connected to the Internet.

terminator — an electrical circuit usually in the form of a blind connector which is attached to each end of a SCSI bus to minimize signal reflections and extraneous noise.

terms and conditions — the legal description for the full execution of an order. Included are terms and conditions for payment, procedures for dispute resolution, assessments and legal expenses, copyright authority and nearly any other routine issue which could arise between customer and vendor. They are usually printed on the back of quotation forms or provided as a separate sheet to be signed by the customer indicating receipt and acceptance.

terrestrial carrier — a telecommunications transmission system using land-based facilities (microwave towers, telephone lines, fiber optic cable), as distinguished from satellite transmission.

tertiary — those colors obtained by mixing two secondary colors. Pronounced TERCH-ee-airy. *Related Term:* primary.

test sheets — sheets run on the press to check points such as ink coverage, image position and halftone dot quality so that any needed adjustments can be made before a press run. *Related Term:* make-ready sheets.

texel texture element — the base unit of a graphic texture map. While pixels are the basic elements in any graphic, texels are their equivalent in a texture map. *Related Terms:* pixel; texture mapping.

text — (1) the written or printed material which forms the main body of a publication. Display matter, headings, and illustrations do not fall into this category. Sometimes used to refer to type sizes between 8 and 14 points, although this is not a good descriptive term. (2) one of the six type classifications; is characterized by a design that attempts to recreate the feeling of medieval scribes. *Related Term:* text type. (3) a data file consisting of alphanumeric characters, defined by a text format such as ASCII or EBDIC. Entries in a text file are available for text searching.

text box — in computing, a screen area where data can be entered.

text file — a data file consisting of alphanumeric characters, defined by a text format such as ASCII or EBCDIC. Entries in a text file are available for text searching.

text paper — a category of printing paper generally used for printing booklets, brochure, menus, etc.

text wrap — to align text on a page or layout, following the contours and shape of another object such as an illustration. Some programs have an automatic text wrap feature that will shorten lines of text when a graphic is encountered; in other systems you need to change the length of lines to go around a graphic. *Related Term:* runaround.

texture — (1) a property of the surface of the substrate. (2) the variations in tonal values that form image detail. *Related Term:* mottling.

texture mapping — in 3D graphics, texture mapping is the process of adding graphics to scenery. Unlike shading, which adds color to the underlying polygons, texture mapping applies simple textured graphics to simulate walls, sky, and so on. *Related Term:* shading.

texturing — a shorthand term for texture mapping. *Related Term:* texture mapping.

TFB (too f- - - ing bad) — often used in flame mail and postings, this acronym indicates sneering, sarcastic "pity." *Related Term:* BFD.

TFT (thin-film transistor) — a technology for building the LCD screens that are commonly found on laptop computers. They are brighter and more readable than dual-scan LCD screens, but consume more power and are generally more expensive. *Related Terms:* active matrix; diode; flat-panel display; passive matrix; transistor.

TGA — *Related Term:* Targa.

The Internet Adapter (TIA) – (1) a product that emulates a SLIP or PPP connection over a serial line, allowing shell users to run a SLIP/PPP session through a UNIX dialup account. (2) also used informally in chat rooms and internet situations as an abbreviation for "Thanks in advance."

thermal — a printing system where dots are selectively heated and cooled on heat-sensitive paper. The paper turns dark in the heated areas.

thermal binding — a type of binding in which a machine applies a strip of heated adhesive material to the spine in a single operation. Similar to perfect binding, except the glue is generally applied as part of a binding tape instead of from a gluing device.

thermal dye sublimation — like thermal transfer, except that pigments are vaporized and their molecules are assimilated into the desired substrate. These inks are highly translucent and produce very high quality. Also called thermal dye diffusion transfer, or D2T2. Thermal printers operate by placing a "transfer" sheet that carries ink in contact with the paper or transparency. A heated print head then touches the transfer sheet to transfer images to the right points on the page. Used for comps and dummies.

thermal ink jet — *Related Terms:* bubble jet; hot melt ink jet.

thermal printer, transfer — printers which operate by placing a "transfer" sheet that carries ink or pigment in contact with the substrate. A heated print head that touches the transfer sheet causes image areas to transfer images to specific points on the substrate. Used for comps and dummies. *Related Terms:* thermal wax; dye sublimation transfer.

thermal printing — any system which utilizes, directly or indirectly, a heat element as the prime activator to cause imaging materials to be deposited on a substrate. Examples of thermal printing would be wax transfer and dye sublimation. Both rely upon heat to activate an imaging material on a carrier sheet and cause its transfer to the substrate.

thermal printing, direct — printing in which the image is created by using selectively heated nibs in a print head to thermally trigger the reaction of components coated on a paper or film. The paper turns dark in the heated areas.

thermal transfer — a printing system like thermal except a one-time ribbon is used and common paper is used as a substrate. Eliminates the problems of fading or changing color inherent in thermal.

thermographic copier — a reflex copying machine that makes copies by using heat reflected from the original.

thermographic paper — (1) a substrate in which an image is formed as a result of irreversible chemical changes that occur when heat is applied. Used in thermal printing applications.

thermography — a post-press finishing process that produces a raised image imitating true engraving. The process takes a previously printed wet image and dusts it with a resinous powder. The application of heat causes the ink and powder to react and a raised image is formed. *Related Term:* Virkotype.

thermo-mechanical pulp — a papermaking pulp which steams the wood chips before and during refining, resulting in a stronger pulp than regular groundwood pulp.

thermoplastic — a condition in which a material is capable of being heated and reformed after hardening and curing.

thesaurus — a (usually) alphabetical or subject order list of synonyms.

thesis paper — *Related Term:* acid-free paper.

thickness — *Related Term:* caliper.

thin negative — a negative with insufficient density due to exposure or processing deficiencies.

thin space — a small fixed space equal in width to the period or comma (usually about one-fifth of an em space).

thinner — a solvent designed to increase the viscosity of another material such as ink or paint and improve its handling characteristics.

third level domain — in a domain name, that portion of the domain name that appears two segments to the left of the top level domain. For example, the williamsport in williamsport.pa.us. *Related Terms:* second level domain; top level domain; domain name system (DNS).

thirty (-30-) — a symbol is used in manuscripts, newspapers, and press releases to signify the end of the story. Derived from the half-hour deadline at newspapers for submission of copy for the morning edition, i.e., 10:30, 11:30, etc.

thirty-two sheet (32-sheet) — a 32-piece poster whose total assembled size measures 120 in. x 160 in. *Related Terms:* eight-sheet; sixteen-sheet.

thread — a chain of postings on a single subject. Most newsreaders include a command that let you follow the thread, that is, jump to the next message on the topic rather than display each message in sequence.

threaded or chained — *Related Term:* pipelining.

three-color patch — the combination of solid yellow, magenta, and cyan areas on a color control strip. This color will normally appear slightly brown.

three-color process printing — a somewhat inferior process printing technique utilizing only cyan, magenta and yellow inks. While it is theoretically possible to reproduce all of the hues found in the original by using only these three subtractive primary inks, dark (black) areas are usually a muddy brown. The fault lies in the impurity of the pigments and colorants used in the inks in that each either reflects or absorbs outside of its theoretical range. *Related Terms:* four-color process; process; process color.

three-cylinder design — a lithographic duplicator or press containing the three standard cylinders: plate, blanket and impression. *Related Term:* three-cylinder principle.

three-em space (3-em) — metal spacing material used in foundry type composition that is 1/3 the width of an em quad. It is the most common word spacing.

three-up — imposition of three items to be printed on the same sheet. *Related Terms:* four-up; two-up; eight-up.

throughput — *Related Terms:* data rate; data throughput, data transfer rate.

thumbnail — (1) in design, a small preliminary sketch of a possible arrangement of elements. (2) in Web publishing, a term that describes the size of an image you frequently find on Web pages. Usually a photo or picture archive will present a thumbnail version of its contents (makes the page load quicker) and when a user clicks on the small image a larger version will appear. Sometimes these links will be to a new page containing the larger graphic and other times right to the image directly, as is the case in the examples below. (3) generally any small photograph less than one column wide used in printed publications. *Related Terms:* rough layout; comp; comprehensive; dummy.

tick marks — *Related Terms:* corner marks; register marks; crop marks; trim marks.

tied letters — *Related Terms:* ligature; dipthong.

TIFF (tagged image file format) — a graphics file format designed to be the universal translator of the graphics world back in the 1980s for sharing graphics across computing platforms. It can handle color depths ranging from one-bit (black and white) to 32-bit photographic images with equal ease. TIFF developed a few inconsistencies along the way. Some graphical software companies estimate that there are more than 50 variations on the TIFF format.

tilde (~) — (Pronounced "tilda") (1) on the Internet, a wavy horizontal line which signifies an individual user's Web site when housed on the server and stands for a path which leads to that person's site on it. (2) in Windows, the substitute character employed as part of the long file name convention. The tilde character is on the top line of your keyboard to the far left.

tile — (1) a single portion of an image or page that has been divided into smaller units so it can be printed. (2) the name of the activity when an image or page is too large to fit onto a standard sheet of paper, to break a page or image into smaller units so that it can be printed.

tilt — in optical reading, the rotation of a bar code symbol about an axis perpendicular to the substrate.

tilt error — another term for rotation error.

time code — the data track which synchronizes all activities in a media event.

time code — value used to number video frames. Usually consists of four two-digit numbers, representing hours, minutes, seconds and frames.

timer — a type of countdown clock used to measure specific amounts of time. Often associated with an electric switching mechanism to turn on (or off) lights, shutters, and other machinery associated with graphic arts production. *Related Term:* integrator.

timesharing — running several jobs on the same press at the same time, sharing common plates, makeup and production staff. A means of accelerating delivery and keeping down costs.

tin — a binding technique which employs a crimped metal strip along an edge of gathered sheets. Commonly used in calendars and cyclical information posters to facilitate removal of the top sheet to reveal the next in the sequence.

tinctorial strength — the concentration of colorant in a printing ink.

tinning — a method of binding by crimping a metal strip along edges of sheets.

tint — (1) a tone area that contains dots of uniform size, that is, no mottling or texture and, when printed, produces less than full strength of a color; usually referred to as a percent of color. An alternate term for screen tint. (2) the mixture of a color with white. The value of color created by adding white to a hue (color). (3) in presswork, the result of adding slight amounts of color to a normally clear varnish or UV coating, producing an light, overall hue on the product. *Related Terms:* screen tint; mechanical tint.

tint block — an area on a printed page produced in a tint, usually with type or art surprinted on it.

tinted-fiber paper — a paper containing thousands of fibers dyed a variety of colors to contrast with the color of the base stock. *Related Term:* Silurian paper.

tinting — an offset press problem identified by slight discoloration over the entire nonimage area, almost like a sprayed mist. Usually caused by contamination of the dampening solution by the inking system.

tip-in/on — (1) gluing an edge of a sheet to another. Can also be used to gather and fasten signatures or to attach loose cards or coupons in magazines and brochures. (2) the separate insertion of a single page or a smaller publication into a book either during or after binding by pasting one edge. To glue one edge of a sheet to another sheet or signature.

tissue — a thin, translucent paper often used in art and design for overlays.

tissue overlay — a translucent covering made of tissue paper that is placed over a page layout for marking instructions for printing or corrections.

title — (1) generally, the name of a book or other published work. (2) in the book publishing industry, the term used to refer to a book, such as "We're publishing five new titles this month."

title page — that part of the front matter of a book that contains the title of the book and the name of the publisher.

TN3270 — a variation of the Telnet program that allows a user to log on to IBM 3270 mainframes and use the computers as if using its terminal.

tock marks — *Related Terms:* tick marks; crop marks.

toe — the lower, concave end of the characteristic curve of a photographic emulsion where exposures are not recorded as proportional densities. *Related Terms:* shoulder; straight line portion.

token-ring — a type of LAN that uses a ring structure with a "token" revolving in the ring. When a workstation needs to send data, it captures the token and attaches information to the token.

tolerance — the acceptable distance registration marks are off.

tombstone — headlines set side by side in similar size, column width and type style that compete for the reader's attention. Two headlines placed next to each other so they seem, at first glance, to be only one.

ton — a measure of weight: (1) short ton = 2,000 pounds (used within the U.S.). (2) long or imperial ton = 2,240 pounds (used within Canada and UK) and (3) a metric ton = 1,000 kilos (used elsewhere in the world). International cargo travelling between countries is usually measured in metric tons, regardless of the system used in the originating country.

tonal compression — because film cannot record all of the tones an eye can see and printing cannot reproduce all tones recorded in a photograph, many tones are eliminated or compressed into larger areas of tones in the reproduction process.

tonal range — a photographic term for density range, or the difference between the maximum highlight density and the maximum shadow density of a photograph as measured by a densitometer or calibrated gray scale.

tone — (1) any variation in lightness or saturation while hue remains constant. The harmonious effect of light and shadows in photographs or art. (2) shading or tinting a printing element.

tone line process — the process of producing line art from a continuous tone original. *Related Term:* posterization.

tone reproduction — a term that relates the density of every reproduced tone to the corresponding original density. This may be enhanced, modified and/or exaggerated by the use of specific graphical techniques.

tone, half — *Related Term:* continuous tone; halftone.

tone, quarter — *Related Terms:* quarter tone; halftone; continuous tone.

tone, three-quarter — a tonal value that is approximately 75 percent of the total dot area.

toner — (1) pigments added to printing inks as supplemental colors, used to get greater tinctorial strength. (2) the powder used to form images in photocopiers and laser printers.

toner cartridge — the replaceable container with the powdered toner used in laser printer imaging.

toning — in photography, the on-purpose process of changing the tone and enhancing the darker tones in a photographic print. The most popular toning agent is a solution of one part selenium to twenty parts water. In a more highly concentrated form it will expand the apparent contrast within a negative. (2) in printing, the undesirable appearance of ink color in non-image areas, often the result of contaminated dampening solution. *Related Terms:* scumming; tinting.

tool — (1) an alternative name for both embossing and debossing. (2) the name of the process used by fine bookbinders when they work with leather covers. The process of decorating the leather.

toolbox — an on-line facility that, with the use of a mouse or other pointing device, allows the user to choose from a selection of "tools" to create simple geometric shapes, lines, boxes, circles etc., to add fill patterns or to select or execute other computer software functions.

tooth — (1) in screen printing, roughening of the threads of monofiliment screen fabrics to increase the stencil's ability to hold on to the fabric; is applied prior to mounting a stencil. (2) in papermaking, the rough surface of paper finishes such as antique or vellum and of papers made for drawing. The characteristic which readily allows it to take printing ink. *Related Term:* bite.

toothiness — the amount of tooth. *Related Term:* tooth.

top level domain (TLD) — in the domain name system (DNS), the highest level of the hierarchy after the root. In a domain name, that portion of the domain name that appears furthest to the right; the com in prof-lyons.com. *Related Terms:* domain name system (DNS); root, domain name.

Total Quality Management (TQM) — the process of making each manufacturing or service step responsive to the customer's needs or expectations.

touch screen — a computer display screen which has infrared (IR) sensors mounted about its perimeter. When a finger or pen is placed against the screen, the intersecting beams are disturbed and cause an action as if it had been entered from a keyboard, mouse or other input device.

touchplate — another plate beyond those used in four-color process printing, usually to strengthen a specific color. *Related Terms:* hexachrome; hi-fi printing; duotone.

toughcheck paper — a strong paperboard made for products such as tickets and tags.

toxicity — the relative poisoning effect of a substance. The higher the toxicity, the less of the substance that would be required to do an individual harm.

TR (terminal ready) — a signal sent during modem communications indicating that the computer is running a communications program.

track — the circular or spiral storage pattern on a diskette, hard drive or CD.

tracking — (1) often a term used by typographers in place of overall letterspacing. (2) in desktop and digital publishing, a feature in some layout and word processing software packages which reduces the set width of all selected characters by a fixed amount, effectively moving every one of them closer together. Usually not a mathematical value which can be specified, but one of several predefined options such as "none," "tight," "very

tight," "loose," "very loose," and "normal." Not to be confused with kerning, which only moves selected characters together for better appearance. (3) in presswork, printed streaks appearing in the paper flow direction when ejected from a printing press. They are usually caused by guide or pullout wheels which have been misplaced and are running over wet heavy ink coverage areas, leaving trace lines over the remainder of the sheet. *Related Terms:* letterspacing; kerning.

tracking number — The reference number assigned to an e-mail message when it is received by Network Solutions at the hostmaster@internic.net address. The format is: NIC-YYMMDD.#, where YYMMDD represent the year, month, and date that the acknowledgment is sent, and # is the unique number assigned to that particular request. *Related Terms:* auto-responder; hostmaster.

tractor feed — the device which moves substrate through a printer by use of prepunched holes manufactured along the edges of the substrate.

trade bindery — a graphics business specializing in trimming, folding, binding, and other finishing operations.

trade camera service — *Related Term:* camera service.

trade customs — business terms and policies followed by businesses in the same field and often codified by a trade association and court opinion. *Related Term:* printing trade customs.

trade paperback — a higher quality paperback or softcover book, as opposed to a mass market paperback, which is printed on less expensive paper.

trade publisher — the classification of publisher who publishes books directly for the book trade, selling books to bookstores and libraries and usually without a retail operation.

trade shop — a printer or other service working primary for other graphic arts professionals.

trademark — a word, phrase, graphic image, or other symbol used to represent a business, commercial, or other organization. Trademarks are used to identify the organization to the public and to consumers and are intended to identify the organization's products and services as well. To be recognized as a trademark, the word, phrase, graphic image, or symbol must be registered with the U.S. Patent and Trademark Office (in the United States) or, in the case of other countries, with the appropriate authority for that country. Registered trademarks should be accompanied by the (™) — a symbol recognizing its official status.

traditional — the creation of documents or publications using mechanical means as opposed to digital means, i.e., hand composition, relief forms, etc., as opposed to modern techniques which employ personal computer terminals and page layout programs.

transaction processing — a system designed for high volume and based on workflow orientation, involving setting up automated procedures for handling commonly repeated tasks, where work is passed automatically from worker to worker in assembly-line fashion.

transceiver — a terminal that can both transmit and receive information.

transfer — in Web publishing, the process of changing the party who is listed as the domain name registrant. The party taking over the domain name is responsible for paying a new registration fee. *Related Terms:* registrant name change agreement; registrant; registration fee.

transfer cylinder — *Related Term:* delivery cylinder.

transfer electrostatic copier — a type of electrostatic copier in which the electrostatic charge is placed on a photo-conductive plate.

Transfer Key — the Imation (formerly 3-M) Corporation trade name for integral color proof.

transfer letter — a letter for printing obtained by rubbing the top of a master sheet to affix a letter or symbol onto the layout sheet. *Related Terms:* dry transfer; Letraset; rub-on letters; cold type.

transfer lithographic plate — a plate formed from a light-sensitive coating on an intermediate carrier; after exposure, an image is transferred from the intermediate carrier to the printing plate.

transfer type —letters, usually printed on plastic or translucent paper, that can be rubbed off onto another surface such as paper using a burnishing tool. *Related Term:* transfer letter; press-on type.

transillumination — passing light through an original for reproduction or display. The opposite of reflective illumination. *Related Terms:* transmissive copy; transparency; transparency camera copy.

transistor — an electronic device that is similar to the electron tube in use and consists of a small piece of a semiconductor such as germanium, and at least three electrodes. It is one of the most important modern inventions and is found in nearly every electronic device. Created in the late 1940s, it is smaller, less expensive, and cooler running than the vacuum tubes then used to amplify current in electronic devices. Today, millions of transistors are packed into single silicon chips in the form of integrated circuits to create the processors used in modern computers.

transition — the flow of a program from one video clip to the next. The simplest transition is a cut; more complex transitions include dissolves and wipes, fades, etc.

transition point — the point in binary scanning where the image changes from black to white or vice versa.

transitional type — a typestyle with serifs that are relatively long and contain smooth, rounded curves. Generally considered a subset of the roman serif classification.

translator — in broadband network operation, a device which is located in a central retransmission facility to filter incoming microwave signals and retransmit them in a higher frequency band. Commonly used television and radio broadcasting to extend service areas.

translite — a piece of glass or plastic lit from behind and on which a photographic image has been reproduced for display. *Related Term:* shadow box.

transmission channel — the medium by which a signal is sent and received between separate locations, i.e., radio, microwave, etc.

transmission copy — *Related Term:* transillumination; reflection copy.

transmission densitometer — an instrument that measures the amount of fight that is transmitted through film from a measured light source. *Related Term:* reflective densitometer.

transmission loss — the decrease in signal along a circuit due to resistance or impedance.

transmissive copy — a slide or transparency used as original art for reproduction. Such copy is viewed by the light transmitted through its surface and can only be photo reproduced with back illumination. *Related Term:* reflection copy; transillumination.

transmittance — the measure of the ability of a material to pass light.

transmitter — a device for transmitting a coded signal when operated by any one of a group of actuating devices.

transparency — (1) a materials characteristic when light passes through it with minimum interference. Clear. (2) a black and white or full color photographically produced image on transparent film; a slide. *Related Terms:* transmission copy; chrome; slide.

transparency scanner — a very expensive, extremely high resolution scanning device designed specifically to scan transparency copy such as 35 mm slides, etc. The optical resolution of these devices commonly is 5-10 times that found in a conventional flatbed scanner.

transparent camera copy — *Related Terms:* transmissive copy; chrome; transillumination.

transparent color proofs — proofs used to check multicolor jobs. Each individual color is carried on a transparent plastic sheet, and all sheets are positioned over each other to give the illusion of the final multicolor job. *Related Terms:* color key; laminated proof.

transparent copy — any color transparency or positive film through which light must pass in order for it to be correctly viewed or reproduced. *Related Terms:* transmissive copy; transillumination.

transparent GIF — a feature of the GIF89a graphics standard, which lets the background show through selected parts of an image. The designer can designate one color in the image's palette as transparent. When the GIF is displayed, areas using that color reveal whatever is underneath. Transparency is most often applied to a GIF's background color to let the page's own background show through, so that images appear to float on the page. Most modern Web browsers support transparent GIFs. Those that don't simply display the images as normal GIFs. *Related Term:* GIF.

transparent ink — a printing ink that contains a vehicle and a pigment with the same refractive index and does not conceal the color underneath. A key feature of process color inks. Excluding the selective color absorption, these inks will allow light to be transmitted through them without loss. *Related Term:* opacity.

transparent-based image — (1) any image carried on a translucent or transparent base and intended to be viewed by light passing through it (trans-illuminated).

transponder — a microwave repeater (receiver and transmitter) in a satellite that amplifies and down-converts the frequency of a received band of signals.

transport mechanism — (1) in computing, a mechanism which moves magnetic or paper tape past reading or recording beads. (2) generally, any mechanism for moving materials at a controlled rate, such as photo paper past a laser in an imagesetter or platesetter, etc.

transpose — commonly used term in both editorial and design to designate that one element (letter, word, picture, etc.) and another should switch relative positions. The instruction is abbreviated tr.

transposition — a common typographic error where letters or words are not correctly placed; "hte" instead of "the," etc.

trap — *Related Term:* trapping.

trapezoid error — on computer monitors, the screen image distorted so that it appears wider at the top of the display than at the bottom, or vice versa. It is a common problem, and most monitors include controls to help correct the image so that it is truly rectangular. *Related Terms:* pincushioning; rotation error; screen geometry.

trapping — a prepress technique which allows for variation in registration during the press run by causing areas which "butt" to actually overprint. In conventional negatives work, it is done by making photographic masks; on the desktop, this is done primarily by allowing an overlap between abutting colors.

trash — in Web browsers, all messages, pages, etc., which are discarded by use of the delete mechanism or the trash can.

trash can — a computer screen icon to which files to be deleted are brought by mouse action. A central repository for the delete operation.

tray agitation method — a chemical tray method used to control the development stage of film processing by washing fresh developer across the emulsion, either constantly or intermittently. In halftone work it makes the black dots smaller and the clear dots larger. *Related Terms:* still development; constant agitation; intermittent agitation; tray development.

tray bath method — conventional tray development. Used in hand developing where the exposed film is placed in four (or more, if the process demands it) successive trays that usually contain developer, short stop solution, fixer (hypo) and water. Opposite of machine processing.

triadic — a color harmony using any three colors at the points of an equilateral triangle on the color wheel, e.g., orange, violet and green.

triangle — a three-sided polygon instrument used to draw right angle lines with a T-square; available in a variety of angles, such as 30 degrees, 45 degrees, and 60 degrees.

trichromatic — the technical name for RGB representation of color to create all the colors in the spectrum. *Related Terms:* additive colors; subtractive colors.

trilinear texture filtering — a complex technique used by 3D graphics cards to make movement through rendered landscapes realistic even in fast-moving games. A much more sophisticated cousin of bilinear texture filtering. *Related Term:* bilinear texture filtering.

trim — the cutting to press size of parent sheets prior to production, or the cutting of the finished product to the correct size. Marks incorporated on the printed sheet show where trimming is to be done.

trim lines — small lines on a mechanical, negative, plate, or press sheet used to guide the paper cutter in trimming the paper pile. *Related Term:* trim marks.

trim marks — lines on a mechanical, negative, printing plate, or press sheet showing where to cut edges off of paper or where to cut paper apart after printing. *Related Term:* trim lines.

trim size — the size of the finished product after last trim is made. When imposing the form for printing, allowance must always be made for final trimming to eliminate unwanted paper areas and make final product the correct size to meet specifications for delivery.

Trinitron — a type of CRT (cathode ray tube) developed by Sony Corporation. It differs from the standard tube types because it employs an aperture grille (wires stretched vertically down the screen) instead of the usual shadow mask (a metal plate with holes in it). Many observers believe that Trinitron tubes generate brighter, clearer images than those using shadow-mask technology. You can tell if you are looking at a Trinitron tube by the presence of one or two very thin black horizontal lines that cross the screen; these lines are the supports for the aperture grille. *Related Terms:* aperture grille; CRT; shadow mask.

tripod — a usually adjustable, three-legged support device for cameras. Often with gimbaled mounting surfaces to facilitate accurate alignment with the field of view.

tristimulus colorimeter — an instrument which measures tristimulus values and converts them to chromaticity components of color.

tristimulus colors — (tri-stim-u-lus) three proportionately matched color stimuli that can closely match any given color. Usually, red, green, and blue lights are used, and their composition can range from monochromatic spectral lines to bands of wavelengths, but each comprises about one-third of the visible spectrum.

tritone — a black and white photo printed with three plates. May produce a multitone effect or extend the tonal range by having each plate concentrate on a particular area of the image, i.e., highlights, midtones, or shadows.

Trojan horse — an unauthorized computer program which, appearing to be a legitimate program, when executed produces undesirable or harmful results on purpose. Most commonly build into unethical games and utilities it might, for instance, delete or erase your hard drive or trash other data files, rendering them unreadable. Sometimes it carries within itself a means to allow the program's creator access to the system using it. *Related Term:* virus; worm.

troll — a term used to define a public message that is posted for the sole purpose of offending people and/or generating an enormous flood of (often) angry replies.

true duotone — reproduction of a (usually) black and white continuous tone image by producing two distinct halftone negatives with different screen angles, exposures and densities. The negatives are printed in register, usually with two different colors, a darker one for the shadows and a lighter one for the highlight information. *Related Terms:* fake duotone; duograph.

TrueSpeech — a Netscape Navigator plug-in which allows real-time audio over the Internet. The codec is as small as possible, and the sound goes through at an eighth the size and rate of the smallest PCM WAV and AIFF files, but has the qualitiy of a mediocre pay-phone. *Related Terms:* ADPCM; AIFF; codec; WAV; PCM.

TrueType — a scalable type technology built into Windows 3.1 and Apple's System 7.

Trumpet Winsock — a popular Windows TCP/IP stack that provides a standard networking layer for many networking applications to use.

truncation — in video compression, the technique of reducing the number of bits per pixel by throwing away some of the least significant bits from each pixel.

trunk — a large-capacity, long-distance channel used by a common carrier to transfer information among its customers.

T-square — a drafting or drawing instrument used to create parallel lines and, in conjunction with a triangle, perpendicular lines or specific angle lines.

T-SQUARE

One traditional tool which has been largely replaced by compter programs and techniques.

TTFN — chat room talk for ta-ta for now.

TTY — an obsolete abbreviation for teletypewriter. Still used as a term for asynchronous transmission.

tungsten — a common artificial light source with a pronounced reddish color, which the human eye is generally unaware of, but is recorded on color film. Some films are balanced to record color naturally under this illumination. Tungsten bulbs with glue glass enclosures emanate daylight color and may be used with daylight films.

Turing experiment (or test) — named for computing pioneer Alan Turing, this is one element of testing the robustness of artificial intelligence. In the 1950s, he believed that, by the millennium, a computer with the right program could engage in a written conversation, and for 5 minutes pass for a real human about 70 percent of the time. The test involves participating in a written conversation via e-mail or online chat with an unseen correspondent who may be a person or a program. If the participant believes it's another person when it's really a program, the software is classified as possessing true artificial intelligence.

turnaround time — the time allotted for completion of individual stages of work or the complete job. Often expressed in days or hours, and referring to the period from customer delivery of copy to pickup of final product.

turning bar — a roller bar that is used to turn a web over to enable printing on the reverse side as it goes through a press.

turnkey — a purchasing specification which makes procurement, installation and satisfactory performance an integrated unit. The process has built-in contractual responsibilities for hardware and software functionality and maintenance. *Related Term:* turnkey supplier.

turnkey supplier — any systems vendor or contractor who supplies all components and installation services required for an operational system. *Related Term:* turnkey.

tusche — (1) in platemaking, a form of greasy ink used to repair holes in solid areas of a presensitized plate. (2) in screen printing, a liquid used to form the stencil image in the tusche-and-glue screen printing method.

tusche-and-glue — an artist's method of preparing a screen printing stencil by drawing directly on the screen fabric with lithographic tusche and then blocking out nonimage areas with a water-based glue material.

TVRO (television receive only) — an earth station capable of receiving satellite TV signals but not of transmitting them. A satellite or direct TV user.

TWAIN (toolkit without an interesting name) — an interface standard that should be on the checklist of anyone buying a scanner, OCR, graphics, or fax software. If your device supports TWAIN, any TWAIN-compliant software can run on it. TWAIN's signature command is Acquire and if the Acquire option appears under a program's File menu, the software is TWAIN-compliant. *Related Term:* OCR.

TWEEZERS

tweezers — a small sprung steel hand tool used by compositors to hold and place small pieces of foundry type and spacing materials.

twin lens reflex camera — *Related Terms:* camera, twin lens; parallax; SLR; medium format camera.

twin wire — (1) in paper manufacture, a paper machine which employs two Fourdrinier wires which provide two smooth sides. (2) in paper merchandising, paper made in a twin wire machine and which has an identical smooth finish on both sides.

twin-wire machine — one relatively new type of papermaking machine that has two Fourdrinier wires instead of one for producing paper with less differences between the a and b sides, i.e., wire vs. felt sides.

twisted pair — telephone companies commonly run twisted pairs of copper wires to each customer's household. The pairs consist of two insulated copper wires twisted into a spiral pattern. The wires are twisted to remove or reduce electromagnetic interference without bulky shielding. Originally designed for plain old telephone service (POTS), these wires can also carry data. New services such as ISDN and ADSL also use twisted-pair copper connections. *Related Terms:* ISDN; ADSL; POTS; two-wire circuit.

two-color — any printed work that is created using only two colors, whether they are spot or process colors.

two-cylinder design — an offset lithographic duplicator configuration that combines the plate and impression functions to form a main cylinder with twice the circumference of a separate blanket cylinder. *Related Terms:* two-cylinder principle; three-cylinder principle.

two-cylinder principle — *Related Term:* two-cylinder design.

two-dimensional space — the design element areas where words, artwork and photographs are to be located and printed.

two-dimensional symbology — a bar code that can, through *x, y* scanning technology, encode more data than the traditional linear symbologies and is designed for use where space is limited or where more data is needed in the same amount of space.

two-dimensional symbols — *Related Term:* stacked codes.

two-sheet detector — a detection device that stops the press or reroutes the paper flow to an auxiliary bin when more than one sheet passes it on the register board. *Related Terms:* two-sheet caliper; double-sheet detector.

two-sided — any type of paper that has a different textures or consistencies on its respective sides. *Related Term:* duplex paper.

two-stage masking — color correction process where a full-range positive premask is made from one separation negative and combined with another separation negative for the purposes of making a final mask that, when returned to the first negative, will correct for some of the unwanted ink absorptions without lowering contrast. A total of three premasks and three final masks are usually made.

two-tone paper — *Related Term:* duplex paper.

two-up — The imposition of two items to be printed on the same. *Related Terms:* three-up; four-up; eight-up.

two-wire circuit — *Related Term:* twisted pair.

tympan sheet — the special, smooth oil-treated paper used as the packing sheet on platen presses that holds the gauge pins and covers the impression cylinder of cylinder letterpress machines; often referred to as a "drawsheet."

type — a distinctive letter design (font), usually produced in a range of sizes (series) and variations (family), including bold, italic, condensed, etc. Includes letters, numerals, punctuation marks and other symbols produced by a machine and that will be reproduced by printing. All the characters used singly or collectively to create words, sentences, paragraphs, etc.

Type 1 — the international type standard for digital type, available on almost every computer platform. Originally invented by Adobe Systems, Type 1 is now the most commonly available digital type format and is used by professional digital graphic designers. More than 30,000 fonts are available in the Type 1 format. *Related Terms:* ATM; PostScript; TrueType.

type bank — a cabinet designed to hold foundry type cases and line spacing material. It often has a sloping top where type cases can be placed during composition. *Related Term:* California job case.

type case — a cabinet or compartmentalized tray used for storing foundry type and from which it is set. *Related Term:* California job case.

TYPE BANK
Type cases were stored as drawers below the sloping bank where they were placed during the hand composition operation.

type casting — (1) molding movable metal (foundry) type or individual types from a molten lead, tin and antimony compound such as in the Monotype casting process; (2) the hot metal formation of full lines of type as is done by a line casting machine such as a Linotype or Intertype.

type classifications — a subjective grouping of alphabet designs by general letter characteristics. Usually, the most basic classifications are roman, serif, sans serif, square serif, gothic, script and occasional or decorative. Some sources further break down these six basic classes to include other definitions such as oldstyle, modern, transitional, etc.

type composition — assembling alphabetical symbols into words, sentences, paragraphs, columns and entire pages.

type face — (1) in relief printing processes, the raised surface carrying the image of a type character cast in metal and which receives ink for letterpress printing. (2) the true name of the types installed in computers, which are often called "fonts." (3) the unique and distinctive design of a font alphabet; all of the letters, figures and punctuation of a specific font. *Related Terms:* type font; font family; font series.

type family — a collection of type faces related in design but differing in character weight, all being variations of one basic style of typeface design. The range of a type family can include roman, italic and boldface, extended, italic, italic bold, etc. Some type families also include condensed, extended and light to extra bold weights and other permutations of the basic design. *Related Terms:* type series; type font.

type font — all of the letters, characters, symbols, and numbers of a particular type face in a specific size, i.e., 8 pt. Times Roman; 12 pt. Times Roman, etc. *Related Terms:* type series; type family.

type form — an assemblage of hand or machine set lines of metal type arranged into paragraphs, columns, paragraphs or pages, usually in a press chase for lockup.

type gauge — a fixed caliper device de-signed to check if type components are .918 in. (type high). Often applied very incorrectly to a line gauge. *Related Terms:* line gauge; printer's measure; type high.

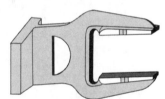

TYPE GAUGE
Often misused name for a line gauge. This device is used to determine that cast metal type is really "type high" or .918", the standard height from base to face of all type and mounted engravings to assure uniform inking and reproduction.

type high — .918 inches. The base-to-face dimension of all standard letterpress image carriers including hot metal composition, hand-set foundry type, mounted engravings, etc. in the United States and Great Britain. *Related Term:* type gauge.

type house — a company that specializes in setting type. *Related Term:* type shop; service bureau; trade shop.

type matter — any part of a form which is type. All of the letters, punctuation, symbols, and other materials used to form words, sentences and paragraphs. Not usually used to refer to illustrations, cuts, etc.

type series — in foundry type, the full range of sizes of a basic typeface design. Customary standard sizes are 5, 6, 8, 9, 10, 12, 14, 18, 24, 30, 36, 42, 48, 54, 60, 72, 84 and 96 points. All may not be included in every type design.

type shop — *Related Terms:* typesetting; type house.

type size — typeface as measured from the top of its ascenders to the bottom of its descenders and expressed in points. *Related Terms:* point size; set size.

type specifications — the directions written on rough layout that inform the compositor about alphabet style, series, size, alignment and amount of leading or space between lines.

type specimen book — a book showing examples of all typefaces available from one type shop, typographer, etc.

type stress — the degree to which a character deviates from the perpendicular of the base line. The more slanted the face, the more stress it is said to possess.

type style — a grouping of alphabet designs related by common characteristics and consistent design. There are essentially six type classifications: serif, sans serif, script, gothic, square serif and display or decorative. All type faces fit into one of these categories.

typescript — *Related Term:* manuscript.

typesetter — (1) a machine that creates phototype or hot type. (2) the person who operates such a machine and applies style specifications such as typeface and point size to raw text.

typesetting — (1) any process which creates or assembles alphabetical symbols, punctuation, and figures into words, sentence, paragraphs, etc., for the purpose of printed communication. *Related Terms:* composition; phototypesetting; imagesetting.

typesetting, digital — imagesetters and third-generation phototypesetting machines that eliminate the need for film fonts by storing digital codes in the computer unit and producing type characters as microscopic dots. *Related Term:* imagesetter.

typewriter composition — type composition for reproduction produced by a typewriter. *Related Terms:* strike-on; direct impression composition.

typo — (1) an unintentional error made in a published work, such as a misspelling or missing letter during keyboarding or by the typesetting machine itself. The latter is often referred to as a machine error. (2) an abbreviation for typographical error.

typographer — (1) in composition, a person able to make typesetting decisions regarding the kind of type, style, size, line lengths, line spacing, etc., and who has particular knowledge in the art of typography. (2) the craftsperson who designs typefaces. *Related Term:* typographic designer.

typographic color — the apparent blackness of a text block. A function of the relative thickness of the strokes of the characters in a font, as well as the width, point size and leading used for setting the text block.

typographic designer — a craftsperson who designs typefaces. *Related Term:* typographer.

typographic errors — errors made in copy while typing. *Related Term:* typo.

typography — the art and process of working with and printing from type. Today's technology, by mechanizing much of the art, is rapidly making typography a science.

Tyvek™ — the trade name for a synthetic paper material made by DuPont. Used extensively for shipping envelopes due to its printability, general toughness and ability to remain waterproof.

U & lc — an abbreviation for upper and lower case letters used to specify text copy that is to be set in capital and lowercase letters.

U interface — an interface that connects NT1 terminated ISDN devices to the telephone company's ISDN network. A device that doesn't have a built-in NT1 has an S/T interface. *Related Term:* S/T interface; NT1.

UART (universal asynchronous receiver transmitter) — the first really successful integrated circuit. It is the key component of the serial communications port. Data is transferred within the computer in units of 1 to 8 bytes, with a separate wire for each bit within the byte. It receives a byte and converts it to a sequence of voltage changes to represent the 0s and 1s on a single wire. The modem takes the signal on this wire and converts it into sound. At the other end, a modem converts the sound back to voltages, and another UART converts the stream of 0s and 1s back into bytes of data. *Related Term:* NS 16550.

UCA — the abbreviation for under color addition.

UCC (Uniform Code Council) — formerly the Uniform Product Code Council; the organization which administers the UPC and other retail standards.

UCR (under color removal) — a technique that can be achieved manually in color separation films or digitally in image-editing applications. The process reduces the amount of magenta, yellow and cyan in neutral areas and replaces them with an appropriate amount of black to improve trapping and reduce ink costs. *Related Term:* gray component replacement (GCR).

UDMA (ultra direct memory access) — a new protocol for the interface protocol between the hard drive and the computer. An improvement of the ATAPI/EIDE standard that doubles data transfer rates, which translates into faster disk reads and writes. For users to take advantage of it, both their system and hard drive must support the protocol. UDMA retains backwards compatibility for previously existing hardware. *Related Terms:* DMA; ATAPI; EIDE.

UDP — *Related Term:* user datagram protocol.

ultraviolet (UV) — invisible, electromagnetic radiation of a shorter wavelength (200-400 nm) than blue. Can create visible fluorescence when used to illuminate some materials. It is used in making exposures to create photopolymer printing plates.

ultraviolet curing ink — specially formulated inks that dried by exposing the printed product to an ultraviolet source. It results in drying at lowered temperatures as opposed to radiated heat and high temperature.

ultraviolet radiation — the range of electromagnetic radiation (light wavelengths of 200 to 400 nanometers) that lies outside the visible spectrum. In the graphic communications, UV rays are used to induce photochemical reactions.

unbalanced folds — folds where the edges do not align normally; products like roadmaps. They are difficult for the binder to create and often are upcharged.

unbleached — a term used to describe papers with a light brown coloring that have been manufactured with unbleached pulp. *Related Term:* kraft paper.

uncalendered paper — a paper unsmoothed by calendering.

uncoated — *Related Term:* uncoated paper.

uncoated paper — a very porous paper that has not been coated with clay. *Related Terms:* advertising paper; offset paper.

under color addition (UCA) — the making of color separations, balanced to increase the amount of the process colors in shadow areas of the image. The effect is to make shadows more dense and increase the apparent tonal range.

undercut of cylinder — the amount a printing press cylinder is made smaller in circumference than its bearers to accept a thicker plate, packing and/or a blanket.

underexposure — a photographic result that appears too dark or lacks shadow detail because too little light reached the film or print.

underlay — pieces of tissue or paper pasted under a letterpress form to increase impression in areas light in impression.

underline — (1) the typographic inclusion of a line under letters and words. This is a sample of underlining. Generally avoided in favor of bold face or italic in better composition. (2) a sometimes used alternate term for a caption.

underrun — a production run of fewer copies than the amount specified. The situation of a printer who manufacturers fewer printed copies than originally specified. *Related Terms:* unders; overs.

unders/overs — the quantity of printed materials that is under or over the originally specified print run.

undertone — a color of ink printed in a thin film.

Uniform Product Code (UPC) — a 12-digit bar code used to identify retail products to automated scanning equipment and computerized merchandising. The codes are assigned by the industry group, the Uniform Code Council.

Uniform Resource Locator (URL) — an absolute location for a given piece of information. URLs are used by Web browsers to locate information. The protocol is: protocol//host/path/filename. For example, the URL for Professor Lyons' homepage is http://www.prof-lyons.com

uniformity — the ideal in a monitor which delivers equal brightness across the entire screen. In practice, many monitors have darker areas or appear patchy, demonstrating a lack of uniformity. These problems are particularly easy to spot on a white background.

union bug — a nickname for labor union identifiers. Placed on work performed by their members to indicate their production of a work.

unit — (1) in typography, a unit of width or an electronic fraction of an em whose value is based on the specific output device. In a 36-unit phototypesetting system, for instance, an em would have 36 units; some systems use more units, which allows greater latitude in programming space between letters and words and in designating character widths. (2) in publishing, an individual printed piece, such as a book or magazine. (3) in presses, a single printing machine. A printing couple. A six-color printing press would be said to have 6 units.

unit cost — the cost of producing a single piece of a production run. It is calculated by adding fixed costs plus variable costs and dividing by the total run length.

unit count — a method of determining whether display type will fit a given area. Often used in calculating headlines and major heads for publications.

unit system — the unit counting method that as-

signs proportional units of width, based on the em, to each typeface character. *Related Term:* unit.

unitized — the unit measuring system used in hot metal, photographic and digital composition. A process that determines the quantity of space (width) that type characters occupy when typeset.

units — self-contained color stations on multicolor presses. Each includes a separate inking, dampening and printing system for a single process or special color, varnish or other coating.

unity — the design characteristic concerned with how all elements of a job fit together as a whole. The tying together of image elements.

Universal Copyright Convention (UCC) — a series of laws which gives protection to authors or originators of text, photographs or illustrations etc., and to prevent their use without permission or acknowledgment. Although still used in the United States, publications should, but are not required, to carry the copyright mark ©, the name of the originator and the year of publication.

universal serial bus (USB) — an easy to use and flexible interconnect specification that enables instant Plug and Play peripheral connectivity. It eliminates all of the ports on the back of your PC, mouse, keyboard, serial, parallel, joystick, etc., with a single port which can daisy-chain as many as 127 peripherals. It allows users to add devices without expensive add-in cards or configuration headaches such as DIP switches and IRQ settings. The port supports data transfer rates up to 12MB/sec, making it suitable for even high-bandwidth applications such as video. A single connector type simplifies connection of all USB-compliant devices, including telephony and broadband adapters, video phones, digital cameras, scanners and monitors in addition to joysticks, keyboards and other peripherals.

UNIX — an operating system developed by the Bell Laboratories and designed to support multiple users in a multitasking environment. It runs on a variety of platforms, which makes its development a subject of widespread discussion. The operating system itself, unlike DOS, is case sensitive. Because it is a somewhat open architecture, many incompatible versions have been developed, creating a large uncontrolled universe which had inhibited its wide acceptance because of these version conflicts.

unjustified — Lines of type set at different lengths

which align on one side (left or right). *Related Terms:* ragged left; ragged right; ragged center; flush left; justification; quad left.

unsharp masking — (1) in conventional darkroom operations, a light-transparent photographic image that is deliberately unsharp. The mask is combined with the sharp image from which it was made, thus enhancing the sharpness of subsequent images made from the combination. (2) in digital imaging, the process of increasing contrast around image edges. Typically by using individual settings which determine the amount or level of masking, the number of pixels around an edge that will be affected and the threshold or contrast level needed between adjacent pixels to be considered an edge.

unwanted colors — colors that should not be in three of the patches of a color reproduction guide; for example, yellow in cyan, blue, and magenta. *Related Term:* white colors.

up — (1) in production, when machines are running, as opposed to "down" when they are not in service. Referred to as up time. (2) in job planning and imposition, a term indicating that multiple copies of a product or piece will be run on the substrate simultaneously by duplicating the image on the image carrier the required number of times. For example 1-up is a single piece, 2-up is two, etc. The technique is utilized to cut press production time and helps improve image uniformity over the total production run. *Related Terms:* duplicate forms; step and repeat; photocomposing machine.

up style — a typographic headline style where every word is capitalized and which ignores the basic rules of grammar.

UPC (Universal Product Code) — the standard bar code symbol for retail food packages in the United States. It encodes a 12-digit number assigned to specific consumer units of general trade items. The first 6 digits are assigned by the UCC. The next 5 digits are assigned by the manufacturer. The final digit is a check digit.

UPCC (Universal Product Carton Code) — a shipping carton encoding standard administered by the UCC.

upgrade fever — the almost uncontrollable, compulsive urge to upgrade hardware and/or software, with little or no consideration extended to a real need or justification.

uplink — (1) an earth station that can transmit signals to a communications satellite. (2) the information which is passed from an earth station to a satellite. The opposite of downlink, when the information passes the opposite direction.

upload — to copy a file from your computer to a server, host system or other computer. The reverse process of download.

upper case — capital letters. *Related Terms:* majuscules; capital letters.

upper rail — a term originally used to designate the upper position on a hot metal linecasting matrix which usually contained italic characters. Opposite of lower rail.

uppercase — the capitalized letters of the alphabet and other symbols produced when the shift key on a typewriter-style keyboard is depressed. Originally called uppercase because the lead type version was located in the upper portion of the California job case. *Related Terms:* minuscules; lowercase; caps; majuscules.

UPS (uninterruptable power supply) — a power source, usually batteries, which keeps your system running in the event of a power outage. Usually designed to take over immediately. On some high-speed systems the short gap between failure of the normal line voltage and resumption from the UPS can cause systems to reboot. Modern techniques include online UPS which continuously routes electricity through the batteries which are continuously recharged by the line current. These system eliminate the lag in power with the older systems and also often provide consistency in electrical current, eliminating spikes and brownouts.

upstairs — the top half of a standard newspaper page, usually containing the flag, main headline, and the most significant stories. It is important because it is usually the only portion of the newspaper which is visible in vending racks and on piles at newsstands.

urban legend — a story, which may have started with a grain of truth, that has been embroidered and retold until it has passed into the realm of myth. On the Internet, such stories spread far, fast and often.

URL (universal resource locator) — a protocol for identifying documents on the World Wide Web, the Internet equivalent of addresses. They describe the location and access method of a resource on the Internet, for example, the URL http://www.prof-lyons.com describes the type of access method being used (http) and the server location which hosts the Web site (www.prof-lyons.com). All Web sites have URLs. Although Web site URLs are sometimes long, hard to read and nearly impossible to remember, browsers have a bookmark feature which gives you the opportunity to save the location (URL) of Web sites so you can return to them later by selecting them from the bookmark list.

USB — *Related Term:* universal serial bus.

Usenet — a collection of thousands of topically named newsgroups, computers which run the protocols and the people who read them and submit Usenet news articles. Not all Internet hosts subscribe to Usenet and not all Usenet hosts are on the Internet. Usenet groups can be "unmoderated" (anyone can post) or "moderated" (submissions are automatically directed to a moderator, who edits or filters and then posts the results).

user — the operator of a particular piece of equipment.

user datagram protocol (UDP) — an Internet standard transport layer protocol. It tends to be an unreliable, connectionless-oriented delivery service, as opposed to TCP.

user friendly — any part of a computer system that is easy for a novice to use.

user interface — the usually visible parts of software which precipitate a user response, i.e., pull-down menus, cancel or proceed radio buttons, etc.

user-defined keys — keys on a terminal that can be programmed to mean any combination or string of characters desired. *Related Terms:* macro; scripts.

userid — a compressed form of "user identification"; the userid always proceeds the @ sign in an e-mail address. *Related Term:* username.

username — a username consists of 1 to 8 characters, and only uses numbers 0 through 9 and the 26 alphabet letters. Usernames do not have spaces.

Usernames are the first part of an e-mail address: yourname@ix.aol.com . You must have a username and a services password to log in to a mailbox or when logging into an access provider or when entering a members only area on the Web. *Related Term:* userid.

utilities — nonessential operations programs which enhance a computer's capability, protect it from viruses, speed up productivity, check spelling, etc. None of the functions are critical to basic operation but all perform work which makes the general computer experience safer, faster and more secure.

UTP (unshielded twisted pair) — a cable type that has one or more pairs of wires twisted together to improve its electrical properties. Unshielded refers to the fact that there is no metal shield around the cable.

UUCP (UNIX to UNIX copy) — a tool for transferring files, sending mail, and executing remote commands.

UUdecoding — the restoration of uuencoded data to its original form.

UUEE (UNIX to UNIX Encode) —one tool for transferring files through e-mail.

UUencoding — the conversion of binary data into a 7-bit ASCII representation using an encoding scheme. Originally implemented to enable users to send such data over UUCP, it is now used to send binary files such as graphics files, user application documents and programs through e-mail and on Usenet. *Related Terms:* UU decoding.

UV coating — a liquid laminate applied to press sheets and which is bonded and cured with ultraviolet light. It is designed to minimize ink chipping, scratching, and other damage from normal use.

UV inks — printing inks used to print on most substrates that can be cured instantly by a high-output ultraviolet light source, and without excessive radiant heat.

V. standards — specifications established by CCITT for modem manufacturers to follow (voluntarily) to ensure compatible speeds, compression, and error correction. *Related Terms:* CCITT; KBPS.

V.120 protocol — a protocol that allows ISDN modems to transfer files using familiar protocols such as X-, Y-, and Zmodem. *Related Terms:* MPPP; PPP; protocol; Xmodem; Ymodem; Zmodem.

V.32 — a modem standard (pronounced "v-dot 32") for error correction and compression at speeds of 9,600 bits per second. *Related Terms:* CCITT; V. standards.

V.32bis — the modem standard (pronounced "v-dot 32 biss") for error correction and compression at speeds of 14.4 kpbs. Bis (French for encore) is an international designation for the first revision to a standard. *Related Terms:* CCITT; KBPS; V. standards.

V.34 — the modem standard (pronounced "v-dot 34") for error correction and compression at speeds of 28.8 KBPS. *Related Terms:* CCITT; KBPS; V. standards.

V.42bis — This is the modem standard (pronounced "v-dot 42 biss") for error correction and compression at speeds of 28.8 KBPS. Bis (French for encore) is an international designation for the first revision of a standard. *Related Terms:* CCITT; KBPS; V. standards.

V.90 — an error correction and compression standard for uploading at 33.6 and downloading at 53.3 KBPS. The system used for 56KB transfer hardware. The U.S. Federal Communications Commission limits the actual speed to 53.3. V.90 requires a digital modem on service provider side.

vacuum frame — usually a flat table device with a glass cover and a perimeter seal and a high light source, often of high intensity. Used to draw materials such as film and plates into tight contact for the purposes of making contact prints, duplicate films, plates or films. Sometimes called a platemaker, although it is not limited to this function. *Related Terms:* contact printing frame; vacuum back.

value – (1) synonym for tone, or the relative tint or shade of a printing element. (2) a term used in the Munsell System to describe lightness.

valve jet — a drop-on-demand ink jet printer in which a nozzle opens and closes to distribute ink to the substrate. *Related Terms:* asynchronous ink jet; bubble ink jet; drop on-demand ink jet.

vanilla — tech talk for the standard version with no extra features.

vanishing — a printing problem that shows up as the gradual disappearance from the press plate image of some lines and halftone dots. *Related Term:* blinding.

vanity press — book publishers whose authors pay, in advance, all of the costs of publishing whatever quantity is sought. *Related Term:* subsidy press.

VAR (value added reseller) — companies which purchase merchandise and enhance its perceived value prior to reselling it. The value added enhancements may be in additional warranty, extras, software bundles, training, or a number of other forms.

variable costs — generally, all of the production costs which are dependent upon quantity. Include such items as paper, ink, production time, binding, packaging, etc. *Related Term:* fixed costs.

variable information printing — electronically driven systems using laser or digital printing techniques which enable changing content with each press revolution under control of a data base. Originally used to place simple black and white messages on products to subscribers or purchasers, it has evolved into a complete printing system which enables the changing of color text and full color process images with each cycle. It can eliminate the need to print overruns to accommodate future need, as most of the systems have nearly no prepress expenses; thus quantities are ordered as required. Sometimes referred to as printing on demand (POD). *Related Term:* variable printing.

variable length code — in bar coding, a code whose number of encoded characters can be within a range, as opposed to a code with a fixed number of encoded characters.

varnish — (1) a usually clear-coat liquid sealer that overprints ink and paper to protect against scratches and scuffing, increase longevity, and enhance image appearance and impact. Can be gloss or dull, and may be lightly tinted for special effect. (2) the major component of an ink vehicle, consisting of solvent plus a resin or drying oil. *Related Term:* vehicle.

varnishing — an over-coating process usually done on press, sometimes even at the same time the piece is being printed. Varnish may be clear or tinted and be dull or gloss in finish. It is simply carried in an ink fountain, in place of ink and applied with a separate image carrier which details the portions of the sheet to be coated. Varnish is applied to printed matter often to enhance halftones, to make blacks "blacker" or denser and to increasing contrast and overall visual impact. Varnish will also protect the printed piece from scuffing and finger marks. Dense blacks and dark colors are unusually susceptible to showing the effects of wear and tear, particularly on glossy paper. Varnishing is usually reserved for high-gloss sheets. Rough-surfaced paper generally doesn't benefit from it.

VAT — a Unix audio teleconferencing tool that enables talk to one or more people over an Internet connection. In most cases, all you need is the Unix VAT program, IP connection, and sound hardware. *Related Terms:* IP; IP address.

VBScript (Visual Basic Scripting) — a programming language developed for creating scripts (miniprograms) that can be embedded in HTML Web pages. These scripts can make Web pages more interactive and works with Microsoft ActiveX Controls, allowing Web site developers to create forms, interactive multimedia, games, and other Web-based programs. VBScript is similar in functionality to JavaScript and is a subset of the widely used Microsoft Visual Basic programming language. *Related Terms:* ActiveX; JavaScript; Visual Basic; applets.

vCalendar — a specification defines a format for exchanging calendaring and scheduling information. vCalendars can be distributed as e-mail attachments or made available for downloading from a Web page. They hold information about event and to-do items that are normally used by personal information managers (PIMs) and group schedulers. Once the recipient has the vCalendar, they can drag and drop it into their vCalendar-compliant electronic organizer. *Related Term:* vCard.

vCard — a specification that defines a sort of electronic business card that can be sent via an e-mail attachment or as a link on a Web page. vCards can store information such as your name, address, tele-phone number, e-mail address, and so on. A recipient of your vCard can easily add your information to their electronic address book (as long as it's vCard-compliant) with a single click. Programs that do not support vCard usually handle the vCard data as an ordinary attachment. *Related Term:* vCalendar.

VCR (video cassette recorder) — an analog or digital magnetic recording and playback machine. Generally used for recording and viewing full motion video. Also useful as a data backup device. *Related Term:* digital video.

VDO — a technology that enables Internet video broadcasting and desktop video conferencing on the Internet and over regular telephone lines and private networks. Its ability to have private point-to-point audio/video contact is currently only available for Windows 95/98 and requires a Pentium processor. The VDOLive player, however, is available for Windows and Power Macs and provides the ability as a Netscape plug-in for viewing and hearing live Internet broadcasts.

VDT — video display terminal. *Related Terms:* cathode ray tube; CRT; computer monitor.

vector — (1) in imaging and computer design, the mathematical description of images and their placement in electronic publishing. Vector graphics information is transferred from a design workstation to a raster image processor (RIP) that interprets all of the page layout information for the marking engine of the imagesetter. PostScript or another page description language serves as an interface between the page layout workstation and the RIP. (2) in mathematics, a line whose length defines force and whose angle defines direction. *Related Terms:* imagesetter; page description language; PostScript; raster; raster image processor; RIP; vector file.

vector digitization — the creation of line drawing by the notation of x and y coordinates and their end points. *Related Terms:* vector; raster; bit mapping.

vector graphics — computer-aided design (CAD) programs and drawing applications such as Adobe Illustrator and CorelDraw produce graphics that don't look blocky when you zoom in on them. They scale up easily because they store geometric information about shapes and lines called vectors. These images are unlike pictures from paint programs or scanners, which are called raster graphics. One-bit-deep raster

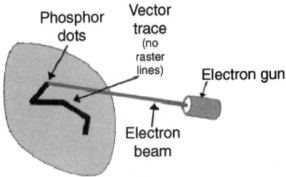

VECTOR TRACKING
Rather than scan in straight lines as in a raster scan, vector lines are determined by mathematical formulas and their actions appear to be generated as more of a freeform scan.

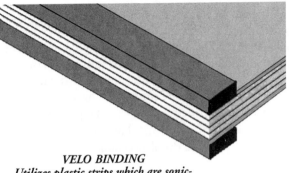

VELO BINDING
Utilizes plastic strips which are sonic-welded together after closely spaced holes are punched in all elements.

graphics are often called bit maps. *Related Terms:* bitmap; vector-based.

vector-based — the type of graphics defined by groups of lines, circles, text, and other objects, in reference to an origin point as opposed to bitmapped graphics, which are defined by pixel positioning. *Related Terms:* object-oriented graphics; vector graphics.

vectoring — *Related Terms:* digitizing; drawing program.

vehicle — a fluid that carries ink pigment in suspension and causes printing ink to adhere to paper or some other substrate. *Related Term:* varnish.

vellum — (1) the treated skin of a calf used as a writing material. (2) a fine cream-colored writing or printing material originally made from split calf skin. It has a fine-grained finish, smoother than antique. (3) the name used to describe a thick creamy book paper. (4) the name applied to artificial paper replacements for the original animal skin material.

vellum bind — binding a book with a cover made of vellum paper. The term is also sometimes applied to books bound with soft leather.

vellum Bristol — Bristol paper with a vellum finish. *Related Terms:* Bristol paper; vellum finish.

vellum finish — a somewhat rough, uncoated, toothy finish; smoother than antique, rougher than English.

Velo binding — a brand name for a binding system which utilizes small pla-stic posts mounted on a common bar. Product is punched with holes the size of the posts. The assembly is passed through the holes and a retainer bar with holes is placed over the pro-

truding posts. A heating system trims off the excess post and welds the elements to the retainer bar, producing a permanent binding.

velour finish — *Related Term:* dull finish.

Velox — (1) a generic term used to describe a photographic print prepared from a halftone negative. It is placed on paste-ups and photographed with line copy, eliminating the need for further screening or stripping operations. (2) the Kodak trade name for a high-contrast photo paper. Often used to refer to any halftone positive print, or interchangeably with PMT. *Related Terms:* photo mechanical transfer; PMT.

vendor — a supplier of goods or services, such as a printer or a service bureau.

Ventura Publisher — a desktop publishing package originally marketed by Xerox and now by Corel. The Ventura approach is a document-oriented one working on the basis that each page will have a similar format. The design lends itself to the production of manuals and directories which have similar page designs throughout. Changes since the Corel acquisition have made it more suitable for publications which have a wider variety of page designs.

verifier — in optical reading, a device that makes measurements of the bars, spaces, quiet zones and optical characteristics of a symbol to determine if the symbol meets the requirements of a specification or standard.

VeriSign server digital ID — electronic identifiers used in much the same way that a business license or articles of incorporation are used in the real world to provide verification of an organization's identity to the people with whom the organization does business. They are used by organizations wanting to authenticate Web sites, establish secure sessions, etc., and make it possible for a Web site to assure visitors of its true identity, enabling users to avoid fraudulent or spoofed Web sites. Digital IDs for servers also enable Web sites to establish secure sessions with visitors, providing a private connection over the Internet that cannot be penetrated by external parties.

vernier wheel — the small knurled wheels used to tighten the plates on the plate cylinders of lithographic duplicators.

Veronica (VEry easy ROdent-oriented Netwide Index to Computerized Archives) — a front end for searching Gopher servers using keywords and subjects. It looks for filenames and produces a menu of items linked to a Gopher data source. It is available at gopher://veronica.scs.unr.edu. *Related Terms:* Archie, FTP, Gopher.

verso — the reverse, back, or left-hand side of a page, folded sheet, book, or cover. *Related Term:* recto.

vertex – where two downward diagonal strokes meet, as in an M or W, or the bottom of a V.

vertical bar code — a bar code pattern presented in such orientation that the symbol from start to stop is perpendicular to the horizon. The individual bars are in an array appearing as rungs of a ladder. *Related Term:* horizontal bar code.

vertical blanking interval — lines 1-21 of video field one and lines 263-284 of video field two, in which frame numbers, picture stops, chapter stops, white flags, closed captions, etc., may be encoded. These lines do not appear on the display screen, but maintain image stability and enhance image access. *Related Term:* horizontal blanking interval.

vertical justification — the automatic adjustment of interline spacing (leading) or the space between lines, in very small amounts so columns on a page can all be made the same depth and end at the same point on the bottom of a page. *Related Term:* scallop.

vertical resolution — in video, the number of scan lines per frame.

VESA (Video Electronic Standards Association) — an industry organization formed to create various personal computer standards, including those for Super VGA video displays and the VLB bus standard. *Related Term:* VLB.

VFW – *Related Term:* video for windows.

VGA (video graphics array) — an IBM video dis-

play standard. Provides medium resolution text and graphics. VGA pixel resolution is 640 x 480. *Related Terms:* SVGA; CGA; XGA.

VI (Visual Interface) — the early and once de facto standard UNIX editor and a nearly undisputed hacker favorite outside of MIT.

video — the descriptive term which applies to all electronic visual techniques, television, Web, etc. The term is derived from the Latin phrase for "I see."

video adapter card — an add-in card that controls the monitor's display characteristics by controlling resolution, number of colors displayed and the rate at which the images are presented. It must be matched to the monitor being used and cannot improve a low-resolution monitor.

video bandwidth — a reference to a monitor's ability to refresh the screen. High bandwidths allow more information to be painted across the display in a given amount of time, which translates into higher resolutions and higher refresh rates. Low bandwidths result in flickering, artifacts and ghosting. To calculate the bandwidth of a monitor (measured in megahertz, or MHz), multiply the horizontal resolution by the vertical resolution, and then multiply the product of the two figures by the refresh rate. For example, 800 x 600 x 75 = 36 MHz. *Related Terms:* artifact; ghosting; latency; refresh rate; resolution.

video board — the device which, when installed in a computer, translates digital values into color information displayed on a monitor. High-end applications can really slow down this process, making accelerator boards a necessity for production environments.

video capture card — an add-in board that is required by computers to acquire video signals for compression.

video compression — a technique used to reduce the bandwidth required for the transmission of video images. It employs the reduction of redundant information within and/or between video frames. *Related Terms:* bandwidth compression; data compression; bit rate reduction.

video conferencing — conducting a conference between two or more participants at different sites by using computer networks or the Internet to transmit fully interactive audio and video data. Software programs such as CUSeeMe have brought video conferencing to the Internet and are easily available and easy to use. *Related Term:* desktop video.

video display terminal (VDT) — a networked computer station's viewing screen. Sometimes includes a keyboard or pen device for input, and may be connected to a printer or other output device. *Related Terms:* cathode ray tube, CRT, Monitor.

Video for Windows (VFW) — the multimedia technology that shipped as part of Windows 95. Its playback files have the extension .avi and can be played using Windows' Media Player. Since the files are large, they are often compressed using a codec. Video for Windows is one of three video technologies used on personal computers. The others are MPEG and QuickTime. *Related Terms:* codec; MPEG; QuickTime.

video hard copy unit — a device that electronically reproduces video images on paper.

video phone — a telephone with two-way video capability, enabling both the caller and the recipient to see one another and/or transmit visual data.

video pointer — an electronic device that produces an arrow or symbol that can be positioned anywhere on a displayed image to point out or highlight information.

videodisc — a high-density reflective disc that stores information in microscopic "pits" indented into the surface; the data is read by a reflected laser beam. It provides a high-capacity storage medium of over 50,000 frames of information; used to store and retrieve video, audio and other information. *Related Terms:* compact disc; DVD; CD-R; CD-RW; CD; CD-ROM; laser disc.

Videotex — a service similar to Teletext except that information is delivered by telephone channels and a user can interact with the data base to select information for viewing. In use internationally but, in France, a national telephone service sponsored by the government.

view camera — *Related Term:* camera, view.

viewable area — the real size of a monitor screen, whose sizes aren't as straightforward as they seem. CRTs are measured diagonally across their glass face. However, the viewable area of the monitor, the diagonal measurement of the largest possible picture the screen can display, is never as large as the actual tube size: A 14-inch monitor typically has a viewable area of 13.1 inches, etc. The exception is the LCD panels used in notebooks. They can display images edge to edge, and their physical and viewable diagonal measurements are the same. *Related Terms:* CRT; LCD.

viewer — any type of application that assists a Web browser by handling files it can't. They are called upon for non-browser files, such as sound files. Some people prefer to call them helpers. *Related Term:* helper applications.

viewing booth — a hooded table or other enclosed space where lighting conditions can be color temperature controlled for proper viewing of transparencies, press sheets, photographic prints, color proofs and separations. Usually the illumination source conforms to standard viewing conditions. *Related Term:* standard viewing conditions.

viewing conditions –*Related Terms:* standard viewing conditions; viewing booth.

vignette — (1) a halftone, drawing, or engraved illustration in which the background gradually fades away from the principal subject until it finally blends into the nonimage areas of the print. (2) an image segment with densities varying from highlight to white. (3) the characteristic of a halftone screen that enables it to make dots of variable size on film based on the vignetted structure of each dot matrix. (4) a small illustration in a book not enclosed in a definite border. (5) an ornamented border on a title page of a book. (6) an alternate, and inaccurate, description sometimes used for graduated or gradient screen tints.

VIM (Vendor Independent Messaging) — a Lotus standard, which allows cc:Mail and Notes to communicate with other applications.

virgin paper — any paper made exclusively from the pulp from new trees or cotton, with no recycled content.

virgule – the common name of the forward slash (/). It is also the shilling mark. Sometimes called a separatrix. *Related Term:* solidus.

virtual — a simulation of the real thing; the same as "almost." The term appears before various computer words to indicate simulation technology which provides the ability to cross boundaries and experience things without needing a physical presence, as in virtual sex, and virtual theme parks.

virtual circuit — a reliable link between a user and an Internet site, even though the two are not communicating over a dedicated phone line.

virtual memory — a part of your hard disk called a swap file, dedicated as a storage area for bits of data in RAM that aren't being used much. You need a virtual memory manager, which is usually a function of the operating system, that can map chunks of data and code onto storage areas that aren't RAM. By freeing up RAM, you're virtually increasing the amount of working memory available to you. *Related Terms:* RAM; swap file.

virtual reality — interactive computer simulations which approximate or imitate the real world. The scenes and their content change based upon user responses in real time.

virus — software of intentional design whose purpose is to cause problems and malfunctions. Most viruses can copy themselves repeatedly, eating up memory, causing unaccountable display problems, be spread throughout networks and cause great damage to systems and software. Many are insidiously designed to destroy large amounts of data or damage a computer system. They can spread rapidly in environments such as schools or businesses where computer users frequently share files on disks or across the network. Antiviral software is available to save data, destroy or remove the offending virus program from systems. Most modern systems continuously monitor data with virus software which eliminates offenders before they can attack the systems. *Related Terms:* Trojan horse; worm.

viscosity — the properties of tack and flow of a printing ink, glue, or other fluid. Viscosity is the opposite of fluidity. *Related Term:* ink viscosity.

visible spectrum — *Related Terms:* prism; spectrum; visible; white light.

visual literacy — knowledge in the use of visual cues and components in conveying messages to an audience. All visually perceptive elements are considered in their total, i.e., color, shape, size, text, art, spatial relationships, photographs, video, etc.

visualization — the act of mentally picturing results of events prior to occurrence. A necessary trait in creative endeavors such as art, design and photography.

VOC — an audio file format used with the earliest Sound Blaster cards under DOS. It has lost much ground to the Windows-native WAV file format. *Related Terms:* AIFF; sound; WAV; waveform.

voice mail — an automated telephone system which answers calls, then records and later retrieves voice messages of callers at the convenience of the user.

voice recognition — the use of speaking to collect data or to input commands. This requires no special printed or encoded symbols, only a device similar to a headset. It provides a hands-free option to enter data into a computer system.

voice switching — an electrical technique for opening and closing a circuit in response to the presence or absence of sound. *Related Term:* voice-actuated.

voice/data terminal — a desktop device that has the combined capability for voice and data communications.

voice-actuated — the ability of a piece of equipment to become activated in response to the sound of a voice. *Related Term:* voice switching.

voice-grade channel — a telephone circuit that carries signals in the voice frequency range of 300 to 3,000 Hertz.

voice-switched microphone — a microphone that is activated by a sound or sufficient amplitude; generally allows only one person to speak at a time.

voice-switched video — refers to a type of videoconference in which the cameras are activated by voice signals to send a picture of a particular person in the group. Not all participants can be seen at any one time, in contrast to continuous presence video.

void — the undesirable absence of ink in an intended print area. *Related Term:* blinding.

volatile organic compounds (VOCs) — chemical substances used in the printing industry that are subject to governmental regulations regarding safety hazards and emissions. Cleaning solvents are an example.

volcano — an infrequent modern-day occurence. The term which describes a condition when evaporating solvents intrude into or through an ink film thickness and cause a visual defect.

volle — the tiny circle over Scandinavian vowels (Å). *Related Term:* ring.

von Neumann architecture — the stored-program concept used in most computers and designed by Hungarian mathematician John von Neumann. With it, you store programs and data in a slow-to-access storage medium, such as a hard disk, and work on them in a fast-access, volatile storage medium (RAM). A flawed concept because it is designed to process instructions sequentially instead of being able to use faster parallel processing. *Related Term:* RAM.

voucher — a free copy of a publication which is given to an advertiser as evidence of appearance and proper execution of advertising material. *Related Term:* tearsheet.

VPN (Virtual Private Network) — a private network of computers that's at least partially connected by public phone lines. They use encryption and secure protocols like PPTP to ensure that data transmissions are not intercepted by unauthorized parties. *Related Terms:* PPTP; LAN; WAN.

VRAM (video RAM) — a type of RAM found on higher quality graphics display adapters. VRAM is much faster than DRAM because it has dual ports and can read and write data at the same time. *Related Terms:* DRAM, RAM.

VRML (virtual reality modeling language) — a 3D specification that enables the creation of 3D Web sites. It is an open, extensible, industry-standard scene description language for 3D scenes, or worlds, on the Internet. With it and certain software tools, you can create and view distributed, interactive 3D worlds that are rich with text, images, animation, sound, music, and even video. *Related Term:* HTML.

VRWeb — a browser for 3D worlds and objects modeled in the VRML. It is the only VRML browser which is freely available in complete source code, does not require commercial packages to work, and which is capable of running on virtually all platforms.

VSAT (very small aperture terminal) — a satellite dish used for reception of high-speed data transmissions. Can also transmit slow-speed data.

VSB (vestigial side band) — a digital frequency modulation technique used to send data over a coaxial cable network.

VSL (virtual software library) — an open standard that has two elements: a back end that updates a database of downloadable files available from shareware, freeware, and corporate software archives on the Internet, and a front end which makes this database available for easy searching and downloading. *Related Terms:* Archie; Veronica.

VTR — videotape recorder; a recording and playback tape deck.

vulcanized — The process of treating rubber or plastic material with chemicals or with heat to give the material useful properties such as elasticity, strength and stability.

Ww

W3 — a shortcut for saying World Wide Web.

wafer seal — gummed sealers which keep folded sheets together while in the mail stream. They may be printed or unprinted.

WAIS (wide area information servers) — a software system used to search indexed databases on remote servers, it returns a ranked list of pages or files that you can retrieve from the server. It also enables you to use so-called natural language input; in other words, you can ask simply "Why is the sky blue?" instead of having to master Boolean AND/OR constructs. *Related Terms:* Archie; FTP; Gopher.

waistline — an invisible or imaginary line which defines the top of the x-height.

walk-off — sometimes, a term used to describe the condition in which a part of an image on a plate deteriorates during printing.

wall — the sides of gravure cylinder wells or cells.

WAN (wide area network) — two local or more area networks hooked together. Wide area networks can be made up of interconnected smaller networks spread throughout a building, a state, or the entire globe. *Related Term:* LAN.

wand — *Related Term:* wand scanner.

wand scanner — a hand-held scanning device used as a contact barcode or OCR reader.

wanted colors — colors that should be in three of the patches of a color reproduction guide, for example, yellow in red, yellow, and green. *Related Term:* black colors.

warehouse management — an application that uses bar codes or other forms of AIDC within the warehouse to keep track of receiving, inspecting, storing, controlling inventory, picking, and shipping of items and locations within the warehouse facility. *Related Term:* inventory management.

warm color — generally a color containing red, orange or yellow.

wash — (1) in photography, usually the final step in film and paper development. A water bath used to remove remaining chemicals from within the layers of photosensitive materials. Failure to remove the residual chemicals degrades permanence and can lead to discoloration. (2) in printing, a term used to indicate cleaning solvents, i.e., blanket wash, etc.

wash drawing — an illustration made with ink mixed with water; watercolors sometimes are also used.

wash up — the process of cleaning ink and dampening solutions from press components such as plates, rollers and fountains, for changing colors or preparing for a new print job.

wash-coated paper — a very thin film coating on paper applied when the sheet goes through the sizing bath rather than during a separate coating process during manufacture. *Related Term:* film-coated paper.

washing out — in screen printing, using running water to remove the image area of an exposed sheet of indirect photo screen film and direct photographic stencils.

washout screen — in screen printing, a process similar to the tusche-and-glue method of preparing a screen printing stencil, painting in the image, blocking out the nonimage area, then opening up (washing out) the image area.

wash-up — the cleaning of ink from rollers, fountains, plates and other components of a press using solvents and other cleaning materials.

waste — (1) paper that cannot be printed for whatever reason, such as being too close to the core of a roll. *Related Terms:* spoilage; white waste.

waste, post-consumer — finished material that is recycled or disposed of as solid waste after its product life span is completed.

waste, post-mill — any waste generated after the paper has left the mill, including pulp substitutes and pre-consumer and post-consumer waste.

waste, secondary — fragments of finished products from a manufacturing process, including printer's trim.

wastepaper — a paper or paper product that has lost its original value and has been discarded. Printing plant waste, waste generated during paper converting and discarded boxes and newspapers fall into this category.

wastepaper, recoverability — the imprecise measure or estimate of the quantity or percentage of a wastepaper grade that could be recovered by intensive collection practice.

water fountain — the reservoir on a press to hold fountain or plate-dampening solution.

water mixer — a device attached to hot and cold water pipes to blend the water to a specific temperature.

water-base ink — printing ink with a vehicle that can be thinned and dissolved with water.

water-bath development — the process of developing the film negative in developer for a determinate amount of time then transferring it to a plain water solution. The tonal values of the shadows will continue to develop, even as the developer in the high values are depleted. The process can be repeated a number of times to accomplish the desired results.

waterless printing — a specialty method of printing in which the plates consist of metal for image areas and rubber for nonimage areas for printing without water. *Related Term:* driography.

watermark — a translucent image in paper created during manufacture by slight embossing from a dandy roll while paper is still more than 60 percent water. *Related Term:* dandy roll.

waterproof paper — a paper substrate containing a relatively high quantity of sizing, causing it to resist wrinkling when moist.

water-soluble film — a screen-printing stencil film that is hand cut and contains an emulsion that will dissolve in water.

WATS (wide area telephone service) — a flat rate or measured bulk rate long distance telephone service provided on an incoming or outgoing basis. WATS permits a customer, by use of an access line, to make telephone calls to any dialable telephone number in a specific zone for a flat or bulk monthly rate using an 800 number. INWATS permits reception of calls from specific zones over an access line in like manner. The U. S. and Canada have been divided into geographical zones of increasingly greater coverage depending on the location of the customer.

WAV — the Windows standard for waveform sound files whose extension is .wav. *Related Terms:* ADPCM; AIFF; AU; PCM; sound; TrueSpeech; waveform.

waveform — a type of sound that works like a tape recording: Speak into a computer's microphone, and your voice becomes a waveform sound file. Waveform sounds play back from disk using sound chips in your computer's system board or on an add-in audio card. System sounds, such as the Windows ding.wav file, and downloadable audio clips are all waveform sounds. *Related Terms:* sound; WAV.

wavelength — a quantitative measurement of radiant energy; the distance between corresponding points on two successive waves of light or sound.

wavetable synthesis — the recreation of sound of an existing instrument or other sound using a synthesizer. An easy way to make a wavetable is to make a digital sample of an existing sound and then modify it to change the pitch. This sample-based synthesis is often called wavetable synthesis because of the way samples are saved and played back. This system sounds better than the alternative, FM synthesis. *Related Term:* FM synthesis.

wax — (1) in paste-up, a paraffin adhesive material used to affix typeset copy and artwork to a paste-up board. (2) in presswork, a material added to ink to prevent setoff.

wax coater — a device used to place a thin layer of adhesive wax over the back of artwork or composition so that the copy can be placed and then repositioned a number of times.

wax holdout — coating materials for type galleys, veloxes, etc., that prevent adhesives from bleeding through one side of a sheet of paper to the other.

waxer — a device used to place a thin layer of semipermanent adhesive wax over the back of composed type and artwork so that the copy can be positioned and repositioned a number of times.

web — (1) a roll of any substrate that passes continuously through a printing press (as opposed to sheet-fed printing). (2) when capitalized, "Web" is short for World Wide Web. *Related Terms:* ribbon; web press.

web break — a break in the paper ribbon or "web" running through a web press, causing production to stop.

Web designer — an individual who is the aesthetic and navigational architect of a Web site, determining how the site "looks" and "feels," etc. Generally, a person on the artistic side of Web site building/developing with an extensive knowledge of Web-based programming, art, and information architecture. *Related Terms:* Web guru; Web master; Web developer.

Web developer — a person who technically architecturally "builds" Web sites, researches and provides through programming the means for a particular Web product to work. Not to be confused with the Web counterpart of Web designer.

Web guru — a flattering title usually given to a person who handles all the Web/Internet needs of an organization or company, or someone with great Internet knowledge and technical prowess and/or Web design skills.

Web jam — a layering of music, media, performers, audience, and the surrounding ecosystem into a rhythmic "jungle" on the Web. Its purpose is to provide a technologically expanded sense of nature and culture. It takes an improvisational approach to cultural, political, and ecological systems.

Web page or World Wide Web page — a single HTML file, which when viewed by a browser on the World Wide Web (WWW) could be several screen-dimensions long; you would "scroll" to view contents that are off screen. Web page sizes vary greatly from system to system and depend largely on monitor resolution. Therefore the contents of a given file, regardless of the number of screens, is considered a single Web page. Large Web sites can have hundreds of pages of information, containing hundreds of separate documents varying in length, each probably with a different topic or subject.

web press — a rotary press that prints on a continu-

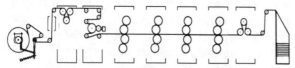

TYPICAL WEB PRESS CONFIGURATION
Note the roll stand on the left which supplies the web of paper, the tensioning units immediately to the right, and the four printing couples which each print a single color. In this example, four color process could be produced on each side of the web. The unit on the right is the delivery system which could be a folder, sheeter or rewinding unit. Also note that this configuration can be multiplied many times in a publication press, and utilize many webs to produce a newspaper or magazine.

ous web, or ribbon, of paper fed from a roll and threaded through the press. *Related Term:* sheet-fed press.

Web site — a specific collection of World Wide Web pages, usually containing a home page plus other pages with a common theme, company or interest.

web tension — the amount of resistance to pull which is applied to a web of paper on a web-fed press.

Webmaster or Webmistress — a fanciful name for anyone in charge of managing the hardware and software that make up a company's Web site. The

person to which all feedback and correspondence for a Web site is usually sent. **Terms:** Internet systems administrator; Web guru; Websmith.

Websmith — a person who builds Web sites. The developer or designer of a Web site.

wedding Bristol — a type of board that is generally two- to four-ply with fancy vellum or antique finish and is used for wedding announcements, menus and programs. Related Term: wedding paper.

wedding paper — *Related Term:* wedding Bristol.

wedge — a device that plugs in between a keyboard and a terminal. It allows data to be entered either by keyboard or by various types of scanners.

wedge spectrogram — a visual representation of a film's reaction to light across the visible spectrum.

weight — (1) the degree of boldness or thickness of a letter or font; how light or dark it appears when set. (2) the common measurement for classifying and grading paper; often a reference to paper thickness within a grade. Weight is calculated by the ream, 500 sheets. A fixed size is assigned to each type or grade of paper (for book grades, for instance, the size is 25 x 38 inches), and the weight corresponds to how much 500 sheets of that fixed size would weigh. Thus, paper stock is "70 pound" if 500 sheets 25 x 38 inches weigh 70 pounds. *Related Terms:* basis weight; gsm; ctw; character weight.

welcome page — the introductory page for a Web site, also referred to as the home page. The first page of a Web site to contain some welcome and/or navigation information about the site. Not to be confused with a buffer page or splash page. *Related Term:* Web site.

welded binding — a method of binding in which sheets of paper are bound together by ultrasonic radio waves and plastic devices. *Related Term:* Velo binding.

well — (1) editorially, the arrangement of advertisements on the right and left sides of a page so editorial matter can be placed between. (2) in intaglio printing, especially gravure, the term which is sometimes interchangeably used with "cell," indicating the etched areas below the surface of the image cylinder.

wet end — the forming section of the paper machine; that area between the head box and the dryer section; the front end.

wet strength — the firmness of a paper after it has been saturated with water for a specified time, as determined by its wet tensile or wet bursting strength.

wet trapping — a method of trapping in which wet ink is printed over previously printed wet ink. *Related Term:* trapping; wet-on-wet printing.

wet-on-wet printing — the process of printing a second color of ink directly over another without waiting for the first ink to dry. *Related Terms:* trapping; wet trapping.

wetting agent — a solution used with continuous-tone photographic film to prevent water spots during processing.

wetting down — the mixing of pigments and vehicle together to make ink.

wf — written as *w.f.* , it is an abbreviation for "wrong font." Used when correcting proofs to indicate a character set in the wrong typeface. *Related Term:* wrong font.

WfMC (Workflow Management Coalition) — a body that provides standards to facilitate communication between various workflow engines and their client applications.

what you see is what you get — *Related Term:* WYSIWYG.

whirler — a machine that contains a platform which turns and is used to evenly distribute light-sensitive coatings on printing plate material.

white — (1) the presence of all colors. (2) the visual perception produced by light in which each wavelength has the same relative intensity in the visible range as sunlight.

white balance — a color adjustment, particularly in motion video, which compensates for off-color light such as fluorescent or sodium vapors and still permits the rendering of white objects as white.

White Book — the fourth major extension to the audio CD (Red Book) standard. Unlike those for CD-ROM, CD-I, and CD-R, it is a very medium-specific format allowing for 74 minutes of video and audio to be laid onto a compact disc. The term is used synonymously with the Sony/Philips trademark Video CD. White Book video and audio are in the MPEG format. *Related Terms:* MPEG; Red Book.

white colors — *Related Term:* unwanted colors.

white light — the visual sensation that results when the wave lengths between 400 and 700 nm are combined in nearly equal proportions. *Related Terms:* prism; spectrum; spectrum, visible.

white space — the area in printed matter that is not covered by type and illustrations. Designer term referring to nonimage area that frames or sets off copy; an important design element. *Related Term:* negative space.

whiteboard — an electronic device used for collaborating on documents. Similar to the conference-room board from which it gets its name, it is a device with driver programs that allow multiple users teleconferencing at their own computers to draw and write comments on the same document.

Whois — a searchable database maintained by Network Solutions, which contains information about networks, networking organizations, domain names, and the contacts associated with them for the com, org, net, edu, and ISO 3166 country code top level domains.

wholesaler — in publishing and graphic supplies, a company that buys goods in quantity for resale to printers, artists, stores, libraries, etc. *Related Term:* distributor.

wicker dryer — an endless belt of lightweight metal racks that hold each screen-printed sheet during the drying travel distance.

wide angle lens — *Related Term:* lens, wide angle.

wide area information servers (WAIS) — a public domain implementation of a distributed information service which offers simple natural language input, indexed searching for fast retrieval and a "relevance feedback" mechanism which allows the results of initial searches to influence future searches. *Related Terms:* Archie; Gopher; Veronica.

widow — (1) any objectionably short line at the end of a paragraph or headline. It may be expressed as anything less than four characters, less than a full word, less than a certain percentage of the line measure, or any other subjective definition. (2) any single line on the top of a column or page. Widows of any definition are generally considered typographically undesirable. *Related Term:* orphan.

width — one of the possible variations of a type face within a type family, such as condensed or extended.

wild card — a placeholder for missing characters in filenames, often a "?", (*) or some other symbol. A wild card in a character string usually means find any character or set of characters. Windows 95 and DOS use the asterisk (*) wild card. All terms beginning with that stem will be found in the same search.

wild formation — the relatively irregular, clumped distribution of fibers in paper.

winding — in bindery work, the process of fanning sheets to get air between them in order to release some cutter dust and to free any edges that may have stuck together during the cutting operation.

window — (1) on mechanical art, a block of masking material that shows position of a photograph or other visual element and, when photographed, creates an opening (window) on the negative for stripping a halftone negative, screen tint, etc. Also, an area cut out of masking material. (2) on a computer screen, rec-tangular portion of the screen within which information appears. (3) in dtp, a graphical user interface (GUI) that allows a rectangular area of a computer screen to display text and graphic output from a program. With a number of programs running at one time, several win-dows can appear on the screen simultaneously, although only the top one is active at any time. Information can be cut and pasted from one window to another among different compatible applications.

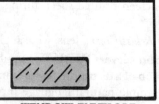

WINDOW ENVELOPE
Note the clear area to permit the address to be shown from the contents, without the additional step of addressing the envelope. The window area may be protected with a clear plastic or glassine or may merely be a cut opening on the envelope face.

window envelope — an envelope with a cut opening on the face to allow addresses and other data from the enclosure to be seen. Generally used by businesses to avoid the necessity of addressing envelopes by designing the inserts to have a name and address panel which coordinates with the position of the window.

WinISDN (Windows ISDN) — an open standard which allows ISDN adapter cards to use a dialer to initiate an Internet connection using PPP and MPPP. It runs on Window 3.1, Windows 95, 98 and Win/OS2 operating systems. *Related Terms:* API; ISDN; PPP; MPPP.

WinSock (Windows socket services) — software that acts as the middleman between Windows applications such as FTP, a Web browser, Telnet, etc., and the Internet protocol. It is a TCP/IP stack that allows you to use your modem to send data to/from the Internet, and is required for Windows internet applications like Netscape, Eudora, Free Agent, and many others. Winsock allows true Internet networking via modem.

WinZip — an independently developed Windows program that can decompress most of the from the Internet. It brings the convenience of Windows to the use of Zip files without requiring PKZip and PKUNZip. The WinZip Wizard makes unzipping easy, and features built-in support for popular Internet file formats, including TAR, Gzip, UNIX compress, UUencode, BinHex, and MIME, ARJ, LZH, and ARC files.

wipe — in videography, a transition between two clips in which the second clip displaces the first. Wipes occur in various shapes and directions.

wipe-on metal surface plate — in offset lithography, a pregrained plate on to which a light-sensitive diazo coating is wiped or applied manually or by a machine to the bare metal image carrier.

wire — (1) in papermaking, the Fourdrinier wire, the wire mesh used at the wet end of the papermaking process. Its fineness determines the texture of the paper. (2) in bindery, the spooled material used to form staples in side stitching and saddle stitching operations. (3) in publishing, the short term used to refer to newsgathering agencies, i.e., the Associated Press wire, etc.

wire frame — in computer-aided design and 3D graphics, the image you see is built on a skeleton called a wire frame. During development, graphics designers use wire frames because they render on screen a lot faster. The images are previewed by covering the frame with shading or a texture map. The viewing mode where the representation of printable attributes is hidden to reveal the mathematical representation of objects. *Related Terms:* shading; texture mapping.

wire services — newsgathering services such as Reuters, Associated Press, etc., that sell information and stories to subscribing members such as newspapers, television, periodical publications, etc.

wire side — the side of a sheet of paper that was formed in contact with the Fourdrinier wire of the paper machine during manufacturing. *Related Term:* felt side.

wire staple binding — the binding category that includes the methods known as saddle wire and side wire or side stitching. *Related Terms:* side stitching; side stabbing; saddle stitching.

wire stitching — *Related Terms:* saddle stitching; side stitching; sewing.

wireless – usually an infrared wavelength connection

from an input device to a computer. Used initially for wireless keyboards, and then printers, it is increasingly used to attach portable laptop computers to cellular phone connections providing literally wireless connection point to point.

Wire-O binding — a brand name double looped wire binding system which used prefabricated binding wire, as opposed to spiral binding, which is produced from a reel of wire.

with the grain — a term used to describe the directional character of paper, often applied to the folding of a sheet of paper parallel to the grain. Paper folds more easily and tears straighter with the grain than against the grain. *Related Term:* grain.

wood block — a relief image carrier or printing plate prepared from a block of wood with special cutting tools. *Related Terms:* woodcut; wood engraving.

wood engraving — *Related Term:* wood block; woodcut; block print.

wood type — blocks of wood into which type characters have been carved in relief. The sizes of wood type were specified in multiples of the pica. Use of wood type predates use of hot metal composition. Usually used for the larger display sizes over 1".

woodcut —illustrations in which lines of varying thickness are cut in relief on plank-grain wood for the purpose of making prints. A similar effect can be achieved by digital techniques. *Related Terms:* wood engraving; wood block.

wood-free paper — made from chemical pulp only, with size added. Available as calendered or supercalendered finish. *Related Term:* free sheet.

word break — the division of a word at the end of a line, usually with a hyphen. *Related Term:* hyphenation.

word processing — a method of typing on a computerized machine in which the keyboarded material is seen on a video screen and recorded on magnetic media so that it can be manipulated before being printed out and stored for future correction and use. Word-processed material can be input directly to computer typesetters for type composition without further keyboarding.

word space — the space between words. Can vary to accomplish justification of lines of type or for other design and aesthetic reasons.

word wrap — in page layout and word processing,

the ability of a program to automatically end a line (with a soft return) and wrap the next words to the following line. The carriage returns set up by this method are termed "soft," as against "hard" carriage returns resulting from the return key being pressed.

word-of-mouth — an informal means of promoting information, a product or a service from one person to another by spoken word.

work and back — when a full press sheet is used without duplication of any pages, one side is printed with one plate then turned end for end, the plate changed and the sheets run through for the second side. *Related Terms:* sheetwise; work and turn; work and flop; work and tumble; work and twist.

work and tumble — a method of printing where pages are imposed in such as way as that the sheet is printed on one side and then turned or tumbled from front to rear to print the opposite side. Work and tumble differs from work-and-turn in that the gripper edge changes, often leading to misregistration unless the stock has been accurately squared. *Related Terms:* work and flop; work and roll.

work and turn — a common printing imposition or layout in which all the images on both sides of a press sheet are placed in such a way that when the sheet is turned over and the same gripper edge is used, one half of the sheet automatically backs up the previously printed half. When the sheet is cut in half parallel to the guide edge, two identical sheets are produced. Work and turn impositions are preferred over work and tumble impositions for accuracy because the same gripper edge and the same side of the press sheet are used to guide the sheet twice through the press.

work and twist — a method of imposition with two different images on a plate, each rotated 180° from the other. After the run, the paper is rotated 180° to produce the second image.

work for hire — any creative work such as illustration, photography, writing, etc., for which the creator acknowledges customer ownership of copyright to the finished product.

work order — a production control form used by printing companies to specify and schedule production of jobs and record the time, materials, and supplies that each job requires to complete.

work station — traditionally, a computer terminal

considered to be more powerful than a PC (personal computer), but less than a mainframe. Current high-speed processors and memory capabilities of desktop personal computers are beginning to rival the capabilities of some work stations.

workaround — manual procedures implemented in order to overcome shortcomings of a program, delivery of materials, equipment, human or natural disasters.

workflow — a program that queues, tracks and otherwise manages production, documents, work items and collections of documents and work items as they progress from entry into the system and through the various individuals or offices in the organization until a business process is completed.

working film — the finished graphic arts negatives which will be used to make final printing plates, but are still loose or not imposed.

working film — the finished film .

working title — the preliminary title used to refer to a written work as it is being written and before the official title is decided upon.

work-in-process — monitoring of production flow during the manufacturing process which tracks an item to the finished goods stage. Also, monitoring processes within the warehouse.

workstation — a PC attached to a local area network.

World Wide Web — also known as the WWW, the W3, or most often simply as the Web. It can be described as a client/server hypertext system for retrieving information across the Internet. On it, everything is represented as hypertext HTML format and is linked to other documents by their URLs. *Related Terms:* browser; FTP; Gopher; HTML; http; hypertext; Telnet; URL.

worm — (1) in computing systems, an unauthorized program. Worms, as opposed to viruses, are meant to spawn in network environments and unlike viruses, don't attach themselves to other programs. Instead, they duplicate themselves over and over until all storage space on the system or network is consumed. The result is paralysis of the operating system. (2) in data storage, WORM (all capitals) meaning Write Once, Read Many. A common type of optical disk that only allows data to be written one time. A term usually applied to digital data disks. *Related Term:* Write Once, Read Many (WORM).

wove finish — a somewhat smooth, finely textured paper without visible wire marks. The finish is accomplished by moderate calendering.

WRAM (window RAM) — a variation on dual-ported memory technology that includes a larger bandwidth and more graphics-handling features than VRAM. Very useful for graphics applications. *Related Terms:* DRAM; EDO RAM; RAM; SRAM; VRAM.

wraparound — (1) in printing, a flexible or molded plate that is wrapped around a printing or plate cylinder of a rotary press and damped in place before being used for printing purposes. (2) in typography, a reference to lines of type which vary in measure to accommodate the shape and size of adjacent artwork without overprinting. *Related Term:* runaround.

wrapper — the paper cover often placed on hardbound or casebound books which contains information about the book, the author and/or his previous efforts, as well as professional reviews of the work.

Wratten filters — a comprehensive range of photographic filters manufactured by Eastman Kodak Company.

wrinkles — (1) the unspecified creases in paper happening during printing or postpress operations. (2) in ink formulation, irregularities in the surface formed during drying.

Write Once, Read Many (WORM) — a common type of optical disk that only allows data to be written directly one time, without mastering, but which cannot be erased or rewritten. Originally a high-cost industrial application, it has successfully migrated to the personal desktop in a variety of WORM and rewriteable formats. *Related Terms:* CD-R; CD-RW; DVD.

write protect — the setting of the appropriate lever or switch to prevent magentic media from being accidentally overwritten.

writing paper — a high-quality paper originally associated with correspondence and record keeping. It is considered the finest classification of paper, except for some specialty items. *Related Term:* bond paper.

writing white — the electrophotographic method in which the image elements remain unexposed while the nonimage, or background, areas are exposed to light, usually laser light.

wrong font — a proofreading mark which usually is written: *w.f.* . It is placed in the margin to indicate type is of a wrong face or font. An error in typesetting in which the letters of different fonts become mixed.

wrong-reading image — in Western countries, printed type that reads from right to left, or an image printed backwards from its normal orientation. Wrong-reading film images are read from right to left when the film is viewed from the base side. *Related Term:* lateral reverse.

WWW (World Wide Web) — a global hypertext network that lets users view text and graphics using a browser. *Related Term:* Web; World Wide Web.

WYSIWYG (what you see is what you get) — (pronounced "wizzy—wig") an accurate screen representation of final output. A term used to describe systems that preview full pages on the screen with text and graphics. It relates to the ability of a monitor to show a representation of the output as it will look when printed, including type styles sizes, colors, complete with correct line breaks, pagination, and other formatting, etc. The term can be misleading due to differences in the resolution of computer screens and that of page printers. Generally, it is a computer screen that displays non-monospaced, approximate true size and true shape typographic characters, rules, tints, and graphics.

X dimension — in code scanning, the dimension of the narrowest bar and narrowest space in a bar code.

X window system — a networked windowing system commonly used on UNIX and VMS systems.

X.25 — a CCITT standard for communicating between computers and specialized peripherals.

X-acto knife — a commercial cutting tool with interchangeable blades, used to cut film and paper galleys, camera film, masking film, etc.

xenon — a relatively inert gaseous chemical used in various gas-discharge lamps largely because it produces a white light. Pulsed xenon lamps are commonly used in graphic arts camerawork, color scanning, and in some optical radiation measuring instruments.

xerocopy paper — *Related Term:* copy paper.

xerography — a nonimpact-class electrostatic technology. A photocopying/printing process developed by Chester Carlson, founder of the Xerox Corporation, in which the image is formed using an electrostatic process. Heat fuses dry ink toner particles to electrically charged areas of the substrate, forming a permanent image. The charged areas of the substrate appear dark on the reproduction, while the uncharged areas remain white. Since the original Carlson patents have expired, most manufacturers of page printers, photocopy machines and laser printers currently use this method of printing.

XGA (eXtended Graphics Adapter) — a more recent IBM graphics standard that includes BGA and supports higher resolutions, up to 1,024 pixels by 768 lines interlaced. *Related Terms:* CGA; VGA; SVGA.

x-height — the vertical measure of a typeface without ascenders and descenders, i.e., a term used to describe the body height of a type character. It is expressed as the total character height without ascenders or descenders. The distance between the mean or waistline and the base line of the letters "x" or "z." They serve as examples of the face body height because they rest on the baseline and vary less in height than curved letters. Typefaces may be designed with small or large x heights relative to the point size. x- height varies with different typeface designs, and the only characters whose x-height is exactly accurate are letters with no top and bottom curves in the letter body, such as the x and z. *Related Terms:* ascender; descender; mean line; waistline; base line; ascender line; x-line; descender line.

x-line — *Related Term:* mean line; waistline.

XML (eXtensible Markup Language) — a pared-down version of SGML, designed especially for Web documents. It enables Web authors and designers to create their own customized tags to provide functionality not available with HTML. XML, for example, supports links that point to multiple documents; HTML links can reference just one. It provides a more powerful set of tools for development of Web applications. XML is gaining in popularity, and may eventually supplant HTML as the standard Web formatting specification as it is supported by all major browsers.

XQL — (not an acronym) The digital convergence of bits of information that form a complete image or story.

Y2K (Year 2000 Problem, or Millennium Bug) — 1,000 is "K" in computing. The "Y2K" generally refers to anticipation of problems which could or might occur in software whenever the millennium changes from "19xx" to "20xx." At the beginning of the computing age in the 1930's and 40's nobody really was thinking that their programming techniques would still be in use 50 or 60 years later. Unfortunately, some legacy programs were designed to assume the first two digits of date data as being 19. When this preface changes to 20 it can throw off the program's ability to calculate dates, times, payments, etc. In some cases it may cause catastrophic consequences such as navigation errors in aircraft or ships at sea, power grid shutdowns, alarm function malfunctions, etc. The exact significance and impact is only a guess until the real event, because nobody knows just where or how many computer programs will be affected.

yapp bind — binding a book with a soft cover which exceeds the trim size of the internal pages. It is designed to protect the edges of the pages, particularly when they are gilded. Often used in dictionaries, bibles and other classic products.

Y-C — the super VHS component video format. *Related Terms:* SVHA; Hi-8.

YCC — the encoding system employed in the Kodak CD format.

yellow — a subtractive primary. In transmitted or additive colors, the combination of the primary colors of green and red. In reflected light or subtractive colors, it is one of the primary colors and is one of the printers used in process color printing.

Yellow Book — the industry standard that defines the format of CD-ROMs, it was the first extension of the audio CD (Red Book) standard, and it enables CDs to contain 650 MB of computer data instead of only digital sound. *Related Terms:* Red Book; Green Book; White Book; Blue Book.

Yellow Pages (YP) — a service used by Unix administrators to manage databases distributed across a network. Now known as NIS (Network Information Services).

yellow printer — in process color printing, the plate used to print the yellow ink image, or the film used to produce the plate that prints the yellow image.

yellowing — a slow change in color of paper as it ages. Often caused by poor chemical and/or lignin removal during the manufacturing process.

YMMV (your mileage may vary) — an acronym often used in e-mail and postings, this disclaimer removes any sense of endorsement from a claim the writer has made. *Related Terms:* IMHO; IMO.

Young-Helmholtz theory — a theory about color vision that was proposed in the early nineteenth century. It suggests that humans perceive color based on the messages received from three receptors in the eye, one of which is particularly sensitive to red, and the others to the green and blue areas of the visible spectrum, respectively.

Yule-Nielson equation — a modification of the Murray-Davies equation to compensate for light scatter within a substrate when measuring printed dot area. This equation calculates the physical dot area.

YUV — in videography, an encoded color video signal; Y is the luminance signal, U and V are chrominance signals.

z-buffer — in a computer graphics card, the section of video memory that keeps track of which on-screen elements can be viewed and which are hidden behind other objects.

zena — a flaming orange/yellow/red hue in an illustration or photograph.

zero lead — in phototypesetting, a line ending with no vertical advancement, similar to a carriage return with no line feed on word processing equipment.

zeroing — setting or calibrating a densitometer or other measuring instrument to a known value.

Z FOLD
A variant of the conventional letter fold. The advantage can be that one panel is exposed after folding and it allows the use of the inside address of letters, for instance, to serve as the recipient addressing when used with a window envelope.

z-fold — an accordion or concertina fold. A very popular folding method used with both letters and letter enclosures because it is easily performed on most desktop folding devices. It works particularly well with window envelopes because one panel of the folded product is exposed and can contain the addressing information.

ZIF (zero insertion force) socket — a simplified chip-mounting and -removal system designed to prevent bending or damaging CPU pins when making a change or upgrading. ZIF sockets use leverage instead of brute force to seat and unseat chips, thus taking the pressure out of chip relocation.

zinc — a metal used in making a photoengraving, sometimes called a "zinc cut" or "cut."

zine (electronic magazine) — sometimes e-zine. A shorthand word for a magazine published in electronic form; an online magazine. It could mean coded in HTML, is available on the World Wide Web, and is updated regularly. *Related Term:* e-journal.

zine.net — a centralized starting point for electronic magazines, which offers users the chance to "try before you buy" or subscribe. A resource on the Web for information on the vast assortment of independent, self-published zines.

zip — an open standard for compression and decompression used widely for PC download archives. Originally a DOS-based program called PKZip, it is now widely used on Windows-based programs such as WinZip and Drag and Zip. The file extension given to ZIP files is .zip. *Related Term:* .sit; WinZip.

ZIP code — abbreviation for Zone Improvement Plan code, originally five numerals that identified every post office and substation in the U.S. Now expanded to ZIP+4 which is five numbers followed by a dash and four more numbers. It provides even more specific delivery data and has been known to enable mail delivery without name, street, apartment number, town or city, although using this approach is not recommended by the postal authorities.

zip drive — a portable diskette storage system from Iomega Corporation which is slightly larger than a standard 3 1/4 in. floppy diskette but which stores either 100 Mb or 250 Mb, de-pending upon model. This com-

ZIP DISKETTE AND CASE

pares with 1.44 or 2.88 Mb per stan-dard floppy disk. It is considered an economic storage medium and is popular as original equipment in newly configured systems and as an add-on item.

Zipatone — a rub-on imaging system based on a transparent sheet containing dot or line patterns that provides a tonal effect similar to that provided by Benday.

zipped files — compressed PC/Windows files you commonly see on the Internet which once downloaded to your computer need to be decompressed by a program like PKUNZip or WinZip.

Zmodem — a first choice file transfer protocol for sending and receiving files using dial-up connections. It has speed, error checking, and it can resume a file transfer after a break in communications. Its name is based on its intention to supersede Xmodem and Ymodem. *Related Terms:* Kermit; Xmodem; Ymodem.

zone file — a portion of the total domain name space that is represented by the data stored on a particular name server. The name server has authority over the zone, or the particular portion of the domain name space, described by that data. *Related Terms:* zone file; name server; domain name space; resolve.

zone system – an approach to black and white photography whose technical objective is the establishment of relation-ships between various tones. The darkest tones are 0 or solid black. X is the whitest area. Proper under-standing of the approach will allow the raising or lowering of certain or selected values in the gray scale, producing a more realistic or surrealistic visual interpretation of the scene.

zone theory – *Related Term:* opponent-process model.

zoom — generally, the process of enlarging all or part of an image for display or print purposes. It may be accomplished in the darkroom, using an enlarger to concentratr on only a portion of the original negative or it may be accomplished in the camera, at the time of exposure by using telephoto lens and attachments. *Related Terms:* pan; tilt.

zoom lens — a camera lens that can be adjusted to various focal lengths along a continuum. *Related Term:* lens, zoom.

zooming — (1) the process of electronically enlarging an image on a video display terminal to facilitate electronic retouching. (2) in photography, the process of using a zoom lens which has variable focal lengths, to frame a subject at some distance.

GRAPHIC ASSOCIATIONS, ORGANIZATIONS, PHONES AND WEB SITES

American Book Producers Association
160 Fifth Avenue
New York, NY 10010
PHONE (212) 645-2368

American Business Press
675 Third Avenue
New York, NY 10017
PHONE (212) 661-6360
FAX (212) 370-0736
www.abp2.com

American Forest and Paper Association
1111 19th Street, NW, Suites 700 & 800
Washington, DC 20036
PHONE (202) 463-2700
www.afandpamg

American Communcation Association
http://www.americancomm.org

American Institute of Graphic Arts
164 5th Avenue
NewYork, NY 10010
PHONE (212) 807-1990
www.ai.ga.org

American Paper Institute
260 Madison Avenue
New York, NY 10016
PHONE (212) 340-0600

Association for Graphic Arts Training
Industrial Technology Department
Arizona State University
Tempe, AZ 85287-6806
PHONE (602) 727-1685

Association of American Publishers
71 Fifth Avenue
New York, NY 10003
PHONE (212) 255-0200
www.publishers.org

Association of Area Business Publications
5820 Wilshire Boulevard, Suite 500
Los Angeles, CA 90036
PHONE (213) 937-5514
wwwbizpubs.com

Binders & Finishers Association
408 Eighth Avenue, Suite 10-A
New York, NY 1000 1 – 1816
PHONE (212) 629-3232

Binding Industries of America
70 E. Lake Street
Chicago, IL 60601
PHONE (312) 372-7606
FAX (312) 704-5025

Bookbuilders West
P. O. Box 7046
San Francisco, CA 94120-9727
PHONE (650) 592-8930
www.bookbuilders.org

Business Forms Management Association
519 S.W. Third Avenue, Suite 712
Portland, OR 97204-2579
(503) 227-3393
FAX (503) 274-7667

Catholic Press Association/U.S. & Canada
3555 Veterans Memorial Highway, Unit 0
Ronkonkoma, NY 11779
PHONE (516) 471-4730

Chicago Association of Direct Marketing
435 N. Michigan Avenue, Suite 1717
Chicago, IL 60611
PHONE (312) 670-2236

Chicago Book Clinic
825 Green Bay Road, Suite 270
Wilmett, IL 60091
PHONE (847) 256-8448
www.chicagobookciinic.org

GRAPHIC ASSOCIATIONS, ORGANIZATIONS, PHONES AND WEB SITES

desktopPublishing.com
http://desktoppublsihing.com/design/html

DesignCafe
http://www.designcafe.com.au

Designer's Guide To The Internet
http://designer.zender.com

designSphere Online
http://www.dsphere.net

Digital Design Communication
http://www.sequel.net/~eagtarap/index.html

Digital Distribution of Advertising for Publications Association
1855 E. Vista Way, Suite 11
Vista, CA 92084

Digital Printing & Imaging Association
10015 Main Street
Fairfax, VA 22031
PHONE (703) 385-1339
hftp://www.dpia.org

Direct Marketing Association
1120 Avenue of the Americas
New York, NY 10036-6700
PHONE (212) 768-7277
www.the-dma.org

Document Management Industries Association
433 E. Monroe Avenue
Alexandria, Va. 22301
PHONE (703) 836-6225

Education Council of the Graphic Arts Industry
1899 Preston White Drive
Reston, VA 22091
PHONE (703) 264-7200

Engraved Stationery Manufacturers Association, Inc.
305 Plus Park Boulevard
Nashville, TN 37217
PHONE (615) 366-1798

Environmental Group
1899 Preston White Drive
Reston, VA 20191-4367
PHONE (703) 648-3218
http://envgroup.org

Fibre Box Association
2850 Gold Road
Rolling Meadows, IL 60008
PHONE (708) 364-9600
FAX (708) 364-9639

Flexographic Technical Association
900 Marconi Avenue
Ronkonkoma, N.Y. 11779-7212
PHONE (516) 737-6020
http://www.fta-ffta.org

GASP Engineering
http://www.gaspnet.com

Graphic Arts Association
1900 Cherry Street
Philadelphia, PA 19103
PHONE (215) 299-3300
www.gaa1900.com

Graphic Arts Education and Research Foundation
1899 Preston White Drive
Reston, Va. 20191-4367
PHONE (703) 264-7200
http://www.npes.org

Graphic Arts Employers of America
100 Daingerfield Road
Alexandria, VA 22314
PHONE (703) 519-8150

GRAPHIC ASSOCIATIONS, ORGANIZATIONS, PHONES AND WEB SITES

Graphic Arts Literacy Alliance
P. O. Box 11712
Santa Ana, CA 92711
PHONE (714) 921-3120
FAX (714) 921-3126

Graphic Arts Marketing Information Service
100 Daingerfield Road
Alexandria, VA 22314
PHONE (703) 519-8179

Graphic Arts Professionals
P. O. Box 3139
New York, NY 10163-3139
PHONE (212) 685-2995

Graphic Arts Sales Foundation
113 E. Evans Street
Matlack Building, Suite A
West Chester, Pa. 19380
PHONE (610) 431-9780

Graphic Arts Show Company
1899 Preston White Drive
Reston, Va. 20191-4367
PHONE (703) 264-7200
http://www.gasc.org

Graphic Arts Suppliers Association
1900 Arch Street
Philadelphia, PA 19103
PHONE (215) 564-3484

Graphic Arts Technical Foundation (GATF)
200 Deer Run Road
Sewickley, PA 15143
PHONE (412) 741-6860
www.gatf.org

Graphic Comm Central
http://teched.edtl.vt.edu/gcc

Graphic Communication Association
100 Daingerfield Road
Alexandria, VA 22314
PHONE (703) 519-8160
www.gca.org

Gravure Association of America
1200-A Scottsville Road
Rochester, NY 14624
PHONE (716) 436-2150
FAX (716) 436-7689
www.gaa.org

Gutenberg Expositions
P.O. Box 11712
Santa Ana, CA 92711
PHONE (714) 921-3120
FAX (714) 921-3126

Hot Stamping Association
40 Melville Park Road
Melville, NY 11747
PHONE (516) 694-7773

IBFI International Association for Document and Information Management Solutions
2111 Wilson Boulevard, Suite 350
Arlington, Va. 22201-3042
PHONE (703) 841-9191
hftp://www.ibfi.org

Graphic Communications Association
100 Daingerfield Road
Alexandria, VA 22314
PHONE (703) 519-8160
FAX (703) 548-2867

International Association of Printing House Craftsmen
7042 Brooklyn Boulevard
Minneapolis, MN 55429
PHONE (612) 560-1620
http://www.iaphc.org

GRAPHIC ASSOCIATIONS, ORGANIZATIONS, PHONES AND WEB SITES

International Business Forms Industries
2111 Wilson Boulevard, Suite 356
Arlington, VA 22201
PHONE (703) 841-9191

International Digital Imaging Association
170 Township Line Road
Belle Mead, NJ 08502
PHONE (908) 359-3924

International Graphic Arts Education Association
200 Deer Run Road
Sewickley, PA 15143
PHONE (412) 74126860
http://wwwigaea.org

International Prepress Association
7200 France Avenue, South, Suite 327
Edina, MN 55435
PHONE (612) 896-1908
www.ipa.org

International Printing Museum
8469 Kass Drive
Buena Park, CA 90621
PHONE (714) 523-4315

International Technology Education Association
http://iteaww.org/index/html

International Thermographers Association
100 Daingerfield Road
Alexandria, VA 22314
PHONE (703) 579-8122

International Association of Business Communicators
1 Hallidie Plaza, Suite 600
San Francisco, CA 94109
PHONE (415) 433-3400

International Publishing Management Association
The IPMA Building
1205 West College Street
Liberty, MO 64068-3733
PHONE (816) 781-1111
hftp://www.ipma.org

International Regional Magazine Association
P. O. Box 125
Annapolis, MD 21404
PHONE (410) 451-2892

IPMA, In-Plant Management Association
The IPMA Building
1205 W. College Avenue
Liberty, MO 64068
PHONE (816) 781-1111

Italian Trade Commission
401 N. Michigan Avenue, Suite 3030
Chicago, IL 60611
PHONE (312) 670-4350

Label Printing Industries of America, PIA
100 Daingerfield Road
Alexandria, VA 22314
PHONE (703) 519-8122

Magazine Printers Section, PIA
100 Daingerfield Road
Alexandria, VA 22314
PHONE (703) 519-8100

Magazine Publishers of America (MPA)
919 3rd Avenue, 22nd Floor
New York, NY 10022
PHONE (212) 872-3700
www.magazine.org

Master Printers of America
100 Daingerfield Road
Alexandria, VA 22314
PHONE (703) 519-8130

GRAPHIC ASSOCIATIONS, ORGANIZATIONS, PHONES AND WEB SITES

The Mining Company Guide to Graphic Design
http://www.graphicdesign.miningco.com
ED. NOTE: The Mining Company is changing to about.com in early 1999. Try URL for about.com if you have difficulty connecting.

National Graphic Arts Dealers Association
116 W. Ottawa
Lansing, MI 48933

National Association of Desktop Publishers
462 Boston Street
Topsfield, MA 01983
PHONE (800) 874-4113

National Association of Quick Printers
401 N. Michigan Avenue
Chicago, 60611
PHONE (312) 644-6610

National Association of Diemakers and Diecutters
P.O. Box 2
Mount Morris, IL 61054
PHONE (815) 734-4178

National Association of Litho Clubs
6550 Donjoy Drive
Cincinnati, Ohio 45242
PHONE (513) 793-2532

National Association of Lithographic Plate Manufacturers, PIA
100 Daingerfield Road
Alexandria, VA 22314
PHONE (703) 519-8100

National Association of Printers and Lithographers
780 Palisade Avenue
Teaneck, NJ 07666
PHONE (201) 342-0700
(800) 642-NAPL
http://www.napl.org

National Association of Printing Ink Manufacturers
581 Main Street
Woodbridge, NJ. 07095-1104
PHONE (732) 855-1525
hftp://www.napim.org

National Association of Quick Printers
401 N. Michigan Avenue
Chicago, IL. 60611
PHONE (312) 3216886
hftp://www.naqp.org

National Business Forms Association
433 E. Monroe Avenue
Alexandria, VA 22301
PHONE (703) 836-6232

National Composition and Prepress Association
100 Daingerfield Road
Alexandria, VA 22314
PHONE (703) 519-8165

National Computer Graphics Association
2722 Merrilee Drive, Suite 200
Fairfax, VA 22031
PHONE (703) 698-9600

National Newspaper Association
1525 Wilson Boulevard, Suite 550
Arlington, VA 22209
PHONE (703) 907-7900
FAX (703) 907-7901

National Paper Trade Association
111 Great Neck Road
Great Neck, NY 11021
PHONE (516) 8293070
www.papertrade.com

GRAPHIC ASSOCIATIONS, ORGANIZATIONS, PHONES AND WEB SITES

**National Printing Equipment
and Supply Association**
1899 Preston White Drive
Reston, VA 22091
PHONE (703) 264-7200

National Soy Ink Association
4554 NW 114th Street
Urbandale, IA 50322-5410
PHONE (515) 251-8640
www.soyink.com

National Association of Independent Publishers
P. O. Box 430
Highland City, FL 33846-0430
PHONE (941) 6484420

Newsletter Publishers Association
1501 Wilson Boulevard, Suite 509
Arlington, VA 22209-2403
PHONE (703) 527-2333
www.newsletters.org

Newspaper Association of America
1921 Gallows Road, Suite 600
Vienna, Va. 22182
PHONE (703) 902-1600
hftp://www.naa.org

Non-Heatset Web Section, PIA
100 Daingerfield Road
Alexandria, VA 22314
PHONE (703) 519-8140

**North American Graphic Arts
Suppliers Association**
1604 New Hampshire Avenue, NW
Washington, D.C. 20009-2512
PHONE (202) 328-8441
http://www.nagasa.org

**NPES: The Association for Suppliers of
Printing Publishing and Converting
Technologies**
1899 Preston White Drive
Reston, VA 20191
PHONE (703) 2647200
www.npes.org

Pacific Press Publishing Association
P. O. Box 5353
Nampa, ID 83653
PHONE (208) 465-2500
www.pacificpress.com

Paper Industry Management Association
2400 E. Oakton Street
Arlington Heights, IL 60005
PHONE (708) 956-0250
FAX (708) 956-0520

Periodical and Book Association of America
475 Park Avenue, South, 8th Floor
NewYork, NY 10016
PHONE (212) 689-4952

Philadelphia Book Clinic
136 Chester Avenue
Yeadon, PA 19050-3831
PHONE (610) 259-7022

PIM Education Foundation
PHONE (810) 354-9200
http://hickory.net-data.com:8001/business/pim/
pimeduc.htf

Pira International
http://www.pira.co.uk:80070/infocentre

**Print Buyers Association
(Printing Industries of   Northern California)**
665 3rd Street, Suite 500
San Francisco, CA 94107-1901
PHONE (415) 495-8242
www.pinc.org

GRAPHIC ASSOCIATIONS, ORGANIZATIONS, PHONES AND WEB SITES

Print/New Jersey
75 Keary Avenue
P.O. Box 6
Keary, NJ 07032
PHONE (201) 99-PRINT

Printing Association of Florida
6250 Hazeltine National Drive, Suite 114
Orlando, FL 32822
PHONE (407) 240-8009

PIA: Printing Industries of America
100 Daingerfield Road
Alexandria, VA 22314
PHONE (703) 5198100
FAX (703) 548-3227
www.printing.org

Printing Industries of the Gulf Coast
1324 W. Clay Street
Houston, TX 77019
PHONE (713) 522-2046
FAX (713) 522-8342

Printing Industries of Wichita
P.O. Box 1377
Wichita, KS 67201
PHONE (316) 264-1363

Printing Industry of New England
10 Tech Circle
Natick, MA 01760
PHONE (508) 804-4119
www.pine.orq

Printing Industry of South Florida
6095 N.W. 167th Street, Suite D-7
Hialeah, FL 33015
PHONE (305) 558-4855

Red Tag News Publications Association
P. O. Box 429
Flossmoor, IL 60422-0429
PHONE (708) 957-5525

Research and Engineering Council of the Graphic Arts Industry
P.O. Box 639
Chadds Ford, PA 19317
PHONE (610) 388-7394
FAX (610) 388-2708

Retail Advertising and Marketing Association International
333 N. Michigan Avenue, Suite 3000
Chicago, IL 60601
PHONE (312) 251-7262
www.ramarac.org

Screen Printing Association International
10015 Main Stt
Fairfax, VA 22031
PHONE (703) 385-1335

Screen Printing Technical Foundation
10015 Main Street
Fairfax, VA 22031
PHONE (703) 385-1417

Screenprinting & Graphic Imaging Association International
10015 Main Street
Fairfax, Va. 22031
PHONE (703) 385-1335
http://-.sqia.org

Society for Scholarly Publishing
10200 W 44th Avenue, Suite 304
Wheat Ridge, CO 80033
PHONE (303) 4223914

Society for Service Professionals in Printing
433 E. Monroe Avenue
Alexandria, VA 22301
PHONE (703) 684-0044

GRAPHIC ASSOCIATIONS, ORGANIZATIONS, PHONES AND WEB SITES

Society of National Association Publications (SNAP)
1650 Tysons Boulevard, Suite 200
McLean, VA 22102
PHONE (703) 506-3285
www-snaponline.org

Suburban Newspapers of America
401 N. Michigan Avenue
Chicago, IL 60611
PHONE (312) 644-5610

Tag and Label Manufacturers Institute, Inc.
104 Wilmot Road, Suite 201
Deerfield, IL 60015-5195
PHONE (708) 940-8800
FAX (708) 940-7218

Technical Association of the Graphic Arts
P.O.Box 9887
Rochester, NY 14623
PHONE (716) 272-0557
FAX (716) 475-2250

The International Association of Printing House Craftsmen
7599 Kenwood Road
Cincinnati, OH 45236
PHONE (513) 891-0611

Typographers International Association
2233 Wisconsin Avenue, NW, # 235
Washington, DC 20007
PHONE (202) 965-3400

Web Offset Section, PIA
100 Daingerfield Road
Alexandria, VA 22314
PHONE (703) 519-8140

Whiskey Creek Document Design
http://www.rrv.net/wcdd

JOURNALS, PUBLICATIONS, MAGAZINES AND OTHER RESOURCES

Adobe Magazine
P. O. Box 24998
Seattle, WA 98124-0998
www.adobe.com

American Printer
650 S. Clark Street
Chicago, IL 60605-9960

AV Video Multimedia Producer
701 Westchester Avenue
New York, NY 10604-3098
PHONE: 914-328-9157
FAX: 914-3228-9093
TOLL FREE: 1-800-800-5474
www.avvideo.com

Color Publishing
Ten Tara Boulevard, 5th Floor
Nashua, NH 03062-2801
PHONE (603) 891-9166
FAX (603) 891-0539

DCC Magazine
270 Madison Avenue, 6th Floor
New York, NY 10016
PHONE 212-951-6000
FAX: 212-951-6717
www.dccmag.com

Desktop Publishers Journal
462 Boston Street
Teopsfield, MA 01983-1232
PHONE: 978-887-7900
FAX: 978-887-9245
www.dtpjournal.com

Digital Design and Production
Cahners Business Information
345 Hudson Street
New York, NY 10014-4504
PHONE: 212-519-7235
FAX: 212-519-7489

Digital Imaging
445 Broad Hollow Road
Melville, NY 11747
PHONE (516) 845-2700 Ext. 278
www.digitalimagingmag.com

Digital Magic
P. O. Box 239
Tulsa, OK 74101-0239

Digital Output
13000 Sawgrass Village Center, Suite 18
Ponte Vedra Beach, FL 32082
PHONE (904) 285-6020
FAX (904) 285-9944
http://www.digitalout.com

e-Business
P. O. Box 469013
Escondido, CA 92046-9964
PHONE: 1-800-336-6060
www.advisor.com

Electronic Publishing
P. O. Box 2709
Tulsa, OK 74101-9689
PHONE (603) 891-9166
FAX (602) 891-0539
pennwell.com

Flexo
P. O. Box 262
Congers, NY 10920-0262

Graphic Arts Monthly
345 Hudson Street
New York, NY 10014-4504
PHONE: 212-519-7235
FAX: 212-519-7489

High Volume Printing
434 W Downer Place
Aurora, IL 60506-9628

JOURNALS, PUBLICATIONS, MAGAZINES AND OTHER RESOURCES

Imaging and Document Solutions
12 West 21 Street
New York, NY 10010
PHONE: 212-691-8215
FAX: 212-691-1191
www.imagingmagazine.com

In-Plant Graphics
P. O. Box 12820
Philadelphia, PA 19108-0820

In-Plant Printer
434 W. Downer Place
Aurora, IL 60506-5080

InternetWeek
P. O. Box 1281
Skokie, IL 60076-9432

Journal of Industrial Technology
3300 Washtenaw Avenue, Suite 220
Ann Arbor, MI 48104-4200
littp://nait.org

Micropublishing News
2340 Plaza del Amo, Suite 100
Torrance, CA 90501
PHONE (310) 212-5802
www.micropubnews.com

NewMedia Magazine
P. O. Box 3039
Northbrook, IL 60065-3039
PHONE: 650-573-5170
www.newmedia.com

Perspectives
1205 West College Street
Liberty, MO 64068-3733

PEI - Photo Electronic Imaging
229 Peachtree Street, NE, Suite 2200
International Tower
Atlanta, GA 30303
PHONE (404) 522-8600
FAX (404) 614-6405
www.peimag.com

Pre
P. O. Box 4949
Stamford, CT 06913-0309

Presentations
50 South Ninth Street
Minneapolis, MN 55402-9973

Printing Impressions
P. O. Box 11571
Riverton, NJ 08076-1571

Print on Demand Business
445 Broad Hollow Road
Melville, NY 11747
PHONE 516-845-2700
FAX: 516-249-5774
TOLL FREE: 1-800-308-6397

Publish
Reader Service Department
12950 SW Pacific Highway, Suite 7
Tigard, OR 97223-9747
www.publish.com

Quick Printing Magazine
Cygnus Publishing, Inc.
445 Broad Hollow Road
Melville, NY 11747-4722
PHONE: 800-547-7377
FAX: 920-563-1704
www.quickprinting.com

Tech Directions
P.O. Box 8623
Ann Arbor, MI 48107-8623
www.techdirections.com

JOURNALS, PUBLICATIONS, MAGAZINES AND OTHER RESOURCES

The Commercial Image
1233 Janesville Avenue
Ft. Atkinson, WI 53538

T.H.E. Journal
150 El Camino Real, Suite 112
Tustin, CA 92780-3670

**The Magazine of Design and
    Technology Education**
103 Armstrong Hall
P.O. Box 7718
Ewing, NJ 08628-0718

Upper & lower Case
866 Second Avenue
New York, NY 100 17
EMAIL: ulc@itc.com
www.itc.com

Video & Multimedia Producer
P.O. Box 3034
Northbrook, IL 60065-3034

Visual Communication Journal
200 Deer Run Road
Sewickley, PA 15143

SHORTHAND REFERENCES

Many of us know terms only by their lettered designations. Sometimes these letters (usually the first one or two of most of the words in the title) form brand new words that often take the place of the words they represent. For instance, MODEM is an acronym for MODulate/DEModulate. Rarely do we ask for a "Modulator/demodulator". We have become accustomed to the shorthand version.

For this reason, the following list of frequently used abbreviations and acronyms is provided to give the user of this dictionary a head start toward locating import terms. There is no need to be embarrassed about not knowing exactly what the QPSK stands for. just look to the page indicated and you will have a short trip to a definitive answer. By the way, it stands for Quadrature Phase Shift Keying. Aren't you glad you stayed long enough to find out?

ABBREVIATIONS AND ACRONYMS

-30-	260	AVK	14	CC	37	CPU	39,59
A/D	1	AVI	14	CCD	37,38,41	CQI	55
A/W	1	AVSS	15	CCITT	38	CRC	59
AA	I	AW	15	CD	38	CRT	37,60
ABA	1	BASIC	18	CD-DA	38	CSC	60
ABI	1	BBL	19	CDF	38	CSLIP	235
ACDSee	2	BBS	19,28	CD-I	38	CSS	60
ACK	2	BCD	19,20	CDMA	38	CT	60
ACPI	2	BCNU	19	CD-R.	38,57	CTP	60
ADB	9	BDR	18	CD-ROM	38,52	CVS	61
ADC	4, 14	BF	20	CD-XA	39	cwt	62
ADF	4	BFD	20	CEPS	40	DASD	63
ADPCM	5	BFT	20	CERN	40	DAT	63
ADSL	5	BIOS	22	CF	40	DBS	65
AGP	1	BITNet	22	CFB	40	DCC	65
AIFF	5	BLOB	24	CGA	40	DCS	65
AIIM	6	BMP	25	CGI	40	DCT	65
AIM	6	BNC	25	CGM	40	DDE	65
ALA	6	BPI	27	CHAP	40	DDN	65
AM	7	BPR	27	CHOP	68	DDS	65
ANSI	7	BPSK	27	CIE	43	DES	67
API	9	BRM	27	CIS	31	DHCP	68
ARC	10	BTW	27	CL	35,43	DHTML	69
ARPA	10	C&lc	31	CLUT	45	DIMM	71
ARPANet	10	C&sc	31,34	CMOs	45	DIP	72
ASA	11	C1S	31	CMYK	45	DMA	74
ASCII	7,11	C2S	31	CNET	45	DMI	74
ASIC	12	CAD	31	COBOL	46	DMS	74
ASN	14	CAM	53	COLD	46	DN	74
AT	13	CAR	35	COM	51	DNS	74, 76
ATAPI	13	CAV	37	CORBA	58	DOM	75
ATM	13	CB	37	CPI	59	DOS	73,76
AU	13	C-Band	37	cps	59	DP	78

ABBREVIATIONS AND ACRONYMS

DPI	79	FIFO	96	ICCF	134	LMP	154
DQPSK	79	FIM	98	ICR	128,134	LOGMARS	154
DRAM	79	FM	102	IDE	128	LOL	155
DRUPA	80	FOB	102	IEEE	128	lpi	153, 156
DS-1	256	FPLL	107	IETF	128	lpm	153, 156
DS-3	256	fpm	96, 107	IFC	128	LS/2000	156
DSL	81	FPO	104,107	IIS	128	LUT	156
DSP	81	FPRR	107	ILD	128	LZW	156
DSSG	81	FTP	108	IMA	129	MAPI	159
DSVD	81	FWIW	109	IMAP	130	Mb	160,161
DTP	68, 81	FYI	104, 109	IMHO	130	MBONE	160
DTV	68, 81	GB	111	IMO	130	MBPS	160
DV	82	GBC	111	IN-ADDR	137	MCA	163
DVD	82,83	GCR	111,115	IP	136, 138	MES	159
DVD-R	83	GDI	111	IPX	138	MG	163
DVD-RAM	83	GEM	111	IR	138	MHZ	163
DVD-ROM	83	GEO	111	IRC	136,138	MICR	163
DVD-VIDEO	83	GHz	112	IRQ	138	MIDI	164
DVI	83	GIF	112,265	ISA	138	MIME	164
EBCDIC	84	GIGO	112	ISAPI	138	MIP	165
ECI	84	GNU	113	ISBN	138	MIPS	165
ECP	84	gsm	116	ISDN	134,138	MIS	165
EDI	84	GUI	114,116	ISO	139	MISF	165
EDO RAM	84	GZip	117	ISP	136,139	M-JPEG	165
EDP	84	H&J	118	ISSN	139	MM	165
EEPROM	84	HAGO	118	ITFS	135	MMX	165
EF	85	HAL	120	ITU-T	139	MODEM	166
EFT	85	HDTV	120	JAR	140	MOPS	167
EGA	85	HFC	122	JDBC	140	MOS	167
EIP	86	HIBC	122	JIT	140	MPC	168
EISA	85	HLS	123	JPEG	141,208	MPEG	168
EMS	88	HS	125	JVM	142	MPEG2	168
EOL	89	HSB	125	K	143	MRD	168
EP	90	HSL	125	KB	143	MRP	168
EPA	90	HSV	125	L8R	146	ms	159, 168
EPP	90	HTML	125	LAN	146,154	MS-DOS	168
EPROM	90	http	125	LATA	147	MSDS	160
EPS	90,201	HTTPS	126	LBA	148	MTBF	168
EPSF	90	Hz	127	lc	148	MTTR	168
F&G	94	I/O	128, 133	LCD	148,153	MUD	168
FACT	94	IANS	128	LCL	148	NAK	170
FAQ	95	IAS	128	LDAP	148	NC	170
FAT	95	IBC	128	LED	148	NCR	170,173,186
FCC	95	IC	128	LIFO	150	NCSA	170

ABBREVIATIONS AND ACRONYMS

NDIS	170	PE	189	RIFF	223	SMIL	241
NFS	172	PEL	190	RIP	216,222,223	SMTP	239,242
NIC	173	PERL	190	RISC	223	SOA	248
NID	173	PGP	191,203	RJ-11	223	sp	243
NIP	173	Pi	194	RJ-45	224	SPARC	243
NOC	173	PICT	194	RNA	224	SPID	245
NREN	174	PICT/PICT2	194	ROFL	224	SPRR	246
NS16550	174	PIF	194	ROM	224	SQL	246
NSA	174	PING	195	ROP	225	SRA	246
NSAPI	174	PKUNZIP	195	ROTFL	225	SRAM	247
NSF	174	PKZIP	195	RS-232	226	SRDS	247
NT-1	174	PLU	197	RSA	226	SS	247
NTFS	175	PLV	198	RTF	226	SSI	247
NTSC	175	PMK	198	RTFM	226	SSL	234,247
OCLC	176	PMS	186, 198	RTP	226	STAC	247
OCR	179	PMT	193, 198	RTS	226	STAC-LZS	247
OCR-A	176	PNC	198	RTV	226	SWOP	254
OCR-B	176	PNG	198	RU	220	SYSOP	255
ODBC	176	PoP	199	RW	227	T1	256
ODMA	176	POP	199,200	S/MIME	229	T3	256
OEM	176,181	POS	200	S/N	229,242	T4S	256
OH	178	POSIX	200	S/NR	229	TA	256
OK	178	POTS	201	S/S	229	TANSTAAFL	257
OLE	178	ppi	185,201	S/T	229	TARGA	257
OMG	178	PPP	191, 201	S-100	229	TCP/IP	257
OMR	180	PPTP	201	SASE	230	TFB	259
OOP	178	PRI	204	SC	230	TFT	259
OOS	178	QA	212	SCA-FM	230	TGA	257,259
OS	181	QPSK	212	S-CDMA	231	TGM	263
OSHA	176	QWERTY	214	SCSI	233	TIA	259
OSP	181	RAD	215	SD	233	TIF	256
PAL	185	RAID	215	SDK	233	TIFF	256,261
PAP	186	RAM	216	SDRAM	236	TLD	263
PC	188	RAMDAC	216	SECAM	233	TN3270	262
PCI	189	RC	216	SGML	236, 248	TQM	263
PCL	205	RCA	216	SGRAM	236	TR	263
PCM	189,210	RD	216	SHTTP	238	TTFN	267
PCMCIA	189	RFCs	221	SIG	239	TTY	267
PCS	189	RFDC	215	SIMM	239	TVRO	267
PCX	189	RFG	223	SIT	240	TWAIN	267
PDF	189	RFI	223	SKU	240	U&lc	271
PDF417	189	RFID	215	SLIP	235,241	UART	271
PDL	184, 189	RFP	223	SLR	33,241	UCA	271
PDP	189	RGB	223	SMATV	241	UCC	272